T0251728

Process Control Engineering

Process Control Engineering

A Textbook for Chemical, Mechanical and Electrical Engineers

Ming Rao

and

Haiming Qiu

University of Alberta
Edmonton, Canada

CRC Press
Taylor & Francis Group
Boca Raton London New York

CRC Press is an imprint of the
Taylor & Francis Group, an **informa** business

First published 1993 by Gordon and Breach Science Publishers

Published 2019 by CRC Press
Taylor & Francis Group
6000 Broken Sound Parkway NW, Suite 300
Boca Raton, FL 33487-2742

© 1993 by Taylor & Francis Group, LLC
CRC Press is an imprint of Taylor & Francis Group, an Informa business

No claim to original U.S. Government works

ISBN 13: 978-2-88124-628-9 (hbk)

Visit the Taylor & Francis Web site at
http://www.taylorandfrancis.com

and the CRC Press Web site at
http://www.crcpress.com

Library of Congress Cataloging-in-Publication Data

Rao, M. (Ming), 1954-
      Process control engineering : a textbook for chemical, mechanical, and electrical engineers / Ming Rao and Haiming Qiu.
          p.   cm.
      Includes index.
      ISBN 2-88124-628-1
      1. Process control.   I. Qiu, Haiming, 1944-   .   II. Title.
TS156.8.R37 1993
670.42 ' 7--dc20   93-33116
                                                                              CIP

CONTENTS

PREFACE

It has been widely recognized that process control is a key element for keeping industries competitive internationally. Interdisciplinary in nature, process control applies the knowledge of control engineering and computer science to industrial processes. Future industrial processes will require managers, engineers and operators who have considerable process control knowledge and experience.

This text was developed for the process control engineering courses offered to chemical, mechanical, and electrical engineering students at the University of Alberta. It emphasizes the interdisciplinary nature and broad scope of control engineering. The first eight chapters introduce process control theory and application for an undergraduate course; the remainder are for graduate and post-graduate study.

The unique features of this book include:

- a succinct presentation of basic concepts, with a synopsis of key information in summary tables;

- integration of state-of-the-art computer-aided instruction and design technology into control engineering education;

- a complete teaching environment that includes a solution manual for hundreds of end-of-chapter problems, examination samples and laboratory simulation projects;

- introduction to the newest development—intelligent control—in process control.

An overview of control is given in chapter 1. In chapter 2 process dynamics, modeling techniques and their mathematical fundamentals are introduced. Laplace transformation and block diagrams are included. Chapter 3 presents system steady state and dynamic performance, analysis and design methods in the time domain, and discusses the *PID* controller in detail. The system stability analysis is given in chapter 4, where the Routh criterion and root locus are discussed. Chapter 5 deals with the analysis and design of control systems in the frequency domain. Nyquist plot and Bode diagram methods are presented. Chapter 6 covers discrete time control systems. Z-transform, discrete system configurations and system analysis and design methods are included.

The next two chapters present practical aspects of process control system implementation. Chapter 7 deals with problems of system design, including the selection of process variables, measurements, controllers and final control elements. Controller tuning is also discussed. Chapter 8 covers advanced control techniques including cascade control, ratio control, feedforward control, internal model control, Smith predictor and selective control. For students in electrical and mechanical engineering, these two chapters may be ignored.

Chapters 9 through 12 can be used for introductory graduate courses, or as material for further study by undergraduate students. Chapter 9 describes linear state space analysis and design techniques. State space representation, solution to state equations, controllability and observability, state feedback and the Lyapunov stability criterion are reviewed. Chapter 10 introduces optimal control theory, including the variational method, the linear regulation problem and Pontryagin's minimum principle. In chapter 11, we present the parameter identification technique, the Kalman filter and adaptive control systems.

Intelligent process control is the new generation of process control. It plays an important role in companywide integration, automation of decision-making, operation support, maintenance automation, safe production, and environmental protection. As an independent section, chapter 12 introduces basic concepts of intelligent control, including computer-integrated manufacturing, expert systems, intelligent system development procedures, neural networks, integrated distributed intelligent systems, and intelligent control applications.

This book is also suitable for self-instruction or professional reference. Included for clarity and ease of use are close to 110 summary tables, which are drawn from the authors' experience in both academia and industry.

Computer-aided instruction technology has been used here as an educational tool. A user-friendly computer software package, "Process Control Engineering Teachware" (PCET), is available on a diskette attached to this book. This carefully designed, PC-based teachware can assist students with educational problems by helping them to gain additional insight into theoretical concepts, and by providing hands-on computer simulation experience. PCET covers a wide spectrum of control engineering, including time domain analyses, stability criteria, root locus techniques, Nyquist plots, Bode diagrams, discrete time system analyses, and linear state space analyses. A case study of an industrial application—a paper machine headbox control system—is included, with which a single loop control system or a cascade control system can be analyzed.

This program is self-instructional, easy to use, and operated under a manual-free environment. *PCET* offers many laboratory simulation exercises which benefit institutions that lack laboratory control equipment. Earlier versions of *PCET* were tested at The State University of New York at Buffalo, the University of Illinois at Chicago, and Rutgers University.

This text and *PCET* have been used for several years in undergraduate and graduate courses for chemical and mechanical engineering students at the University of Alberta. The instructor can flexibly select the order in which the material from the text is presented. For example, the authors have used chapters 1 through 4 (with a brief introduction to chapter 5), and chapters 7, 8 and 12, in a one-semester senior undergraduate course.

Many individuals in universities and corporations have offered valuable assistance during the preparation of this book. We would like to acknowledge Clarence de Silva (University of British Columbia), Mengchu Zhou (New Jersey Institute of Technology), Michel Perrier (Pulp and Paper Research Institute of Canada), Mike Folley (Sunoco Inc.), Weiping Lu (Cominco Corp.), Jim Zurcher (Weyerhaeuser Pulp Ltd.), and Jules Thibault (Laval University). We would also like to thank our colleagues Lixin Peng, Xuemin Shen, Qun Wang, Qijun Xia, and Dahai Wang for their contributions and encouragement. Heon Chang Kim, Randy Dong, Munawar Saudagar, Yiqun Ying, and Hong Zhou were instrumental in the development of *PCET*. Finally, we gratefully acknowledge our graduate and undergraduate students, who gave us much important feedback.

Ming Rao

Haiming Qiu

CHAPTER 1

INTRODUCTION

1.1. COMPOSITION OF CONTROL SYSTEMS

A control system is an interconnection of components that act together to satisfy a common control objective. The famous Watt steam engine is often used to mark the beginning of the industrial revolution in Great Britain. Invented in 1769, it was one of the earliest examples of a commercial device utilizing a built-in control system. Control systems currently play an important role in manufacturing, production, and, indeed, in human life generally.

Control engineering is concerned with the analysis and design of control systems.

A device, a plant or a system under control is called a process. More practically, a process is a set of material equipment along with the physical or chemical operations that take place within it.

Examples of feedback control systems

Feedback control is the most popular type of control system configuration. To understand the concept of feedback control, let us consider two illustrative examples.

The steering of a car can be considered as a control system. When one drives a car, the direction and the speed of the car must be controlled. The eyes of the driver observe the car's direction and speed. This information is sent to the driver's brain, where decisions are made. Decisions are executed by the arms and feet of the driver in turning the steering wheel and/or adjusting the pedals. If the car deviates from its desired direction or speed, the driver's eyes sense this, and a corrective command is sent from the brain to the arms and feet, which then steer the car to reduce the error. There are two signal paths in the system: the command signals are sent from the brain to the arms and feet to control the car's direction and speed, 'measurements' of which are fed back to the brain via the eyes.

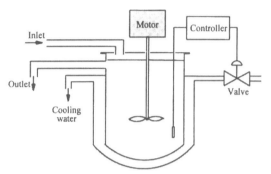

FIGURE 1.1.1. Temperature control of a continuous, stirred-tank reactor.

In contrast to the steering system example, an industrial process is usually controlled by automatic means, rather than by human operators. This is called process automation. Automation is very important for increasing productivity and ensuring high product quality. A simple reactor example is shown in Fig. 1.1.1. The temperature of the reactor is to be controlled to a prespecified value. Cooling water runs through the cooling jacket to remove heat generated by the reaction. The mixed raw material fed to the reactor may be of variable flow rate and temperature. The generated product is withdrawn at such a rate as to keep the amount of material in the reactor unchanged. The temperature of the reactor is measured by a thermocouple, and the measurement signal is transmitted to a comparator, where it is compared with the desired value. The difference between them is calculated, amplified, and used to adjust the control valve, thereby changing the flow rate of cooling water to the jacket. In this way, the temperature of the reactor may be kept at or near its target value.

Composition of systems
From the two examples above, the following hardware elements of a control system may be distinguished:
    1. *Process* (Object): This represents the material equipment along with the physical and/or chemical operations which take place. The car and its physical operation compose the object of the steering system. The heat transfer between the reactor and cooling jacket constitutes the process in the second example.
    2. *Measuring instrument* or *sensor*: Sensors are used to measure physical and/or chemical variables, and are the main sources of information about what is going on in the process. In the above examples, the driver's eyes measure the direction and speed of the car, and the thermocouple measures the temperature of the reactor.
    3. *Controller*: The controller receives information from the measuring devices and after calculation decides what action should be taken. In other

TABLE 1.1. Composition of a control system.

Components	Definition	Steering system	Temperature control system
Process (Object)	Material equipment along with physical and/or chemical operations	Car	Reactor and cooler
Measuring instrument	Device which measures physical and/or chemical variables	Eyes	Thermocouple
Controller	Control decision-maker	Brain	Comparator and amplifier
Final control element	Device which manipulates physical or chemical variable of a process	Arms, feet	Valve

words, it is a decision-maker which implements a control law.

In a steering system, the driver's brain makes decisions and is therefore a controller. The comparator and the amplifier would be included in the controller of the second example.

4. *Final control element*: This device (also called an actuator or an executive component) manipulates a process variable. The flow control valve in the engineering example is the final control element. In the steering system, the driver's arms and feet are final control elements.

From the viewpoint of control engineering, then, a system is composed of four basic parts: a process or object, measuring instrument(s), a controller and final control element(s). In practice, however, there may exist additional elements such as recording devices, measurement filters and protective gear.

The composition of a control system is summarized in Table 1.1.

General concepts and terminology

An intuitively appealing tool used to represent the configuration of a control system is the process block diagram. The block diagram is especially helpful in illustrating the relationship among the process variables. A functional block diagram of the temperature control system in the reactor example is shown in Fig. 1.1.2. Each block represents an element or combination of elements. (In Chapter 2, transfer functions describing the dynamic relationship between signals will replace the functional words in the blocks.) Each line represents a signal, and its arrowhead indicates the direction of signal flow. The circles indicate addition or subtraction of signals.

Before proceeding further, some general concepts and terminology should be introduced.

1. *Manipulated variable*. This is a variable in a control system that we can change in order to keep the system in a desired operating region. In the steering

example, the angle of the steering wheel and the displacements of the brake pedal and gas pedal are manipulated variables. The manipulated variable in the reactor example is the cooling water flow rate.

2. *Controlled variable* or *Output*. This is the variable in a control system that we will attempt to control, either by trying to keep it constant or by having it follow an assigned signal. The direction and the speed of the car, and the temperature of the reactor in the examples are all controlled variables.

3. *Input*. This refers to any variable which influences the system when it changes in value. The desired direction and speed, as well as random wind gusts, can be considered inputs in the steering system. In the reactor example, the prespecified temperature in the reactor, the flow rate and temperature of mixed raw material at the inlet, and the ambient air temperature are all input variables.

4. *Setpoint* (*Reference input*). A setpoint is an input representing the desired value of a system output. In a feedback system, the measured output is compared with the setpoint and the difference between them is used to actuate the controller. The desired direction and speed in the car steering system are both setpoints. The target value of the reactor temperature is the setpoint in the engineering example.

5. *Actuating error*. This is the difference between the setpoint and measured output, and serves as input to the feedback controller. In the car steering system, the actuating error is the difference between the desired speed (direction) and the measured car speed (direction). In the other example, it is the difference between the prespecified temperature and the measured reactor temperature.

6. *Disturbance*. Any input that cannot be manipulated is classified as a disturbance variable. For example, bad road conditions will interfere with the direction and speed of the car, and are therefore disturbances. The ambient temperature is a disturbance in the reactor example. The temperature and flow rate of the mixed raw material must also be considered as disturbance variables,

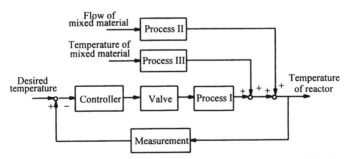

FIGURE 1.1.2.   Block diagram of the reactor temperature control system.

TABLE 1.2. Control system terminology.

Terminology	Definition	Steering system	Temperature control system
Manipulated variable	Variable being changed in order to control the system	Angle of steering wheel, displacement of pedal	Flow rate of cooling water
System output	Variable we try to control	Direction and speed of the car	Temperature of the reactor
External input	Variable influencing the system	Desired speed and direction, etc.	Prespecified reactor temperature, etc.
Setpoint	Desired value of system output	Desired speed and direction	Prespecified reactor temperature
Actuating error	Difference between setpoint and measured system output	Difference between desired and measured speeds, etc.	Difference between prespecified and measured temperatures
Disturbance	Any input which cannot be manipulated	Bad road conditions, random wind gusts, etc.	Temperature of mixed raw material, etc.

although it is entirely possible that they might be utilized as manipulated variables. This would, however, require hardware modifications to the control system configuration of Fig. 1.1.2.

The above terminology are summarized in Table 1.2.

Single loop feedback control configuration
Single loop feedback control configuration is a basic control strategy. The block diagram of a single loop control system is shown in Fig. 1.1.3. There are two paths in a single loop feedback control system: feedforward path and feedback path. The output of the system is fed back by the feedback path to compare with the reference input (setpoint), and the difference, i.e., the actuating error, is fed

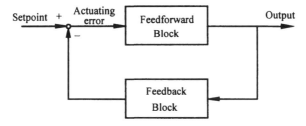

FIGURE 1.1.3. A single loop feedback control system.

to the feedforward path. By comparing reference input with output and using the difference as a means of control, the feedback control system tends to maintain a prescribed relationship between the output and the reference input of the system.

Ignoring the disturbances and combining the controller, valve and process I into a single block in Fig. 1.1.2, we can find the reactor temperature control system is a single loop feedback control system.

1.2. CONTROL SYSTEM DEVELOPMENT AND SIMULATION

For some simple processes, the control strategy and control system hardware can be selected based on process knowledge and operating experience. After a control system is installed, the control law can be modified using a trial and error procedure. In many industrial situations it becomes necessary to obtain good initial estimates from controller 'tuning' formulas. For this we require a mathematical model of the process.

Steps in control system development
The initial step in developing a control system is to clarify the control objectives, i.e., what we specifically desire from the control system. Usually, the control objectives will involve some combination of the following factors:

    i) Suppression of the effect of external disturbances,

    ii) System stability,

    iii) Optimization of steady state and/or dynamic system performance.

After the control objectives have been formulated, a mathematical model of the process should be developed so that the control system can be properly designed, and an improved understanding of the process can be obtained. One approach is to build the process model on a theoretical basis, involving, for example, physical and chemical principles such as conservation laws and rates of reaction. The alternative is an empirical approach, in which the model is constructed using experimental data.

An input-output model has the following general form:

$$output = f(input\ variables) \qquad (1.2.1)$$

This is a convenient means of expressing cause-and-effect relationships between the input and output variables of a processing system.

Another kind of mathematical model is the state variable model. The state variables are a set of variables which completely describe the state of the process at a given time. A state variable model specifies the relationships among state variables, inputs and outputs, and is of the form:

$$state\ variables = f(input\ variables) \qquad (1.2.2)$$

$$output\ variables = g(state\ variables) \tag{1.2.3}$$

The next step in developing a control system is to design a control law that helps us to achieve the control objectives, given the system constraints. Indeed, the main objective of this textbook is to present a complete exposition of this stage of the design process. It is of course necessary to obtain a measurement of the process output, or of an auxiliary variable from which a value of the output may be inferred. We also need to select manipulated variables which can be used to control the process. After the measured and manipulated variables have been selected, a control configuration may be determined and a control law developed based on automatic control theory.

Simulation is needed after a tentative control law has been developed. Computer simulation is often the only feasible or safe technique by which to analyze a system and evaluate the controller design. It is certainly much cheaper and easier to change the conditions or parameters of a simulated process than to make modifications to the real plant. Computer simulation is currently a widely used design tool, because it allows the engineer to examine the dynamic behavior of the system under a variety of proposed control laws.

It should always be remembered that the integrity of the simulation is critically dependent upon the validity of the model. Simulation results obtained using grossly inadequate process models are meaningless.

Analog computers were the first means of simulating the dynamics of control systems. They played an important historical role in control system analysis and design, and are still used today to a limited extent. However, most simulation work is presently carried out with the help of digital computers, which do not suffer from the hardware problems associated with their analog precursors. Digital computer simulation allows engineers to anticipate the behavior of a process not only qualitatively, but also quantitatively, and to deal with more sophisticated control systems. If simulation results indicate that the primary control objectives cannot be met, even after adjustment of the controller parameters, then redesign of the control law and/or the control loop configuration must be considered.

Following the simulation stage is the selection of control hardware, e.g., transducers (sensors and transmitters), final control elements and controllers. It is standard design practice to neglect or otherwise idealize the dynamics of various hardware components of the control system. However, if this assumption leads to erroneous results, it may prove necessary to model these devices more rigorously in the computer simulation and redesign the controller.

The final step in developing a control system is installation and adjustment. On-line controller adjustment, or tuning, is usually necessary to account for model inaccuracies, noisy measurement signals, etc. Such fine-tuning of the controller design is generally conducted in a trial and error fashion.

TABLE 1.3. Stages of control system development.

Step	Engineering activity	Explanation
1	Formulation of control objective	Obtain process information. Identify constraints and primary control objectives.
2	Development of process model	Formulate mathematical model based on first principles and/or empirical basis.
3	Control law design	Using automatic control theory, design control law which fulfills control objectives while satisfying process constraints.
4	Computer simulation	Test performance of proposed control law design using analog or digital simulation. If simulation results don't meet control objectives, return to Step 3.
5	Selection of control hardware	If models of control hardware elements do not reflect vendor hardware, go back to Step 2. Step 5 is sometimes executed before Step 3.
6	Installation and adjustment	Install the system and adjust the parameters, a necessary step since some model mismatch is inevitable. A trial and error tuning procedure is often used.

The major steps involved in designing and installing a control system are summarized in Table 1.3.

Computer-aided instruction for control engineering education
Computer simulation is very important for the analysis and design of control systems. It is also a convenient, low-cost tool for process control education. Simulation studies help engineering students gain clear insight into theoretical concepts, solve educational problems, obtain practical experience and, hopefully, reduce stress through increased efficiency.

A copy of a computer-aided instruction package called *PCET* (Process Control Engineering Teachware) is included with this book and can be used for analyzing and designing control systems, as well as enhancing one's understanding of fundamental concepts. *PCET* is a menu-driven program, and provides a manual-free operating environment. The user needs only to follow on-line instructions and select items displayed on the screen. Its calculation and graphic capabilities allow students to obtain results in numerical or graphical form, either on-screen or through a hard copy print-out.

The functions of *PCET* are listed in Table 1.4.

TABLE 1.4. *PCET* software options.

Acronym	Explanation	Functions	Related chapters
TDA	Time domain analysis	Open loop system response Closed loop system response PID control algorithm Controller tuning	2, 3, 7, 8
RSC	Routh stability criterion	Routh array display Pole calculation	4
RLT	Root locus technique	Root locus plots Stability analysis	4, 7, 8
FDA	Frequency domain analysis	Nyquist plot Bode diagram Controller tuning Stability analysis	5, 7, 8
DSA	Discrete time system analysis	Open loop system response Closed loop system response PID control algorithm Controller tuning	6
LSA	Linear state space analysis	Controllability analysis Observability analysis Eigenstructure analysis System matrix diagnostics	9
IAC	Industrial application case	Single loop system design Cascade system design	8

1.3. CLASSIFICATION OF CONTROL SYSTEMS

Control systems may be classified in a number of different ways. Some common classifications are described below.

Linear systems and nonlinear systems
If a system can be described with a linear differential equation, or with a linear difference equation, then it is said to be a linear system. Otherwise, it is called a nonlinear system. The outputs of a linear system obey the superposition principle. In reality, perfectly linear systems do not exist; most so-called linear systems possess linear characteristics only within a limited range of operation.

Because analytical solutions are available for linear difference/differential equations, many significant results have been achieved in the area of linear systems theory. Linearization is often used when modeling the behavior of nonlinear systems in the neighborhood of a specified operating point. If the process is 'mildly' nonlinear, then this local model may be used to develop a control law based on linear control theory.

Time varying systems and time invariant systems

A time varying system is defined as a system in which one or more parameters vary with time. For instance, the design of a missile guidance system requires a time varying process model, because the mass of the missile decreases as fuel is consumed during the flight. Most chemical processes are also time varying. Changes in system parameters commonly result from the decay of catalyst activity in a reactor, or the decrease due to fouling of the overall heat transfer coefficient in a heat exchanger.

By contrast, the parameters of a time invariant system remain unchanged during system operation. The analysis and design of a time invariant system are much easier than those of a time varying one. Many analysis and design methods for time varying systems are extensions of methods developed for time invariant systems. For example, an engineer may design the control law assuming the process is time invariant, and allow the controller to update its parameters on-line according to a preprogrammed schedule of operation.

Open loop systems and closed loop systems

A closed loop system is a system operating under feedback control. As described earlier, the feedback signal is compared with a desired value, and the difference between them actuates the controller. The path from the comparator to the output is called the forward path, and the path from the output to the comparator is called the feedback path. An open loop system is one in which the manipulated variable is not being adjusted to keep the output at setpoint.

An example of a closed loop system was given previously in Fig. 1.1.1 of

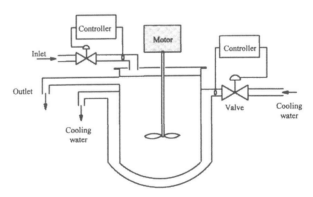

FIGURE 1.3.1. Reconfigured reactor control system.

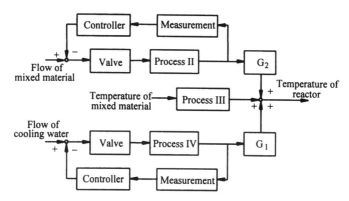

FIGURE 1.3.2. Block diagram of modified reactor control system.

Section 1.1. The configuration of this control system can of course be altered if desired. The variation of temperature in the reactor is mainly caused by variation of the flow rate of mixed raw material in the inlet and by variations in the cooling water flow rate. If these two flow rates could be kept constant, then the reactor temperature would hold constant as well (neglecting the effect of feed temperature   fluctuations). With this understanding, a new control configuration is obtained as shown in Fig. 1.3.1, with corresponding block diagram Fig. 1.3.2. There are two feedback loops operating in the system, but neither includes the output of the system. This implies that an open loop strategy has been adopted in the new configuration.

Continuous time systems and discrete time systems
A continuous time control system, or equivalently, a continuous data control system, is one in which all signals are continuous functions of time. Most conventional control theory was developed originally for continuous time systems. It is primarily for this reason that we restrict our analyses to continuous time systems in this text.

Discrete time control systems, or discrete data control systems, possess one or more signals that are sampled signals. Such systems are also called sampled data systems. In general, a discrete time system receives data or information only intermittently, at specific instants of time called sampling instants. Any system that includes digital codes generated by a digital computer or digital controller is referred to as a digital control system.

Servo systems and regulating systems
In the study of feedback control systems, we may distinguish between two types of problems: servo and regulatory. In a servo problem, the controller is designed

to keep the system output as close as possible to a time varying setpoint signal. It can also be considered a "tracking" problem. The regulator problem results when the setpoint is assumed to remain constant. The control objective is then to keep the system output steady at setpoint despite variations in the disturbance variables.

SISO systems and MIMO systems

A single input single output (*SISO*) system is one in which a single manipulated input and a single controlled variable are considered in the analysis and design of the control strategy. In the example shown in Fig. 1.1.1, the temperature of the reactor is the only output, and the cooling water flow rate is the only manipulated variable. Because the temperature and flow rate of the mixed raw material are considered to be disturbances, the system is *SISO*.

As its name suggests, a multi-input multi-output (*MIMO*) system possesses more than one manipulated variable and more than one output. The car steering example we discussed in Section 1.1 is a *MIMO* system. The direction and speed of the car are controlled variables, hence there are two outputs. The steering wheel angle, and the displacements of the brake and gas pedal are manipulated inputs.

Another *MIMO* system is shown in Fig. 1.3.3. The control objectives are to keep the water in the tank at a prescribed level and temperature. The water level and temperature are the two outputs of the system; the temperatures and flow rates of the two inlet streams are the manipulated variables. This is an example of a MIMO process with four inputs and two outputs.

Deterministic systems and stochastic systems

A deterministic system can be described by deterministic variable equations. Each variable in such a system has a definite value at a given time if the initial conditions and inputs of the system are completely specified.

Since most disturbances acting on a real process are unmeasurable, their combined effect on the system output is often modeled as a random process. A system in which some or all of the variables are described by random processes

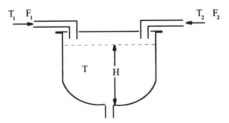

FIGURE 1.3.3. A four-input two-output *MIMO* system.

is called a stochastic system. Modern control algorithms such as minimum variance control and linear quadratic Gaussian control are based on a stochastic system model.

Lumped parameter systems and distributed parameter systems
If time is the only independent variable of a system, i.e., if every dependent variable in the system can be assumed to be a function only of time, then it is termed a lumped parameter system. For example, in the reactor control system discussed earlier, we assumed for simplicity that spatial variations of temperature in the reactor could be neglected. Lumped parameter models are commonly used to describe the dynamic behavior of chemical processes, and form the basis of the approach adopted in this textbook.

A distributed parameter system is one in which both time and spatial independent variables exist. Such process models will contain one or more partial differential equations. The distributed parameter approach is appropriate for certain chemical engineering applications, such as shell-and-tube heat exchangers, packed columns, and long pipelines carrying compressible gases.

Continuous process systems and batch process systems
Batch processes are commonly encountered in process industry in the form of processes with discontinuous feed and product stream flows. Intermittent, rather than continuous, modes of operation are primarily involved. A typical cycle consists of start-up, operation, shutdown, cleanup, and change-over phases. Conventional control theory for continuous processes may not be applicable to batch systems without some modification. A batch switch is often used to activate/deactivate controllers required for different stages in a cycle.

Many other frames of reference exist by which we might categorize process control systems. For example, a control scheme can be classified as adaptive

TABLE 1.5. Classification of control systems.

Viewpoint	Classification
Nature of system differential equation	Linear vs. nonlinear
Number of inputs and outputs	SISO vs. MIMO
Variability of reference input	Servo vs. regulatory
Variability of system parameters	Time invariant vs. time varying
Control loop configuration	Open loop vs. closed loop
Nature of signals	Continuous vs. discrete time
Nature of disturbance modeling	Deterministic vs. stochastic
Nature of parameters	Lumped vs. distributed
Mode of operation	Continuous vs. batch

versus nonadaptive, optimal versus suboptimal, modulated versus unmodulated, and so on. Mixed terminology can also be used; e.g., we may describe a system as being time invariant linear MIMO, or as a discrete nonlinear stochastic control system, etc. A number of common control system classifications are listed in Table 1.5.

Artificial intelligence techniques have been incorporated into control systems. This development marked a new generation of process automation. Intelligent control systems will be discussed in Chapter 12, where computer integrated manufacturing systems, computer integrated process systems, real time discrete event systems and so on are included.

Problems

1.1. Fig. 1.1.2 is a functional block diagram for an engineering control example. Construct a similar block diagram for the car steering system discussed in Section 1.1.

1.2. Describe the water level control system in a home toilet tank as shown in Fig. 1.P.1, and draw its functional block diagram.

1.3. A missile is controlled in the following process: The desired flying direction information is stored in a computer. The desired direction signal as a reference input is sent to an autopilot. The signal from the autopilot drives the rudder that makes the missile direction change. A gyroscope in the missile detects the missile direction and sends the direction signal back for comparison with the desired direction signal. The difference between the desired direction and the detected direction is sent to the autopilot. Draw a functional block diagram for this missile flying system. The missile direction is designated as the system output.

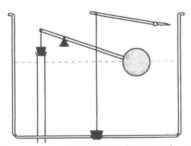

FIGURE 1.P.1.   A water level control system in a toilet tank.

CHAPTER 2

MATHEMATICAL MODELS
AND TRANSFER FUNCTIONS

2.1. MATHEMATICAL MODELS

A process can be described using a number of different mathematical modeling techniques, involving differential equations, transfer functions, state-space models, etc. A mathematical model may be derived theoretically or obtained empirically. Alternatively, a semi-empirical model could be obtained from a theoretical model with one or more parameters evaluated from empirical data.

This section is devoted to theoretical differential equation modeling, based upon fundamental physical and/or chemical principles. We shall restrict our discussion to ordinary differential equation models, although partial differential equations are sometimes needed to describe a process precisely.

Electrical systems

Differential equations modeling electrical systems often comprise the following three components:

$$\text{Resistance} \qquad u = iR$$

$$\text{Capacitance} \qquad i = C\frac{du}{dt} \quad \text{or} \quad u = \frac{1}{C}\int idt$$

$$\text{Inductance} \qquad u = L\frac{di}{dt}$$

where R, C and L represent resistance, capacitance and inductance, respectively. u denotes the voltage drop across a particular element, and i the current passing through it.

Kirchhoff's voltage law and current law are the most important principles for the analysis of electrical networks. The voltage law states that the sum of all voltage drops around a closed loop is zero. The current law points out that the sum of currents into a node must equal the sum of currents out of the node.

Example 2.1.1

An electrical network is shown in Fig. 2.1.1. We need to set up a differential equation model relating the output u_2 to the input u_1.

The following differential equations can be obtained by application of Kirchhoff's laws:

$$i_3 = i_1 - i_2$$

$$u_1 = i_1 R_1 + \frac{1}{C_1} \int i_3 dt$$

$$\frac{1}{C_1} \int i_3 dt = i_2 R_2 + \frac{1}{C_2} \int i_2 dt$$

$$u_2 = \frac{1}{C_2} \int i_2 dt$$

Eliminating the intermediate variables i_1, i_2 and i_3 yields

$$R_1 C_1 R_2 C_2 \frac{d^2 u_2}{dt^2} + (R_1 C_1 + R_2 C_2 + R_1 C_2) \frac{d u_2}{dt} + u_2 = u_1$$

Translational mechanical systems

The motion equations of mechanical elements are always directly or indirectly related to Newton's laws of motion, e.g.,

$$F = ma = m \frac{dv}{dt} \qquad (2.1.1)$$

where F is the vector sum of forces acting on a moving rigid body, m is the mass of the body, a its acceleration in the direction of F, and v the velocity of the body.

A differential equation describing the motion of a spring is

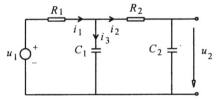

FIGURE 2.1.1. An electrical network.

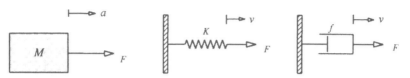

a. Force-mass system   b. Force-spring system   c. Viscous friction syste

FIGURE 2.1.2. Some basic mechanical systems.

$$v = \frac{1}{K}\frac{dF}{dt} \qquad (2.1.2)$$

where F is the force acting on or exerted by the spring, K is the Hooke's law constant, and v is the spring velocity.

Viscous friction represents a retarding force for which a linear relationship exists between the applied force and the resultant velocity:

$$F = fv \qquad (2.1.3)$$

where F is the retarding force, v is the velocity of the body on which the force acts and f is a frictional constant.

The force-mass system, force-spring system and viscous friction element are illustrated in Fig. 2.1.2.

Example 2.1.2

A mechanical system is depicted in Fig. 2.1.3. The forces acting on the mass M are spring force Ky in the direction of $-y$, a viscous force as expressed in (2.1.3) and an external force r. The output Y represents displacement from rest position. Using (2.1.1), the following differential equation is easily obtained:

$$M\frac{d^2y}{dt^2} + f\frac{dy}{dt} + Ky = r \qquad (2.1.4)$$

Rotational mechanical systems

The rotational motion of a body can be defined as its motion about a fixed axis.

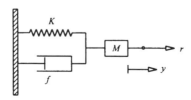

FIGURE 2.1 3. A composite mechanical system.

A rotational mechanical system is analogous to a translational one with the replacement of mass, force, velocity and distance by moment of inertia, torque, angular velocity and angle, respectively.

The torque equation corresponding to (2.1.1) is

$$T = J\frac{d\omega}{dt} \tag{2.1.5}$$

where T is the torque acting on a body that spins around a fixed axis. J is its moment of inertia about the axis, and ω is its angular velocity.

The moment of inertia is a measurement of inertia of a rotating body, which can be calculated using the following formula:

$$J = \sum m_i r_i^2 \tag{2.1.6}$$

where m_i represents the mass of a small element of a rigid body, and r_i the distance between this element and the fixed axis around which the body rotates.

With (2.1.6) the moment of inertia of a circular disk about its geometric axis is given by (see Problem 2.1)

$$J = \frac{1}{2} MR^2 \tag{2.1.7}$$

where M is the mass of the disk, and R is its radius.

Example 2.1.3

A simple pendulum composed of a rod of length l and mass M fixed at one end is shown in Fig. 2.1.4. The moment of inertia of the pendulum can be calculated as

$$J = \int_0^l \rho r^2 dr = \frac{1}{3}\rho l^3 = \frac{1}{3} M l^2$$

where ρ is the mass per unit length of the rod (i.e., $M = \rho l$). The torque exerted on the rod by its own weight is

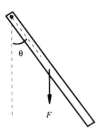

FIGURE 2.1.4. A simple pendulum.

$$T = \frac{1}{2} Mgl \sin\theta$$

with g denoting gravitational acceleration, and θ the vertical displacement angle. The system differential equation is then

$$\frac{1}{2} Mgl \sin\theta = \frac{1}{3} Ml^2 \frac{d^2\theta}{dt^2}$$

If θ is small, the approximation $\sin\theta \approx \theta$ becomes valid, leading to

$$\frac{d^2\theta}{dt^2} - \frac{3g}{2l}\theta = 0$$

Chemical processes

For chemical processes such as stirred tank heaters, batch reactors, distillation columns, heat exchangers, etc., the principles of conservation of mass, energy and momentum are used to obtain system differential equations. The conservation principle of a quantity states that

$$\begin{pmatrix} rate\ of \\ quantity \\ accumulation \end{pmatrix} = \begin{pmatrix} rate\ of \\ quantity \\ in \end{pmatrix} - \begin{pmatrix} rate\ of \\ quantity \\ out \end{pmatrix} + \begin{pmatrix} rate\ of \\ quantity \\ generated \end{pmatrix} \qquad (2.1.8)$$

where the quantity may represent mass, energy or momentum.

Mass balance equations can be written as

$$\frac{d(\rho V)}{dt} = \sum_{i:inlet} \rho_i F_i - \sum_{j:outlet} \rho_j F_j \qquad (2.1.9)$$

where

ρ = density of the material in the system
V = total volume of the system
ρ_i = density of the material in ith inlet stream
ρ_j = density of the material in jth outlet stream
F_i = volumetric flow rate of the ith inlet stream
F_j = volumetric flow rate of the jth outlet stream

Energy balance equations typically assume the form

$$\frac{dE}{dt} = \sum_{i:inlet} \rho_i F_i h_i - \sum_{j:outlet} \rho_j F_j h_j + Q \qquad (2.1.10)$$

where ρ_i, ρ_j, F_i, F_j are as defined above and
E = total energy of the system

h_i = specific enthalpy of the material in ith inlet stream
h_j = specific enthalpy of the material in jth outlet stream
Q = rate of heat exchange between the system and its surroundings per unit of time

Example 2.1.4

Consider a stirred tank heater as shown in Fig. 2.1.5. The liquid level and temperature in the tank are to be controlled, and thus represent the two outputs of the system. The total mass M in the tank is

$$M = \rho V = \rho A h$$

where ρ, V, and h represent the liquid density, volume and level, and A is the cross-sectional area of the tank. The mass balance equation reduces to

$$\frac{d(\rho A h)}{dt} = \rho F_i - \rho F \qquad (2.1.11)$$

where F_i and F are the volumetric flow rates of the inlet and outlet streams, respectively. If liquid density can be assumed to remain constant, it follows that

$$A \frac{dh}{dt} = F_i - F \qquad (2.1.12)$$

The total energy of the system is given by

$$E = \rho A h C_p (T - T_{ref}) \qquad (2.1.13)$$

with ρ, A, h as stated above. C_p is defined as the heat capacity of the liquid in the tank, and is assumed to be independent of liquid temperature T. T_{ref} is a reference temperature corresponding to a zero datum level for liquid enthalpy, and is henceforth taken to be zero. The energy balance equation (2.1.10) becomes

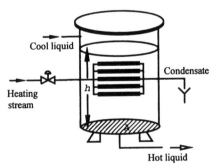

FIGURE 2.1.5.   A stirred tank heater.

$$\frac{d(\rho A h C_p T)}{dt} = \rho F_i C_p T_i - \rho F C_p T + Q \tag{2.1.14}$$

where T_i is the temperature of the liquid in the inlet stream, and Q is the amount of heat supplied to the tank by the heating stream per unit of time. In (2.1.14), C_p is assumed to be a constant. Substitution of (2.1.12) into (2.1.14) leads to

$$Ah \frac{dT}{dt} = F_i (T_i - T) + \frac{Q}{\rho C_P} \tag{2.1.15}$$

Eqs. (2.1.12) and (2.1.15) constitute the system differential equations.

Pure time delay

For some systems, it is observed that when the input variable changes, there exists a time interval during which no effect is observed in the system output. This 'waiting period' is referred to as a time delay or dead time, and is frequently encountered in process control problems. A time delay will occur whenever material or energy is physically transported from one part of the plant to another.

This phenomenon is illustrated in Fig. 2.1.6, where fluid is traveling through a pipe in plug flow fashion. The transportation time between point A and point B is given by

$$\tau = L/v \tag{2.1.16}$$

where L is the distance between A and B, and v is the fluid velocity. From this equation it is apparent why a time delay is often referred to as a distance velocity lag, or as a transportation lag. If the temperature (or other physical variable) of the fluid at point A is denoted as $r(t)$, then the temperature at point B, denoted as $y(t)$, will be given by

$$y(t) = r(t - \tau) \tag{2.1.17}$$

This relationship is illustrated in Fig. 2.1.7.

Besides the physical movement of liquid or solid material, there are several other sources of time delay in engineering applications. For example, some

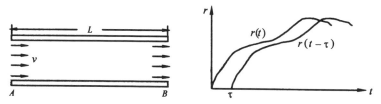

FIGURE 2.1.6. Plug flow transportation. FIGURE 2.1.7. Effect of time delay.

TABLE 2.1. Elements of system differential equations.

Systems	Fundamental formulas	Elements
Electrical system	$u = iR$ for resistance $i = C(du/dt)$ for capacitance $u = L(di/dt)$ for inductance	R - resistance C - capacitance L - inductance u - voltage i - current
Translational mechanical system	$F = ma = m(dv/dt)$ for motion $v = (1/K)(dF/dt)$ for spring $F = fv$ for viscous fiction	m - mass a - acceleration F - force v - velocity K - spring constant f - frictional constant
Rotational mechanical system	$T = J\dfrac{d\omega}{dt}$	T - torque J - moment of inertia ω - angular velocity
Tank system	$A\dfrac{dh}{dt} = F_i - F$ $Ah\dfrac{dT}{dt} = F_i(T_i - T) + \dfrac{Q}{\rho C_P}$	A - area, h - level F_i - inlet flow rate F - outlet flow rate T - temperature in tank T_i - inlet temperature Q - supplied heat per unit of time ρ - density C_P - heat capacity
Pure delay system	$y(t) = r(t - \tau)$	y - output r - input τ - time delay

measurement instruments such as gas chromatographs, which are used to obtain on-line concentration measurements, require a fixed amount of time (e.g., 3~5 minutes) to process a sample.

Systems with time delay are inherently difficult to control, because the outputs do not describe the present state of the process. Clearly, every effort should be made at the process design stage to minimize sources of time delay in the plant.

A summary of terms commonly appearing in differential equation models can be found in Table 2.1.

Deviation variables
The steady state value of a system variable refers to its limiting value as time approaches infinity. The difference between a variable and its steady state value

is called a deviation or perturbation variable, which is a very useful concept for process control purposes. When the control objective is regulatory, we seek to maintain process variables such as temperature, pressure, flow rate, etc., at desired values which specify the system operating point. Deviation variables may be introduced to describe the deviation of the process variables from the operating point.

Consider a linear system described by

$$y = kx + b \qquad (2.1.18)$$

with operating point (x_0, y_0). That is, at steady state $x = x_0$, and $y = y_0$, where

$$y_0 = kx_0 + b \qquad (2.1.19)$$

Define the deviation variables

$$\Delta x = x - x_0 \qquad (2.1.20)$$

$$\Delta y = y - y_0 \qquad (2.1.21)$$

Subtracting (2.1.19) from (2.1.18) yields

$$y - y_0 = k(x - x_0) \qquad (2.1.22)$$

or

$$\Delta y = k\Delta x \qquad (2.1.23)$$

This is the deviation variable equivalent of Eq. (2.1.18).

Linearization of nonlinear models
Mathematically, a linear differential or algebraic equation is one for which the following superposition properties hold:

1. If $y = f(x)$, then $cy = f(cx)$.

2. $y = f(x_1 + x_2) = f(x_1) + f(x_2)$

Many mechanical and electrical elements can be assumed to be linear over a reasonably wide range of operation [Pease, 1987]. Most chemical processes, on the other hand, behave in a nonlinear fashion. In order to apply linear control theory to these systems, we will require some means of approximating their true dynamic behavior with a set of linear differential equations.

Consider a nonlinear differential equation

$$\frac{dx}{dt} = f(x) \qquad (2.1.24)$$

Expanding the nonlinear function $f(x)$ into a Taylor series about the operating point $(x_0, f(x_0))$,

$$f(x) = f(x_0) + \left(\frac{df}{dx}\right)_{x_0} \frac{x - x_0}{1!} + \left(\frac{d^2f}{dx^2}\right)_{x_0} \frac{(x - x_0)^2}{2!} + \cdots \qquad (2.1.25)$$

If $(x - x_0)$ is small, i.e., $(x - x_0) \ll 1$, then all terms of order two and higher can be neglected. It follows that

$$f(x) \approx f(x_0) + \left(\frac{df}{dx}\right)_{x_0} (x - x_0) \qquad (2.1.26)$$

which is an approximation to (2.1.25), and is linear in x. The initial differential equation (2.1.24) becomes

$$\frac{dx}{dt} = f(x_0) + \left(\frac{df}{dx}\right)_{x_0} (x - x_0) \qquad (2.1.27)$$

The above linearization technique is depicted graphically in Fig. 2.1.8. The behavior of the curves $f_1(x)$ and $f_2(x)$ at the point $(x_0, f(x_0))$ is well approximated by (2.1.26) when the deviation $x - x_0$ is small.

In fact, we have already used the linearization technique in Example 2.1.3, where $\sin\theta \approx \theta$ was assumed for small θ. This approximation can be obtained by setting $x_0 = \theta_0 = 0°$ and $(df/dx)_{x_0} = \cos 0° = 1$ in (2.1.26).

When dealing with a multivariable system, the linearization procedure is conducted in a similar manner. For example, the system

$$\frac{dx_1}{dt} = f_1(x_1, x_2) \qquad (2.1.28)$$

$$\frac{dx_2}{dt} = f_2(x_1, x_2) \qquad (2.1.29)$$

has a linear approximation at the point (x_{10}, x_{20}):

$$\frac{dx_1}{dt} = f_1(x_{10}, x_{20}) + \left(\frac{\partial f_1}{\partial x_1}\right)_{(x_{10}, x_{20})} (x_1 - x_{10}) + \left(\frac{\partial f_1}{\partial x_2}\right)_{(x_{10}, x_{20})} (x_2 - x_{20}) \quad (2.1.30)$$

$$\frac{dx_2}{dt} = f_2(x_{10}, x_{20}) + \left(\frac{\partial f_2}{\partial x_1}\right)_{(x_{10}, x_{20})} (x_1 - x_{10}) + \left(\frac{\partial f_2}{\partial x_2}\right)_{(x_{10}, x_{20})} (x_2 - x_{20}) \quad (2.1.31)$$

Remarks

1. The linearization is computed around an operating point $(x_0, f(x_0))$. Choice of different operating points will lead to different approximation results.

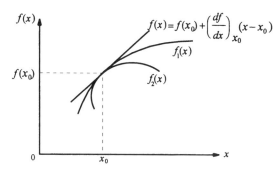

FIGURE 2.1.8. Linearization via Taylor series approximation.

In practice, the operating point is usually defined in terms of nominal or steady state values of the process variables.

2. The validity of the linear approximation will be restricted to a narrow range of $x - x_0$. However, its accuracy depends not only on $x - x_0$, but also upon the characteristics of the function being approximated. For instance, the nonlinear functions $f_1(x)$ and $f_2(x)$ in Fig. 2.1.8 have the same linear approximation at the point $(x_0, f(x_0))$, but are not equally well represented by $f(x)$ when $x \neq x_0$.

3. Some essentially nonlinear models cannot be linearized. For example, the relay function

$$f(x) = \begin{cases} 1 & x \geq 0 \\ -1 & x < 0 \end{cases} \tag{2.1.32}$$

cannot be linearized at the point $x = 0$.

4. A process may be described by several nonlinear differential equations, which can be difficult to combine in a single equation. In this case, linearize the equations individually first, making easier the task of combining them if needed.

Example 2.1.5

Returning to the stirred tank heater of Example 2.1.4, we have the total mass balance differential equation

$$A \frac{dh}{dt} = F_i - F \tag{2.1.33}$$

If the exit flow rate F follows the formula

$$F = \beta \sqrt{h} \tag{2.1.34}$$

then a linear approximation to F at the nominal point (F_{i0}, h_0, F_0) is

$$F = \beta\sqrt{h_0} + \frac{\beta}{2\sqrt{h_0}}(h - h_0) \tag{2.1.35}$$

or

$$\Delta F = \frac{\beta}{2\sqrt{h_0}}\Delta h = \frac{\Delta h}{R} \tag{2.1.36}$$

where $R = 2\sqrt{h_0}/\beta$ is referred to as a flow resistance. Substituting (2.1.36) into the deviation variable form of (2.1.33), we have

$$A\frac{d\Delta h}{dt} + \frac{1}{R}\Delta h = \Delta F_i \tag{2.1.37}$$

The procedure involved in constructing a linear input-output model is summarized in Table 2.2.

Degrees of freedom
We might ask ourselves whether each of the differential equations derived in the previous examples has a unique solution. The answer to this question hinges on whether the number of variables involved is equal to the number of independent equations. We define the degrees of freedom of a set of equations, N_f, as

$$N_f = N_v - N_e \tag{2.1.38}$$

where N_v is the total number of variables, including outputs and unspecified inputs, and N_e is the number of independent equations (differential and/or algebraic). If $N_f = 0$, the set of equations has a unique solution, and the process is exactly specified. If $N_f > 0$, i.e., the number of process variables is greater than the number of independent equations, then an infinite number of solutions exists, and the process is underspecified. If there are fewer process variables than independent equations, i.e., if $N_f < 0$, then no solutions exist, and the

TABLE 2.2. Development of a linear input-output model.

Step	Content
1	Designate input and output variables consistent with control objectives and process analysis.
2	Find physical and/or chemical relationship between inputs, intermediate variables, and outputs.
3	Delete intermediate variables to obtain differential equations involving inputs and outputs only.
4	If there are nonlinear equations, linearize them.
5	If necessary, determine the deviation variable equation.

TABLE 2.3. Degrees of freedom analysis.

Degree of freedom	$N_f = 0$	$N_f > 0$	$N_f < 0$
Nature of process	Exactly specified	Underspecified	Overspecified
Number of solutions	One	Infinity	Zero

process is overspecified. These three cases are summarized in Table 2.3.

In Example 2.1.4, we had two differential equations ((2.1.12) and (2.1.15)) in six variables: the height h of liquid, the heat supply rate Q, the temperatures T and T_i in the tank and inlet, and the volumetric flow rates F_i and F in the inlet and outlet streams. The degrees of freedom $N_f = 4$, and thus an infinite number of solutions exist. If we specify the values of four variables, e.g., F_i, F, T_i and Q, then h and T can be uniquely determined, as N_f would be zero. Furthermore, if two feedback loops were introduced so that

$$F_i = f_1(h, T) \tag{2.1.39}$$

$$F = f_2(h, T) \tag{2.1.40}$$

i.e., the volumetric flow rates of the inlet and outlet depend upon the variables h and T, then $N_e = 4$ while $N_v = 6$, and only two process variables T_i and Q would remain to be specified.

Modeling is an important and often critical step in the synthesis of a control system. Experience has proved that control engineers cannot restrict their activities to the derivation of sophisticated control algorithms based on purely empirical plant models. The design of a practical control system demands that all available process knowledge be utilized in the development of the process model.

As space permits only a cursory description of the modeling process, we refer the reader to the literature for further details [see, for example, Kuo 1991, Luyben 1990, Seborg et al. 1989, and Stephanopoulos 1984].

2.2. LAPLACE TRANSFORMS AND TRANSFER FUNCTIONS

The Laplace transform is a very powerful tool for analysis of linear systems, since it can transform a time domain function to the complex s-domain, where linear differential equations become algebraic.

The Laplace transformation of a time-dependent function $f(t)$ is defined as

$$F(s) = \mathcal{L}[f(t)] = \int_0^\infty f(t)e^{-st}dt \tag{2.2.1}$$

(It should perhaps be mentioned here that not all time domain functions are Laplace-transformable, but all physically realizable signals are. For this reason, we will ignore any questions of existence which may arise in the following development.) The inverse Laplace transform is defined as

$$f(t) = \mathcal{L}^{-1}[F(s)] = \frac{1}{2\pi j}\int_{\sigma-j\infty}^{\sigma+j\infty} F(s)e^{st}ds \qquad (2.2.2)$$

where σ is a real constant that is greater than the real part of all singularities of $F(s)$.

Equation (2.2.2) converts $F(s)$, the Laplace transformation of a time function $f(t)$, back to the original function $f(t)$. This complex variable inversion formula is somewhat unwieldy, and so in practice we make use of tables of Laplace transform pairs such as that provided in Appendix A.

Basic Laplace transform functions
The Laplace transforms of some common time functions are given in Table 2.4. Proofs for selected entries are developed in the following examples.

Example 2.2.1
Consider a step function $R(t) = Ku(t)$, where $u(t)$ is the unit step function defined as

$$u(t) = \begin{cases} 1 & t \geq 0 \\ 0 & t < 0 \end{cases} \qquad (2.2.3)$$

The Laplace transformation of $R(t)$ is K/s.
Proof:

TABLE 2.4. Basic Laplace transform pairs.

$f(t)$	$F(s)$	$f(t)$	$F(s)$
Step function $u(t)$	$1/s$	t^n	$n!/s^{n+1}$
Impulse function $\delta(t)$	1	e^{-at}	$\dfrac{1}{s+a}$
$\sin\omega t$	$\dfrac{\omega}{s^2+\omega^2}$	$\cos\omega t$	$\dfrac{s}{s^2+\omega^2}$
$e^{-at}f(t)$	$F(s+a)$	$f'(t)$	$sF(s)-f(0)$
$f^{(n)}(t)$	$s^nF(s)-s^{n-1}f(0)-\cdots$ $-sf^{(n-2)}(0)-f^{(n-1)}(0)$	$\int_0^t f(t)dt$	$F(s)/s$

$$\mathcal{L}[R(t)] = \int_0^\infty R(t)e^{-st}dt = K\int_0^\infty e^{-st}dt = \left[-\frac{K}{s}e^{-st}\right]_0^\infty = \frac{K}{s}$$

Example 2.2.2

The Laplace transformation of a sinusoidal function $f(t) = A\sin\omega t$ is

$$\mathcal{L}[A\sin\omega t] = \frac{A\omega}{s^2 + \omega^2} \tag{2.2.4}$$

Proof:

$$\mathcal{L}[A\sin\omega t] = \int_0^\infty (A\sin\omega t)e^{-st}dt = A\int_0^\infty \frac{e^{j\omega t} - e^{-j\omega t}}{2j}e^{-st}dt$$

$$= \frac{A}{2j}\int_0^\infty \left[e^{-(s-j\omega)t} - e^{-(s+j\omega)t}\right]dt$$

$$= \frac{A}{2j}\left[-\frac{e^{-(s-j\omega)t}}{s-j\omega} + \frac{e^{-(s+j\omega)t}}{s+j\omega}\right]_0^\infty$$

$$= \frac{A}{2j}\left[-\frac{1}{s-j\omega} - \frac{1}{s+j\omega}\right]$$

$$= \frac{A\omega}{s^2 + \omega^2}$$

Example 2.2.3

The unit pulse function is defined as

$$f(t) = \begin{cases} 0 & t < 0 \\ 1/h & 0 \le t < h \\ 0 & t \ge h \end{cases} \tag{2.2.5}$$

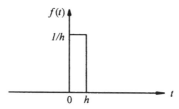

FIGURE 2.2 1. Unit pulse function.

and is graphed in Fig. 2.2.1. Its Laplace transform is $(1 - e^{-hs})/hs$.
Proof:

$$\mathcal{L}[f(t)] = \int_0^\infty f(t)e^{-st}dt = \int_0^h (1/h)e^{-st}dt = \left[\frac{-1}{hs}e^{-st}\right]_0^h = \frac{1}{hs}(1-e^{-hs})$$

Example 2.2.4

A unit impulse function is obtained as a limiting case of the unit pulse function when h approaches zero, i.e.,

$$\delta(t) = \lim_{h \to 0} f(t) \tag{2.2.6}$$

where $f(t)$ denotes the function (2.2.5). The value of $\delta(t)$ at $t = 0$ is infinity, but the integral

$$\int_0^\varepsilon \delta(t)dt = 1 \tag{2.2.7}$$

for all positive ε. The Laplace transform of $\delta(t)$ may be derived by considering the limit of $F(s)$ as h approaches zero, i.e.,

$$\mathcal{L}[\delta(t)] = \mathcal{L}[\lim_{h \to 0} f(t)] = \lim_{h \to 0}\{\mathcal{L}[f(t)]\} = \lim_{h \to 0}\frac{1}{hs}(1-e^{-hs})$$

By application of L'Hospital's rule, it follows that

$$\mathcal{L}[\delta(t)] = \lim_{h \to 0}\frac{se^{hs}}{s} = 1 \tag{2.2.8}$$

Properties of the Laplace transform

1. *Linearity*

The Laplace transformation is a linear operator, i.e.,

$$\mathcal{L}[\alpha f(t) \pm \beta g(t)] = \alpha\mathcal{L}[f(t)] \pm \beta\mathcal{L}[g(t)] \tag{2.2.9}$$

where α and β are constants. The proof is straightforward and is left to the reader.

2. *Translation*

If $\mathcal{L}[f(t)] = F(s)$, then $f(t - a)$, the translated function of $f(t)$, has the following Laplace transformation:

$$\mathcal{L}[f(t-a)] = e^{-sa}F(s) \tag{2.2.10}$$

This result is proven as follows:

$$\mathcal{L}[f(t-a)] = \int_0^\infty f(t-a)e^{-st}dt = e^{-sa}\int_0^\infty f(t-a)e^{-s(t-a)}d(t-a)$$

Letting $\tau = t - a$, we have

$$\mathcal{L}[f(t-a)] = e^{-sa}\int_{-a}^\infty f(\tau)e^{-s\tau}d\tau$$

$$= e^{-sa}\int_0^\infty f(\tau)e^{-s\tau}d\tau \qquad (f(\tau) = 0 \text{ if } \tau < 0)$$

$$= e^{-sa}F(s)$$

Example 2.2.5

Recalling Example 2.2.3, the unit pulse function $f(t)$ of (2.2.5) can be described as the difference of two unit step functions:

$$f(t) = (1/h)u(t) - (1/h)u(t - h) \tag{2.2.11}$$

where $u(t)$ is as defined in (2.2.3). From Example 2.2.1 and the linearity property,

$$\mathcal{L}[f(t)] = (1/h)\mathcal{L}[u(t)] - (1/h)\mathcal{L}[u(t - h)]$$

$$= (1/h)\{\mathcal{L}[u(t)] - e^{-sh}\mathcal{L}[u(t)]\}$$

$$= (1 - e^{-sh})/hs \tag{2.2.12}$$

This result is identical to that arrived at in Example 2.2.3.

3. *Differentiation with respect to time*

Integrating by parts, the Laplace transformation of a derivative can be obtained as

$$L\left[\frac{df}{dt}\right] = \int_0^\infty \frac{df(t)}{dt}e^{-st}dt = \int_0^\infty e^{-st}df(t)$$

$$= f(t)e^{-st}\Big|_0^\infty + \int_0^\infty f(t)e^{-st}s\,dt = sF(s) - f(0) \tag{2.2.13}$$

This formula shows that the Laplace transform of a derivative depends upon the initial value of the time function. When constructing Laplace transform models for process control purposes, we generally deal with functions having the property $f(0) = 0$, in which case (2.2.13) reduces to $\mathcal{L}[df/dt] = sF(s)$.

Similarly, it can be proved that

$$\mathcal{L}\left[\frac{d^2 f}{dt^2}\right] = s^2 F(s) - sf(0) - f'(0) \tag{2.2.14}$$

and

$$\mathcal{L}\left[\frac{d^n f}{dt^n}\right] = s^n F(s) - s^{n-1} f(0) - s^{n-2} f'(0) - \cdots - sf^{n-2}(0) - f^{n-1}(0) \tag{2.2.15}$$

where $f'(0)$ is the derivative of $f(t)$ at $t = 0$ and $f^{(i)}(0)$ is the ith derivative at $t = 0$.

Example 2.2.6

Consider the third order linear differential equation

$$\frac{d^3}{dt^3} y(t) + a\frac{d^2}{dt^2} y(t) + b\frac{d}{dt} y(t) + cy(t) = Kr(t) \tag{2.2.16}$$

The Laplace transformation of (2.2.16) is

$$s^3 Y(s) + as^2 Y(s) + bs Y(s) + c Y(s) = KR(s) \tag{2.2.17}$$

when the initial conditions are described by

$$y(0) = y'(0) = y^{(2)}(0) = 0 \tag{2.2.18}$$

4. *Integration with respect to time*

The Laplace transform of the integral of a function with respect to time is

$$\mathcal{L}\left[\int_0^t f(t)dt\right] = \int_0^\infty \left[\int_0^t f(t)dt\right] e^{-st} dt$$

$$= -\frac{1}{s}\left[\int_0^t f(t)dt e^{-st}\right]_0^\infty + \frac{1}{s}\int_0^\infty f(t)e^{-st} dt$$

$$= (1/s)(0 - 0) + (1/s)F(s)$$

$$= F(s)/s \tag{2.2.19}$$

which can be interpreted as having the 'inverse' effect of differentiation in the s-domain.

5. *Final value theorem*

The final value theorem states that

$$\lim_{t\to\infty} f(t) = \lim_{s\to 0} sF(s) \tag{2.2.20}$$

This assertion may be proved by observing the behavior of (2.2.13) as s approaches zero:

$$\lim_{s \to 0} \int_0^\infty (df/dt)e^{-st}dt = \lim_{s \to 0}[sF(s) - f(0)]$$

or

$$\int_0^\infty (df/dt)dt = f(\infty) - f(0) = \lim_{s \to 0}[sF(s) - f(0)]$$

Clearly,

$$f(\infty) = \lim_{s \to 0} sF(s)$$

which establishes (2.2.20).

It should be noted that we have assumed the existence of $f(\infty)$ in the above proof. For some functions, e.g., $f(t) = \sin\omega t$ or $f(t) = e^t$, final values do not exist, and the final value theorem is invalid. To be precise, the theorem holds true only when $sF(s)$ is analytic on the imaginary axis and in the right half s-plane.

6. *Initial value theorem*

Following arguments similar to those employed above for the final value theorem, we may derive the initial value theorem, which states that

$$\lim_{t \to 0} f(t) = \lim_{s \to \infty} sF(s) \tag{2.2.21}$$

The initial and final value theorems permit the computation of initial and final values of a time function from its Laplace transformation.

Example 2.2.7

Consider a time function $f(t)$ with Laplace transform

$$F(s) = \frac{(s+1)(s+2)}{s(s+3)(s+4)}$$

The initial value of $f(t)$ is

$$f(0) = \lim_{s \to \infty} s\frac{(s+1)(s+2)}{s(s+3)(s+4)} = 1$$

and its final value is

$$\lim_{s \to \infty} f(t) = \lim_{s \to 0} s\frac{(s+1)(s+2)}{s(s+3)(s+4)} = \frac{1}{6}$$

Example 2.2.8

Consider the time function $f(t)$ whose Laplace transformation is given by

$$F(s) = \frac{(s-1)(s+2)}{s(s+3)(s-4)}$$

Application of the final value theorem leads to the result

$$f(\infty) = \lim_{t \to \infty} f(t) = \lim_{s \to 0} s \frac{(s-1)(s+2)}{s(s+3)(s-4)} = \frac{1}{6}$$

which is incorrect. This is because the s-domain function $sF(s)$ is not analytic in the right half s-plane and the theorem is invalid. In actuality, $f(\infty) = \infty$, i.e., the final value of $f(t)$ does not exist.

7. Complex shifting

It is an easy matter to prove that the Laplace transform of a time function $f(t)$ multiplied by e^{-at} (where a is a constant) is equal to the Laplace transform $F(s)$ with s replaced by $s + a$, i.e.,

$$\mathcal{L}[e^{\mp at} f(t)] = F(s \pm a) \tag{2.2.22}$$

8. Convolution theorem

It can be shown that

$$F_1(s)F_2(s) = \mathcal{L}\left[\int_0^t f_1(\tau)f_2(t-\tau)d\tau\right] = \mathcal{L}[f_1(t) * f_2(t)] \tag{2.2.23}$$

TABLE 2.5.   Useful properties of the Laplace transform.

Properties	Formulas
Linearity	$\mathcal{L}[\alpha f(t) \pm \beta g(t)] = \alpha\mathcal{L}[f(t)] \pm \beta\mathcal{L}[g(t)]$
Translation	$\mathcal{L}[f(t \mp a)] = e^{\pm sa} F(s)$
Differentiation	$\mathcal{L}[df/dt] = sF(s) - f'(0)$
Integration	$\mathcal{L}\left[\int_0^t f(t)dt\right] = \frac{1}{s}F(s)$
Final value theorem	$\lim_{t \to \infty} f(t) = \lim_{s \to 0} sF(s)$
Initial value theorem	$\lim_{t \to 0} f(t) = \lim_{s \to \infty} sF(s)$
Complex shifting	$\mathcal{L}[e^{\mp at} f(t)] = F(s \pm a)$
Convolution	$F_1(s)F_2(s) = \mathcal{L}\left[\int_0^t f_1(\tau)f_2(t-\tau)d\tau\right] = \mathcal{L}[f_1(t) * f_2(t)]$

where $F_1(s)$ and $F_2(s)$ represent the Laplace transforms of $f_1(t)$ and $f_2(t)$, and the symbol "*" denotes convolution in the time domain. This result does not imply that the Laplace transform of a product of time functions, e.g., $f_1(t)$ and $f_2(t)$, equals the product of their individual transforms.

Properties 1-8 of the Laplace transform are summarized in Table 2.5.

Partial fraction expansion and residues

Partial fraction expansion is the most commonly used method of inverting Laplace transforms. Consider the rational function

$$F(s) = \frac{Z(s)}{Q(s)} = \frac{Z(s)}{(s-p_1)(s-p_2)\cdots(s-p_N)} \tag{2.2.24}$$

where $Q(s)$ and $Z(s)$ are polynomials in s, and p_i ($i = 1, 2, \cdots, N$) are the distinct roots of $Q(s)$. The order of $Z(s)$ is assumed to be less than that of $Q(s)$. If all p_i ($i = 1, 2, \cdots, N$) are distinct, then $F(s)$ can be expressed as the sum of N terms:

$$F(s) = \frac{C_1}{s-p_1} + \frac{C_2}{s-p_2} + \cdots + \frac{C_N}{s-p_N} \tag{2.2.25}$$

The constants $C_1, C_2, \cdots, C_N$ are called the residues of $F(s)$, and can be evaluated from

$$\begin{aligned} C_1 &= \lim_{s \to p_1} [(s-p_1)F(s)] \\ C_2 &= \lim_{s \to p_2} [(s-p_2)F(s)] \\ &\cdots \qquad \cdots \\ C_N &= \lim_{s \to p_N} [(s-p_N)F(s)] \end{aligned} \tag{2.2.26}$$

Another method of finding $C_1, C_2, \cdots, C_N$ is to compare the coefficients of the numerator of (2.2.24) with those of (2.2.25) after combining all terms on the right-hand side of (2.2.25) to obtain a common denominator.

Because all terms of (2.2.25) are simple, the inverse Laplace transform of each is readily seen to be an exponential function, and the inverse Laplace transform of $F(s)$ is simply the sum of these, i.e.,

$$f(t) = C_1 e^{p_1 t} + C_2 e^{p_2 t} + \cdots + C_N e^{p_N t} \tag{2.2.27}$$

This formula is a special case of the residue theorem of complex analysis.

If repeated roots exist in the denominator of (2.2.24), $p_1 = p_2$, say, then the expansion of $F(s)$ will be

$$F(s) = \frac{C_{12}}{(s - p_1)^2} + \frac{C_{11}}{s - p_1} + \frac{C_3}{s - p_3} + \cdots + \frac{C_N}{s - p_N} \qquad (2.2.28)$$

where C_{11} and C_{12} are calculated as follows:

$$C_{12} = \lim_{s \to p_1} [(s - p_1)^2 F(s)]$$

$$C_{11} = \lim_{s \to p_1} \{\frac{d}{ds}[(s - p_1)^2 F(s)]\} \qquad (2.2.29)$$

The time function corresponding to (2.2.28) is

$$f(t) = C_{12} t e^{p_1 t} + C_{11} e^{p_1 t} + C_3 e^{p_3 t} + \cdots + C_N e^{p_N t} \qquad (2.2.30)$$

More generally, if the denominator of $F(s)$ has a factor of multiplicity r at p_1, then $F(s)$ is of the form

$$F(s) = \frac{Z(s)}{(s - p_1)^r (s - p_{r+1}) \cdots (s - p_N)} \qquad (2.2.31)$$

which can be expanded as

$$F(s) = \frac{C_{1r}}{(s - p_1)^r} + \frac{C_{1,r-1}}{(s - p_1)^{r-1}} + \cdots + \frac{C_{11}}{s - p_1} + \frac{C_{r+1}}{s - p_{r+1}} + \frac{C_{r+2}}{s - p_{r+2}} + \cdots + \frac{C_N}{s - p_N}$$
$$(2.2.32)$$

The constants C_{r+1}, C_{r+2}, $\cdots$, C_N, corresponding to the $N-r$ simple roots may be evaluated using (2.2.26). The C_{1i} are determined by

$$C_{1r} = \lim_{s \to p_1} [(s - p_1)^r F(s)]$$

$$C_{1,r-1} = \lim_{s \to p_1} \{\frac{d}{ds}[(s - p_1)^r F(s)]\}$$

$$C_{1,r-2} = \lim_{s \to p_1} \{\frac{1}{2!} \frac{d^2}{ds^2}[(s - p_1)^r F(s)]\} \qquad (2.2.33)$$

$$\cdots \qquad \cdots$$

$$C_{11} = \lim_{s \to p_1} \{\frac{1}{(r-1)!} \frac{d^{r-1}}{ds^{r-1}}[(s - p_1)^r F(s)]\}$$

or alternatively, from the coefficient comparison method discussed above. The original time function is readily obtained:

$$f(t) = \frac{C_{1r}}{(r-1)!} t^{r-1} e^{p_1 t} + \cdots + C_{11} e^{p_1 t} + C_{r+1} e^{p_{r+1} t} + C_{r+1} e^{p_{r+2} t} + \cdots + C_N e^{p_N t} \quad (2.2.34)$$

TABLE 2.6. Partial fraction expansion of a third order function.

Roots of Q	Partial fraction expansion	Solution of A, B, C
Real $p_1 \neq p_2 \neq p_3$	$F(s) = \dfrac{C_1}{s - p_1} + \dfrac{C_2}{s - p_2} + \dfrac{C_3}{s - p_3}$	Coefficient comparison method or Eq. (2.2.26)
Real $p_1 = p_2 \neq p_3$	$F(s) = \dfrac{C_{12}}{(s - p_1)^2} + \dfrac{C_{11}}{s - p_1} + \dfrac{C_3}{s - p_3}$	Coefficient comparison method or Eq. (2.2.29)
Real $p_1 = p_2 = p_3$	$F(s) = \dfrac{C_{13}}{(s - p_1)^3} + \dfrac{C_{12}}{(s - p_1)^2} + \dfrac{C_{11}}{s - p_1}$	Coefficient comparison method or Eq. (2.2.33)
Complex conjugates p_1 and p_2, real p_3	$F(s) = \dfrac{As + B}{s^2 + fs + g} + \dfrac{C}{s - p_3}$ where $f = -(P_1 + P_2)$ and $g = P_1 P_2$	Coefficient comparison method

When $Q(s)$ contains irreducible factors, the expansion procedure becomes a little more complicated. For example, if $Q(s)$ includes the pair of complex conjugate roots $p_1 = -\alpha + j\beta$ and $p_2 = -\alpha - j\beta$, then the partial fraction expansion of $F(s)$ will be given by

$$F(s) = \frac{As + B}{s^2 + fs + g} + \frac{C_3}{s - p_3} + \cdots + \frac{C_N}{s - p_N} \qquad (2.2.35)$$

with $f = 2\alpha$ and $g = \alpha^2 + \beta^2$. The constants in this expression are best obtained by comparison of coefficients, as illustrated in the following section.

Table 2.6 outlines four possible strategies for the expansion of a third order rational function. Their inverse Laplace transforms can be found in Appendix A.

Example 2.2.9
Find the inverse Laplace transform of

$$F(s) = \frac{3}{s(2s+1)^2} = \frac{0.75}{s(s+0.5)^2} = \frac{C_{12}}{(s+0.5)^2} + \frac{C_{11}}{s+0.5} + \frac{C_3}{s}$$

C_{11} and C_{12} may be calculated from (2.2.33), and C_3 from (2.2.26):

$$C_{12} = \lim_{s \to -0.5} [(s+0.5)^2 \frac{0.75}{s(s+0.5)^2}] = -1.5$$

$$C_{11} = \lim_{s \to -0.5} \{\frac{d}{ds}[(s+0.5)^2 \frac{0.75}{s(s+0.5)^2}]\} = -3$$

$$C_3 = \lim_{s \to 0} s \frac{0.75}{s(s+0.5)^2} = 3$$

giving

$$F(s) = \frac{-1.5}{(s+0.5)^2} + \frac{-3}{s+0.5} + \frac{3}{s}$$

Hence, the original time function is

$$f(t) = -1.5\, te^{-0.5t} - 3e^{-0.5t} + 3$$

Transfer function, its poles and zeros
A transfer function is an s-domain algebraic representation of the relationship between the input and output variables of a system. It is typically obtained by taking the Laplace transform of a system differential equation with zero initial conditions. A transfer function completely determines both the dynamic and steady state behaviors of the process output when the input signal is specified.

For physically realizable systems, the dégree of the denominator polynomial of the corresponding transfer function always exceeds that of the numerator. If the order of the numerator is larger than that of the denominator, the transfer function is said to be unrealizable.

The roots of the numerator polynomial are called the zeros of the transfer function. If s assumes the value of any zero, the transfer function vanishes. The roots of the denominator are known as the poles of the transfer function. When s approaches a pole, the transfer function becomes infinite.

The poles and zeros of the process transfer function play an important role in the analysis and design of control systems; they will be discussed in detail in Chapter 4.

Example 2.2.10
In Example 2.2.6, we applied the Laplace transform to the system differential equation (2.2.16) to obtain (2.2.17) with zero initial condition (2.1.18). The transfer function between the system input $R(s)$ and output $Y(s)$ is

$$\frac{Y(s)}{R(s)} = \frac{K}{s^3 + as^2 + bs + c}$$

which may be represented in block diagram form as shown in Fig. 2.2.2. The block diagram transformation technique will be discussed in Section 2.4.

FIGURE 2.2.2.   Block diagram with transfer function expression.

2.3. TRANSFER FUNCTIONS OF ELEMENTARY SYSTEMS

As mentioned earlier, the Laplace transform is an important analytical tool for the solution of linear differential equations. The general procedure may be summarized as follows:

i) Apply the Laplace transform to both sides of the differential equation.

ii) Resolve the s-domain algebraic equations by partial fraction expansion.

iii) Take the term-by-term inverse Laplace transform of the expression obtained in Step ii).

We now employ this strategy to analyze some elementary dynamic systems.

First order systems

Consider a process described by the first order differential equation

$$a\frac{dy(t)}{dt} + by(t) = K_0 r(t) \qquad (2.3.1)$$

with y denoting the system output and r the input. For obvious reasons, we call this a first order system.

The Laplace transform of (2.3.1) with zero initial conditions is

$$asY(s) + bY(s) = K_0 R(s) \qquad (2.3.2)$$

and the transfer function of the system is given by

$$\frac{Y(s)}{R(s)} = \frac{K_0}{as+b} = \frac{K}{Ts+1} \qquad (2.3.3)$$

where $K = K_0/b$ is called the steady state or static gain of the process, and $T = T_0/b$ is known as the time constant.

A general time domain solution for $y(t)$ may be found once the structure of the input $r(t)$ has been specified. For example, if the input is chosen to be a step function of magnitude A, then

$$Y(s) = \frac{K}{Ts+1}\frac{A}{s} \qquad (2.3.4)$$

and after partial fraction expansion, we obtain

$$Y(s) = \frac{KA}{s} - \frac{KAT}{Ts+1} \qquad (2.3.5)$$

Taking the inverse Laplace transform then reveals the response of the system output to a step change in input:

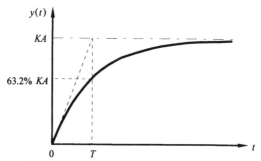

FIGURE 2.3.1 Response of a first order system to a step input.

$$y(t) = KA(1 - e^{-t/T}) \qquad (2.3.6)$$

which is illustrated in Fig. 2.3.1.

The following features of a first order system subjected to a step input may be extracted from the figure:

i) The output $y(t)$ rises to a new steady state value of KA asymptotically. Its rate of change is fastest initially, and becomes progressively slower as t increases. Values of the response at selected times are listed below:

Elapsed time	T	$2T$	$3T$	$4T$
$y(t)/y(\infty)$	63.2%	86.5%	95%	98%

ii) The speed of response is determined by the time constant T. Larger T gives slower response. If the initial rate of change of $y(t)$ was maintained, then $y(t)$ would assume its final value of $y(\infty)$ at time $t = T$ (see Problem 2.13).

iii) The new steady value of the system output is K times the magnitude of the input change. This is readily observed by setting $t = \infty$ in (2.3.6) or $s = 0$ in (2.3.4) (cf. final value theorem), and justifies the designation of K as the steady state gain of the process.

Second order systems
A second order system is one which can be described by a second order, ordinary differential equation of the form:

$$a\frac{d^2}{dt^2}y(t) + b\frac{d}{dt}y(t) + cy(t) = K_0 r(t) \qquad (2.3.7)$$

The process transfer function is clearly

$$\frac{Y(s)}{R(s)} = \frac{K_0}{as^2 + bs + c} = \frac{K}{T^2 s^2 + 2\zeta Ts + 1} \qquad (2.3.8)$$

The rightmost part of this expression illustrates the accepted standard form of a second order transfer function. As before, $K = K_0/c$ represents the steady state gain of the process. $T = \sqrt{a/c}$ is termed the natural period of oscillation, but is also referred to in some textbooks [e.g., Luyben 1990] as the process time constant by analogy with a first order system. ζ is called the damping factor. T and ζ are observed to be nonnegative for all physical systems, thereby guaranteeing stability (a topic to be discussed further in Chapter 4). In an alternative version of the standard form, the transfer function is written in terms of $\omega_n = 1/T$, i.e.,

$$\frac{Y(s)}{R(s)} = \frac{K_1}{s^2 + 2\zeta\omega_n s + \omega_n^2} \qquad (2.3.9)$$

with $K_1 = K/T^2$. ω_n is called the undamped natural frequency because if $\zeta = 0$, then the system oscillates indefinitely with angular frequency ω_n.

Once again, let us specify the input as a step function of magnitude A. The s-domain output is then given by

$$Y(s) = \frac{K}{T^2 s^2 + 2\zeta Ts + 1} \frac{A}{s} \qquad (2.3.10)$$

The two poles of the transfer function are

$$s_{1,2} = -\frac{\zeta}{T} \pm \frac{1}{T}\sqrt{\zeta^2 - 1} \qquad (2.3.11)$$

and (2.3.10) can be rewritten in these terms as

$$Y(s) = \frac{KA/T^2}{s(s - s_1)(s - s_2)} \qquad (2.3.12)$$

The character of the system output is dependent upon the magnitude of the damping factor ζ. There are three cases of practical interest:

Case I. $\zeta > 1$ (Overdamped system)
Rewrite the pole expression as $s_{1,2} = -\alpha \pm \beta$, where

$$\alpha = \zeta/T, \qquad \beta = \frac{1}{T}\sqrt{\zeta^2 - 1}$$

Because $\zeta > 1$, the two poles are real. Partial fraction expansion of (2.3.12) gives

$$Y(s) = \frac{E}{s} + \frac{F}{s - s_1} + \frac{G}{s - s_2} \qquad (2.3.13)$$

FIGURE 2.3.2. Step response of an overdamped second order system.

The coefficient comparison method can be used to find parameters E, F and G. Eq. (2.3.13) can be rewritten as

$$Y(s) = \frac{s^2(E + F + G) + s(-s_1 E - s_2 E - s_2 F - s_1 G) + s_1 s_2 E}{s(s - s_1)(s - s_2)}$$

Comparing its coefficients with those of (2.3.12) yields

$$E + F + G = 0$$

$$-s_1 E - s_2 E - s_2 F - s_1 G = 0$$

$$s_1 s_2 E = KA / T^2$$

The solution to these equations is

$$E = KA, \qquad F = \frac{KA / T^2}{2\beta s_1} \qquad \text{and} \qquad G = -\frac{KA / T^2}{2\beta s_2}$$

By taking the inverse Laplace transform we obtain

$$y(t) = KA[1 + \frac{1/T^2}{2\beta s_1} e^{s_1 t} - \frac{1/T^2}{2\beta s_2} e^{s_2 t}] \qquad (2.3.14)$$

This response is plotted in Fig. 2.3.2, and appears similar to that of a first order system (cf. Fig. 2.3.1). Indeed, a second order system with $\zeta > 1$ may be constructed by combining two first order systems in series. However, unlike the first order response, the maximum rate of change does not occur at $t = 0$. Larger values of ζ imply a slower response, as the name 'damping factor' would suggest.

Case II $\zeta = 1$ (Critically damped system)
Setting $\zeta = 1$ in (2.3.11) yields

$$s_1 = s_2 = -1/T$$

and the partial fraction expansion of (2.3.12) leads to an expression of the form:

$$Y(s) = \frac{KA}{s} + \frac{KAs_1}{(s-s_1)^2} - \frac{KA}{s-s_1} \qquad (2.3.15)$$

The inverse Laplace transform is then

$$y(t) = KA(1 - \frac{1}{T}te^{-t/T} - e^{-t/T}) \qquad (2.3.16)$$

For a given T of a second order system, $\zeta = 1$ corresponds to the fastest possible response, one with no oscillation.

Case III $\zeta < 1$ (Underdamped system)
For this case, the poles of the transfer function are expressed as

$$s_{1,2} = -\alpha \pm j\beta \qquad (2.3.17)$$

with

$$\alpha = \zeta / T, \qquad \beta = \frac{1}{T}\sqrt{1-\zeta^2} \qquad (2.3.18)$$

Following (2.2.35), a partial fraction expansion of (2.3.12) leads to

$$Y(s) = \frac{KA/T^2}{s[(s+\alpha)^2 + \beta^2]} = \frac{KA}{s} - \frac{KA(s+2\alpha)}{(s+\alpha)^2 + \beta^2} \qquad (2.3.19)$$

or

$$Y(s) = \frac{KA}{s} - \frac{KA(s+\alpha)}{(s+\alpha)^2 + \beta^2} - \frac{KA\alpha}{\beta} \frac{\beta}{(s+\alpha)^2 + \beta^2} \qquad (2.3.20)$$

which has inverse Laplace transform

$$y(t) = KA[1 - e^{-\alpha t}(\cos\beta t + \frac{\alpha}{\beta}\sin\beta t)] \qquad (2.3.21)$$

Substitution of (2.3.18) into (2.3.21) and further rearrangement yields

$$y(t) = KA[1 - e^{-\zeta t/T} \frac{1}{\sqrt{1-\zeta^2}} \sin(\frac{\sqrt{1-\zeta^2}}{T}t + \phi)] \qquad (2.3.22)$$

where

$$\phi = \text{arctg}(\sqrt{1-\zeta^2}/\zeta) \qquad (2.3.23)$$

From Fig. 2.3.3, we observe the oscillatory character of this response, which becomes more pronounced as the damping factor ζ is decreased. In all three cases, the final value of the output is KA, i.e., K times the magnitude of the input

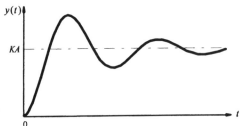

FIGURE 2.3.3. Step response of an underdamped second order system.

change. This result can be confirmed by taking the limit as t approaches infinity in (2.3.14), (2.3.16) and (2.3.22), or alternatively from (2.3.13), (2.3.15) and (2.3.20) by means of the final value theorem.

The effect of ζ on the nature of a second order response is summarized in Table 2.7. The first two rows of the table were ignored in the preceding discussion, but are easily verified. Consideration of the third column may be postponed until the reader has studied the concept of system stability, which will be presented in Chapter 4.

Integrating process
The dynamics of an integrating process (or pure integrator) is described by the differential equation

$$\frac{dy}{dt} = Ku(t) \qquad (2.3.24)$$

or equivalently, by the transfer function

$$\frac{Y(s)}{U(s)} = \frac{K}{s} \qquad (2.3.25)$$

TABLE 2.7. Effect of ζ on response of a second order system.

	Dynamics	Location of poles	Stability of system
$\zeta < 0$	Divergent	At least one pole in right half s-plane	Unstable
$\zeta = 0$	Oscillatory	Poles on imaginary axis	Marginally stable
$0 < \zeta < 1$	Underdamped	Pair of complex poles in left half s-plane	Stable
$\zeta = 1$	Critically damped	Two coincident, negative real poles	Stable
$\zeta > 1$	Overdamped	Two negative real poles	Stable

For step inputs of magnitude A, (2.3.25) becomes

$$Y(s) = \frac{KA}{s^2} \tag{2.3.26}$$

or

$$y(t) = KAt \tag{2.3.27}$$

upon taking the inverse Laplace transform. Notice that a pure integrator permits no definition of a steady state gain, since the output continually increases in a linear fashion with slope KA.

Time delay
In Section 2.1, a pure time delay model was expressed in the form

$$y(t) = r(t - \tau) \tag{2.3.28}$$

where r and y were the system input and output, respectively.

Using the translation property of the Laplace transform (cf. Eq. (2.2.10)), we obtain

$$Y(s) = e^{-\tau s} R(s) \tag{2.3.29}$$

That is, the transfer function of a τ-unit time delay is $e^{-\tau s}$.

The salient features of the transfer functions studied thus far are summarized in Table 2.8. The parameters of each of these systems have differing effects on

TABLE 2.8.   Dynamics of elementary systems.

Process	Transfer function	No. of poles	Parameters
First order system	$K/(Ts + 1)$	1	K, T
Second order system	$K/(T^2s^2 + 2\zeta Ts + 1)$	2	K, T, ζ
Integrating process	K/s	1	K
Pure time delay	$e^{-\tau s}$	0	τ

TABLE 2.9.   Influence of system parameters.

Parameter	Symbol	Output characteristics affected	Effect on performance
Static gain	K	Steady state value	$K\uparrow$, Steady state value$\uparrow$
Time constant	T	Speed of response	$T\uparrow$, Stability$\downarrow$
Damping factor	ζ	Speed of response Presence of oscillations	$\zeta\uparrow$, Stability$\uparrow$
Time delay	τ	Time coordinate of response	$\tau\uparrow$, Stability$\downarrow$

the time response of the process output, as indicated in Table 2.9.

2.4. BLOCK DIAGRAMS

The block diagram concept was introduced in Chapter 1 as a convenient means of representing the dynamic structure of feedback control loops. The purpose of the present section is to discuss some important quantitative aspects of this technique.

Composition of a block diagram
Block diagrams consist of a series of interconnected blocks representing the dynamic relationship between system variables. Each such representation is unidirectional, or causal. By this we mean that the sense of the arrows connecting the various blocks may not be reversed. Because of their more quantitative nature, we might also describe the block diagrams to be presented here as operational, in contrast to the functional approach adopted in Chapter 1. The single block shown in Fig. 2.4.1a, for example, represents the operational relationship of a transfer function $G(s)$.

$$Y(s) = G(s)R(s) \qquad (2.4.1)$$

Each interconnecting line in a block diagram represents a system variable or signal. The summation of two system variables is indicated by means of a summing junction. The particular relationship depicted in Fig. 2.4.1b is

$$E(s) = R(s) - Y(s) \qquad (2.4.2)$$

Additionally, the value of any signal in the block diagram may be forwarded to two different locations using a 'pickoff point', as illustrated in Fig. 2.4.1c.

Block connection modes
There are three basic modes of block connection: series, parallel and feedback. The relationship among system variables in each case is easily deduced. The series and parallel block connections are depicted in Fig. 2.4.2. For the series connection in Fig. 2.4.2a,

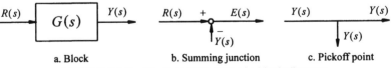

a. Block   b. Summing junction   c. Pickoff point
FIGURE 2.4.1.   Basic components of a block diagram.

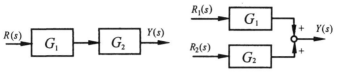

a. Series connection   b. Parallel connection

FIGURE 2.4.2.   Block diagram configurations.

$$Y(s) = G_2(s)G_1(s)R(s) \tag{2.4.3}$$

Indeed, the overall transfer function of any number of blocks connected in series is simply the product of the individual transfer functions. For the parallel connection of Fig. 2.4.2b, we have

$$Y(s) = G_1(s)R_1(s) + G_2(s)R_2(s) \tag{2.4.4}$$

When $R_1(s) = R_2(s)$, it follows that

$$Y(s) = [G_1(s) + G_2(s)]R_2(s) \tag{2.4.5}$$

In this case, the overall transfer function is the sum of the individual transfer functions.

A feedback block connection is shown in Fig. 2.4.3a. This is a single loop feedback control system, discussed earlier in Chapter 1. The feedforward block transfer function is denoted as $G(s)$, while the feedback block transfer function is $H(s)$. Based on the figure, the following relationships are true:

$$Y(s) = G(s)E(s) \tag{2.4.6}$$

and

$$E(s) = R(s) - H(s)Y(s) \tag{2.4.7}$$

Substituting (2.4.7) into (2.4.6), it follows that

$$Y(s) = \frac{G(s)}{1 + G(s)H(s)} R(s) \tag{2.4.8}$$

This result implies that the block diagram of a feedback system may be replaced

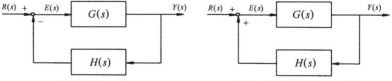

a. A feedback control system   b. A positive feedback control system

FIGURE 2.4.3. Feedback control configuration.

by a single block containing the closed loop transfer function. For the unity feedback case, i.e., when $H(s) = 1$, Eq. (2.4.8) reduces further to

$$Y(s) = \frac{G(s)}{1+G(s)} R(s) \qquad (2.4.9)$$

The feedback we discussed above is called negative feedback, since the negative output signal is fed back to the summing junction. If the feedback signal is positive, it is called positive feedback. A single loop positive feedback system is shown in Fig. 2.4.3b. The Laplace transform of the actuating error of this system is

$$E(s) = R(s) + H(s)Y(s) \qquad (2.4.10)$$

and the relationship between input and output is

$$Y(s) = \frac{G(s)}{1-G(s)H(s)} R(s) \qquad (2.4.11)$$

Block diagram transformation

Block diagrams are in no way limited in structure to the simple configurations portrayed in Fig. 2.4.2 and 2.4.3. However, more complex arrangements can often be reduced to this level using the techniques illustrated in Table 2.10. Generally speaking, the simplification of a block diagram is achieved at the expense of more complicated transfer functions in the new diagram.

TABLE 2.10.   Block diagram manipulation.

TABLE 2.10.   Block diagram manipulation (Continued).

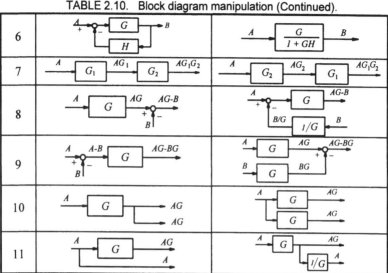

In block diagram manipulation, care must be taken when exchanging the connection points. As shown in Table 2.10, the same kind of points can be exchanged. Method 1 in the table indicates that the adjacent summing junctions can exchange their locations. Method 3 in the table relates to the pickoff points. Attention must be paid to the fact that simply exchanging a pickoff point and an adjacent summing junction will bring about a wrong result.   Table 2.11 summarizes the exchange rules of elements in a block diagram. To reduce a block diagram, we can simplify three basic mode connections in the block diagram. For a complicated block diagram, it is usually necessary to move a summing junction (or pickoff point) to another summing junction (or pickoff point) for exchange to construct basic mode connections. In this way, a complicated block diagram can be reduced into a single block, yielding a system transfer function.

TABLE 2.11.   Exchange of elements in a block diagram.

Adjacent elements exchanged	Overall result	Relevant items in Table 2.10
Summing junctions	No change	1
Pickoff points	No change	3
Block and summing junction	Extra block must be introduced	8 and 9
Block and pickoff point	Extra block must be introduced	10 and 11
Two blocks in series	No change	4

Example 2.4.1

Assume that we wish to simplify the block diagram of Fig. 2.4.4a to the form of Fig. 2.4.1a. Using methods 1 and 8 in Table 2.10, we first move the summing junction of feedback loop H_2 to the outside of feedback loop H_1 as shown in Fig. 2.4.4b. Successive application of method 6 of the table then leads to the single block shown in Fig. 2.4.4e.

Remarks

1. All signals appearing in a block diagram represent Laplace transforms of system variables.

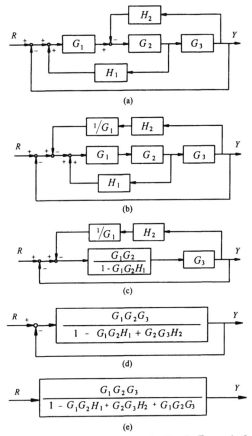

FIGURE 2.4.4.   Block diagram reduction in Example 2.4.1.

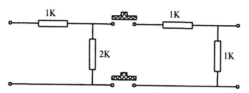

FIGURE 2.4.5. Circuit loading problem.

2. When deriving transfer functions to model the dynamics of electrical circuits, care must be taken to ensure that the loading problem is well-defined. For example, consider the two circuits shown in Fig. 2.4.5. The transfer functions relating the input and output voltages of each circuit are $G_1 = 2/3$ and $G_2 = 1/2$. If the circuits are then connected in series, the overall transfer function is $1/4$, clearly not the product of the two. However, if the loading problem is accounted for in the derivation of G_1, we obtain $G_1 = 1/2$ and the multiplicative property is restored.

3. As explained previously, transfer functions are obtained by taking the Laplace transform of the system differential equation under the assumption of zero initial conditions (cf. Eq. (2.2.18)). If this is not the case, then the effect of any nonzero initial conditions may be incorporated into the block diagram by means of additional external inputs.

Problems

2.1. Prove Eq. (2.1.7); i.e., prove that the moment of inertia of a circular disk about its geometric axis is given by

$$J = \frac{1}{2} MR^2$$

where M is the mass of the disk, and R is its radius.

2.2. Write the differential equations with the input variables u_1 and output

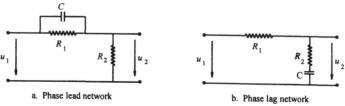

a. Phase lead network   b. Phase lag network

FIGURE 2.P.1. Two electrical networks.

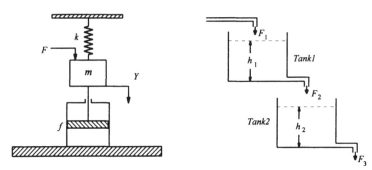

FIGURE 2.P.2. A spring-mass system.   FIGURE 2.P.3. Two tank system.

variable u_2 for the networks shown in Fig. 2.P.1.

2.3. Set up the differential equation model for the mechanical spring-mass system shown in Fig. 2.P.2. In this figure, the displacement Y is the output of the system, F is a force acting on the body (which has mass m), k denotes the spring constant, and f is the frictional constant.

2.4. Consider the system depicted in Fig. 2.P.3, where the output of Tank 1 is the input to Tank 2. Assume that the cross-sectional area of Tank 1 is A_1 , and that of Tank 2 is A_2. Obtain the system differential equations of Tank 1 and 2, selecting the corresponding flow rates as inputs and outputs. Linearize these equations and find their deviation variable equivalents. Combine the linearized equations into a single differential equation.

2.5. Compute the Laplace transforms of the differential equations obtained in Problems 2.2 and 2.3.

2.6. Recalling Problem 2.4 and Fig. 2.P.3, what is the transfer function between h_2 and h_1? between h_2 and F_1?

2.7. Prove that

$$\mathcal{L}[A\cos\omega t] = \frac{As}{s^2 + \omega^2}$$

2.8. Find the Laplace transforms of the time functions shown in Fig 2.P.4.

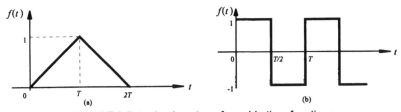

FIGURE 2.P.4. Laplace transformable time functions.

The first figure is a triangle, and the second one is a square wave valid from $t = 0$ to ∞.

2.9. Calculate the initial and final values of the function $f(t)$, whose Laplace transform is given by

a. $\dfrac{s+b}{s(s+a)}$

b. $\dfrac{s+1}{(s+1)^2 + 1}$

c. $\dfrac{s+2}{s(s^2 + 3s + 1)}$

d. $\dfrac{s\omega}{s^2 + \omega^2}$

2.10. Invert the following Laplace transforms:

a. $\dfrac{3}{(s+1)(s+2)}$

b. $\dfrac{s^2 + 2s + 6}{(s+1)^3}$

c. $\dfrac{5\omega}{(s+1)^2 + \omega^2}$

d. $\dfrac{3s+6}{s^2 + 2s + 5}$

2.11. Assume that all initial conditions are zero and solve the following differential equation with Laplace transforms:

$$x'' + 3x' + 2x = 20$$

2.12. Solve the differential equation of Problem 2.11, with initial conditions taken as

$$x(0) = 5, \qquad\qquad x'(0) = 15$$

2.13. Prove that for a first order system, the time constant T is the time at which the system response to a step input would reach its steady state value if the initial speed of the response was maintained indefinitely.

2.14. For a tank as shown in Fig. 2.P.5, the cross-sectional area of the tank A $= 1.2m^2$ and the flow resistance $R = 0.2m/(m^3pm)$ are given. What is the transfer function of the system if

a. the inlet flow rate is assigned as the input and the height level in the tank as

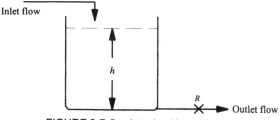

FIGURE 2.P.5.   A tank with resistance.

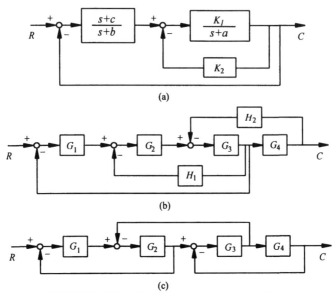

FIGURE 2.P.6.   Block diagrams to be simplified.

the output.

b. the inlet flow rate is assigned as the input and the outlet flow rate as the output.

2.15. Plot the unit step response of the transfer functions:

a. $G(s) = \dfrac{s + 0.5}{s + 1}$ 　　　　　 b. $G(s) = \dfrac{5}{8s^2 + 3.2s + 2}$

2.16. Simplify in step-by-step fashion the block diagrams shown in Fig. 2.P.6.

CHAPTER 3

TIME DOMAIN ANALYSIS AND DESIGN

3.1. TEST SIGNALS

When designing process control systems, the engineer requires an objective means of evaluating the performance of the control schemes under consideration. Performance criteria are quantitative measures of the system response to a specific input or test signal. The present section establishes the mathematical properties of some common test signals that may be used in the definition of various performance criteria.

Step input
As discussed in the previous chapter, a step input is a sudden and sustained change in input defined mathematically as

$$u(t) = \begin{cases} M & t \geq 0 \\ 0 & t < 0 \end{cases} \tag{3.1.1}$$

(See Fig. 3.1.1a). The constant M is known as the magnitude or size of the step function. $u(t)$ is called a unit step function if $M = 1$. The Laplace transform of a step function was derived in Example 2.2.1:

$$\mathcal{L}[u(t)] = M/s \tag{3.1.2}$$

The response of a system to a step input is referred to as its step response, and conveys information regarding the dynamic and steady state behavior of the system. The step function has been used extensively in evaluating the performance of control systems for the following reasons:

    i) It is simple and easy to produce.

    ii) It is considered to be a serious disturbance, since it has a very fast change at the initial moment. Many other kinds of bounded disturbances can be overcome if a step disturbance can be overcome.

iii) The response of the system to other types of disturbances can be inferred from the process step response.

iv) The step response is easy to measure, and thus get an approximated transfer function of the system.

Ramp input

A ramp input is a function that varies linearly with time in the following manner:

$$r(t) = \begin{cases} Rt & t \geq 0 \\ 0 & t < 0 \end{cases} \qquad (3.1.3)$$

where R is a constant determining the slope of the ramp. Ramp inputs arise in chemical processes as batch reactor temperature or pressure setpoints, and in mechanical systems in the guise of angular displacement of shafts rotating at constant speed. The Laplace transform of a ramp function is

$$\mathcal{L}[r(t)] = R/s^2 \qquad (3.1.4)$$

and its time domain behavior is as illustrated in Fig. 3.1.1b.

Parabolic input

A parabolic input function is described mathematically as:

$$r(t) = \begin{cases} Rt^2 & t \geq 0 \\ 0 & t < 0 \end{cases} \qquad (3.1.5)$$

which is represented graphically in Fig. 3.1.1c. The Laplace transformation of a parabolic function is

$$\mathcal{L}[r(t)] = 2R/s^3 \qquad (3.1.6)$$

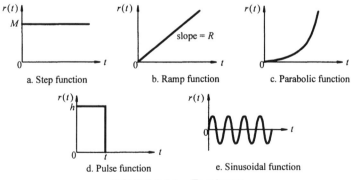

a. Step function   b. Ramp function   c. Parabolic function

d. Pulse function   e. Sinusoidal function

FIGURE 3.1.1.   Test inputs.

Pulse and impulse inputs

A typical pulse function is the rectangular pulse:

$$f(t) = \begin{cases} 0 & t \geq \tau \\ h & 0 \leq t < \tau \\ 0 & t < 0 \end{cases} \qquad (3.1.7)$$

This can be viewed as the superposition of a positive step function followed by a negative step function of the same size (see Fig. 3.1.1d). The area $h \times \tau$ is called the strength of the pulse; a unit pulse is one whose area $h \times \tau$ is unity.

If the area is held constant as the time t approaches its limiting value of zero, then the magnitude h approaches infinity and the resulting function is called an impulse. If $h \times \tau = 1$, we obtain a unit impulse, or Dirac delta function, denoted by $\delta(t)$.

The Laplace transform of $\delta(t)$ was previously obtained as Eq. (2.2.8) of the last chapter, i.e.,

$$\mathcal{L}[\delta(t)] = 1 \qquad (3.1.8)$$

An ideal impulse function can never be realized in practice, since it requires an infinite amount of energy to be injected into the process during an infinitesimal length of time. However, the delta function and the concept of the system impulse response are of great theoretical significance.

Sinusoidal input

The sinusoidal input displayed in Fig. 3.1.1e is generated by the expression

TABLE 3.1.   Typical test signals.

Test signal	Definition	Transfer function	Parameters
Step input	$u(t) = \begin{cases} M & t \geq 0 \\ 0 & t < 0 \end{cases}$	$\mathcal{L}[u(t)] = M/s$	M
Ramp input	$r(t) = \begin{cases} Rt & t \geq 0 \\ 0 & t < 0 \end{cases}$	$\mathcal{L}[r(t)] = R/s^2$	R
Parabolic input	$r(t) = \begin{cases} Rt^2 & t \geq 0 \\ 0 & t < 0 \end{cases}$	$\mathcal{L}[r(t)] = 2R/s^3$	R
Pulse or impulse	$f(t) = \begin{cases} 0 & t \geq \tau \\ h & 0 \leq t < \tau \\ 0 & t < 0 \end{cases}$	$\mathcal{L}[\delta(t)] = 1$	h, τ
Sinusoidal input	$f(t) = \begin{cases} A\sin\omega t & t \geq 0 \\ 0 & t < 0 \end{cases}$	$\mathcal{L}[f(t)] = \dfrac{A\omega}{s^2 + \omega^2}$	A, ω

$$f(t) = \begin{cases} A\sin\omega t & t \geq 0 \\ 0 & t < 0 \end{cases} \qquad (3.1.9)$$

The parameters A and ω fix the amplitude and angular frequency of the oscillation. (Recall that the period T of a sine wave is related to its angular frequency by $T = 2\pi/\omega$.) The Laplace transform of a sinusoidal function was obtained in Example 2.2.2:

$$\mathit{l}[f(t)] = \frac{A\omega}{s^2 + \omega^2} \qquad (3.1.10)$$

When a linear system is subjected to a sinusoidal input, its ultimate response is also a sustained wave. This important fact forms the basis of frequency response analysis, which will be studied in Chapter 5.

Table 3.1 summarizes the key features of the test inputs introduced in this section.

3.2. STEADY STATE ERRORS

Steady state error or offset is defined as the error between the output and its desired value as time approaches infinity. Offset is an important measure of control system performance because it represents the final accuracy of the system in response to a specific test signal.

Computation of steady state error
In a single loop feedback system shown in Fig. 3.2.1, $E(s)$ denotes the Laplace transform of the actuating error $e(t)$, defined as the difference between the setpoint and the measured output. If we make the (physically justifiable) assumption that the gain of the sensor transfer function $H(s)$ is unity, then for the purposes of steady state analysis there is no need to distinguish between actuating error and steady state error.

From Fig. 3.2.1, we can write

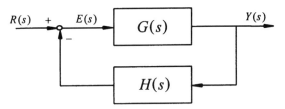

FIGURE 3.2.1.   Feedback control system.

$$E(s) = R(s) - H(s)Y(s) \tag{3.2.1}$$

Substituting $Y(s) = G(s)E(s)$ into this expression yields

$$E(s) = \frac{R(s)}{1 + G(s)H(s)} \tag{3.2.2}$$

Using the final value theorem, the steady state error e_{ss} is given by

$$e_{ss} = \lim_{s \to 0} sE(s) = \lim_{s \to 0} \frac{sR(s)}{1 + G(s)H(s)} \tag{3.2.3}$$

which depends upon the reference input $R(s)$ and the open loop transfer function $G(s)H(s)$.

System type

The general form of an open loop transfer function is:

$$G(s)H(s) = \frac{K(C_m s^m + \cdots + C_1 s + 1)}{s^p (T_n s^n + \cdots + T_1 s + 1)} \tag{3.2.4}$$

where C_i and T_j ($i = 1, \cdots, m; j = 1, \cdots, n$) are real constants. The parameter p defines the type of a feedback control system, and plays an important role in its steady state behavior; the transfer function GH is said to be a 'type-p system'.

Step inputs and the step error constant

If the reference input is specified as a step input of magnitude R, then using Table 3.1, the steady state error becomes

$$e_{ss} = \frac{R}{1 + \lim_{s \to 0} G(s)H(s)} = \frac{R}{1 + K_p} \tag{3.2.5}$$

where the parameter

$$K_p = \lim_{s \to 0} G(s)H(s) = G(0)H(0) \tag{3.2.6}$$

is defined as the step error constant. The larger the step error constant is, the smaller the steady state error. However, it is clear that an infinite value of K_p is required to reduce e_{ss} to zero. From (3.2.4) we see that $K_p = \infty$ whenever $p \geq 1$ (i.e., for all type-1 or higher systems).

Ramp inputs and the ramp error constant

The steady state error obtained for a ramp of slope R is

$$e_{ss} = \lim_{s \to 0} \frac{sR/s^2}{1 + G(s)H(s)} = \frac{R}{\lim_{s \to 0} sG(s)H(s)} = \frac{R}{K_v} \tag{3.2.7}$$

and the ramp error constant K_v is defined as

$$K_v = \lim_{s \to 0} sG(s)H(s) \tag{3.2.8}$$

For a system of type 2 or higher, the ramp error constant is infinite, which implies a zero steady state error. On the other hand, for type-0 systems, the ramp error constant is zero, and infinite steady state error results.

Parabolic inputs and the parabolic error constant
When the parabolic input $0.5Rt^2$ is injected into a system, the steady state error

$$e_{ss} = \lim_{s \to 0} \frac{sR/s^3}{1 + G(s)H(s)} = \frac{R}{\lim_{s \to 0} s^2[1 + G(s)H(s)]} = \frac{R}{K_a} \tag{3.2.9}$$

results, where

$$K_a = \lim_{s \to 0} s^2 G(s)H(s) \tag{3.2.10}$$

is designated the parabolic error constant.

For systems of type 0 or 1, it is apparent that $K_a = 0$ and $e_{ss} = \infty$. For systems of type 3 or higher, the parabolic error constant is infinite, and the steady state error is therefore zero.

Values of the three error constants and their related steady state errors are listed in Tables 3.2 and 3.3, respectively.

Remarks
1. The error constants K_p, K_v and K_a describe the ability of a system to reduce or eliminate steady state actuating error, and may therefore serve as numerical

TABLE 3.2.   Step, ramp and parabolic error constants.

System type	Error Constants		
	K_p	K_v	K_a
0	$K_p = \lim_{s \to 0} G(s)H(s)$	0	0
1	∞	$K_v = \lim_{s \to 0} sG(s)H(s)$	0
2	∞	∞	$K_a = \lim_{s \to 0} s^2 G(s)H(s)$
3	∞	∞	∞

TABLE 3.3. System accuracy for typical input signals.

System type	Steady state error e_{ss}		
	Step input $r(t) = R$	Ramp input $r(t) = Rt$	Parabolic input $r(t) = Rt^2/2$
0	$R/(1+K_p)$	∞	∞
1	0	R/K_v	∞
2	0	0	R/K_a
3	0	0	0

measures of steady state performance.

2. Error constants are determined by the open loop transfer function of the system, and are independent of the structure of the input signal.

3. The steady state error of a linear system perturbed by more than one input is the sum of the errors due to the individual inputs.

4. It is evident from the preceding analysis that offset may be reduced/removed by increasing the open loop gain or the type of the system. In doing so, we must be careful to preserve system stability, the most fundamental of all performance criteria.

Example 3.2.1

Given the open loop transfer function

$$G(s)H(s) = \frac{10s}{(s+5)(s+8)}$$

with input $r(t) = Ce^{\lambda t}$, evaluate the error constants and steady state actuating error.

Because no integration terms are present in the denominator (i.e., $p = 0$ in (3.2.4)), the system is of type zero. Using Table 3.2, the three error constants K_p, K_v and K_a are found to be zero. The steady state error cannot, however, be found from Table 3.3 because the input $r(t)$ does not appear as a column heading. Hence, we refer instead to the general equation for e_{ss} in (3.2.3).

The Laplace transform of $r(t)$ may be obtained from Table 2.4 as

$$\mathcal{L}[Ce^{\nu}] = \frac{C}{s-\lambda}$$

and we also have

$$1+G(s)H(s) = \frac{(s+5)(s+8)+10s}{(s+5)(s+8)}$$

Substituting these expressions into (3.2.3),

$$e_{ss} = \lim_{s \to 0} \frac{s(s+5)(s+8)}{(s+5)(s+8)+10s} \frac{C}{s-\lambda} = 0$$

Example 3.2.2

Consider the system shown in Fig. 3.2.2, where

$$G_1(s) = \frac{K_1}{1+T_1 s} \qquad \text{and} \qquad G_2(s) = \frac{K_2}{s(1+T_2 s)}$$

The Laplace transforms of the input and disturbance are specified as follows:

$$R(s) = R_1 / s \qquad\qquad N(s) = R_n / s$$

What is the steady state error of the system?

The input and disturbance are both step functions, and they are considered separately. In other words, when considering $R(s)$, let $N(s) = 0$; and when considering $N(s)$, let $R(s) = 0$. Since the open loop transfer function $G_1(s)G_2(s)$ has a pole at zero, the system is of type one. K_p is therefore infinite, and the steady state error due to the reference input is zero, i.e.,

$$e_{ss,r} = 0$$

Now, consider the disturbance. The transfer function from $N(s)$ to $E(s)$, $G_{e,n}(s)$, can be found as follows. From the signal relationship in the block diagram and $R(s) = 0$, we have

$$Y(s) = G_2(s)C_2(s)$$
$$C_2(s) = N(s) - C_1(s)$$
$$C_1(s) = G_1(s)E(s)$$
$$E(s) = R(s) - Y(s) = -Y(s)$$

Eliminating $C_1(s)$, $C_2(s)$ and $Y(s)$ yields

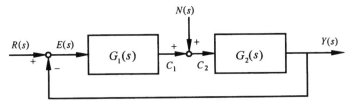

FIGURE 3.2.2. Control system of Example 3.2.2.

$$G_{e,n}(s) = \frac{-G_2(s)}{1 + G_1(s)G_2(s)}$$

Therefore, the steady state error caused by the disturbance signal is

$$e_{ss,n} = \lim_{s \to 0} \frac{-sG_2(s)}{1 + G_1(s)G_2(s)} N(s) = \lim_{s \to 0} \frac{-s\dfrac{K_2}{s(1 + T_2 s)}}{1 + \dfrac{K_1 K_2}{s(1 + T_1 s)(1 + T_2 s)}} \frac{R_n}{s}$$

$$= \lim_{s \to 0} \frac{-K_2(1 + T_1 s)R_n}{s(1 + T_1 s)(1 + T_2 s) + K_1 K_2} = -\frac{R_n}{K_1}$$

Following Remark 3 above, the steady state error of the system is found to be

$$e_{ss} = e_{ss,r} + e_{ss,n} = -\frac{R_n}{K_1}$$

3.3. TRANSIENT RESPONSES

The transient response of a dynamic system refers to the behavior of the process output in response to a change in input. The nature of the transient response depends upon the structure of the input used to perturb the process. We restrict our attention in this section to the step input, because it is by far the most common form of test signal used in practice.

Transient response performance criteria

A number of different performance criteria may be used to characterize the transient response of a system to a step change in input. Figure 3.3.1 portrays the response of a control system to a unit step change in setpoint. With reference to this figure, we may introduce the following standard performance criteria:

1. *Overshoot*. In general, the term overshoot refers to the deviation of the response beyond its steady state value. More commonly, however, it is used to denote the maximum deviation represented by the symbol 'a' in Fig. 3.3.1. Percent overshoot expresses this quantity as a percentage of the final value of the response, i.e.,

$$P.O. = \frac{overshoot}{steady\ state\ value} \times 100\% = \frac{a}{b} \times 100\% \tag{3.3.1}$$

2. *Rise time*. This is the time required for the response to initially attain its steady state value, as represented by 't_r' in the figure. An alternative definition refers to the time for the response to rise from 10% to 90% of its final value.

3. *Settling time*. The settling time, t_s , is the time taken by the output before settling within a band of arbitrary width (e.g., 5%) centered at its steady state value.

4. *Peak time*. Denoted by the symbol 't_p' in Fig. 3.3.1, this is the time at which the output attains its maximum value.

5. *Decay ratio*. The decay ratio (*D.R.*) is defined as the ratio of the height of second peak to that of the first (c/a in Fig. 3.3.1).

6. *Period of oscillation*. The time interval between successive peaks or valleys of the response is referred to as the period of oscillation, denoted by 'P' in Fig. 3.3.1.

It is important to remember that the above performance criteria are defined with reference to the characteristics of an underdamped response. With the exception of settling time, these criteria become meaningless when applied to systems which behave in a different (e.g., overdamped) manner.

Performance criteria for a second order system
The standard form of the transfer function for a second order process was given in Section 2.3 as

$$\frac{Y(s)}{R(s)} = \frac{1}{T^2 s^2 + 2T\zeta s + 1} \qquad (3.3.2)$$

or, equivalently,

$$\frac{Y(s)}{R(s)} = \frac{\omega_n^2}{s^2 + 2\zeta\omega_n s + \omega_n^2} \qquad (3.3.3)$$

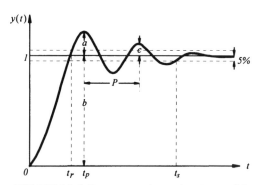

FIGURE 3.3.1 Response of a system to a unit input

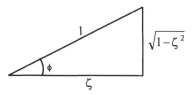

FIGURE 3.3.2. Relationship between φ and ζ.

with $\omega_n = 1/T$ defined as the natural frequency of oscillation. The time response of an underdamped system ($\zeta < 1$) to a unit step input was found to be

$$y(t) = 1 - e^{-\zeta\omega_n t}\frac{1}{\sqrt{1-\zeta^2}}\sin(\omega_n\sqrt{1-\zeta^2}\,t + \phi) \qquad (3.3.4)$$

where

$$\phi = arctg(\sqrt{1-\zeta^2}\,/\zeta) \qquad (3.3.5)$$

as readily observed in Fig. 3.3.2.

Using Eq. (3.3.4), we may derive mathematical expressions of the six performance criteria in terms of the system parameters ω_n and ζ. The derivative of (3.3.4) is

$$\frac{dy(t)}{dt} = \zeta\omega_n e^{-\zeta\omega_n t}\frac{1}{\sqrt{1-\zeta^2}}\sin(\omega_n\sqrt{1-\zeta^2}\,t + \phi) - \omega_n e^{-\zeta\omega_n t}\cos(\omega_n\sqrt{1-\zeta^2}\,t + \phi)$$

$$(3.3.6)$$

and from Fig. 3.3.2, we can obtain

$$\sin\phi = \sqrt{1-\zeta^2} \qquad \text{and} \qquad \cos\phi = \zeta$$

Application of the trigonometric identity

$$\sin\alpha\,\cos\beta \pm \cos\alpha\,\sin\beta = \sin(\alpha \pm \beta)$$

yields

$$\frac{dy(t)}{dt} = \omega_n e^{-\zeta\omega_n t}\frac{1}{\sqrt{1-\zeta^2}}\sin(\omega_n\sqrt{1-\zeta^2}\,t) \qquad (3.3.7)$$

The stationary points of $y(t)$ are obtained when its derivative is zero, i.e., when

$$\omega_n\sqrt{1-\zeta^2}\,t = n\pi \qquad (3.3.8)$$

Maximum overshoot occurs at $n = 1$, and so the peak time of the response is given by

TABLE 3.4. Performance criteria underdamped system.

Item	Symbol in Fig.3.3.1	Mathematical expression	Effect of ζ
Maximum overshoot ($M.O.$)	a	$\exp(-\zeta\pi/\sqrt{1-\zeta^2})$	$\zeta\uparrow$, $M.O.\downarrow$
Percent overshoot ($P.O.$)	a/b	$100\exp(-\zeta\pi/\sqrt{1-\zeta^2})\%$	$\zeta\uparrow$, $P.O.\downarrow$
Rise time	t_r	$\approx(0.8+2.5\zeta)/\omega_n$	$\zeta\uparrow$, $t_r\uparrow$
Settling time	t_s	$\approx 3/\zeta\omega_n$	$\zeta\uparrow$, $t_s\downarrow$
Peak time	t_p	$\pi/(\omega_n\sqrt{1-\zeta^2})$	$\zeta\uparrow$, $t_p\uparrow$
Period of oscillation	P	$2\pi/(\omega_n\sqrt{1-\zeta^2})$	$\zeta\uparrow$, $P\uparrow$
Decay ratio ($D.R.$)	c/a	$\exp(-2\pi\zeta/\sqrt{1-\zeta^2})$	$\zeta\uparrow$, $D.R.\downarrow$

$$t_p = \pi/(\omega_n\sqrt{1-\zeta^2}) \qquad (3.3.9)$$

and the maximum overshoot ($M.O.$) by

$$M.O.= y(t_p)-1 = \frac{-e^{-\zeta\omega_n t_p}}{\sqrt{1-\zeta^2}}\sin(\pi+\phi) = \frac{-e^{-\zeta\omega_n t_p}}{\sin\phi}(-\sin\phi)$$

$$= \exp(-\zeta\omega_n t_p) = \exp(-\zeta\pi/\sqrt{1-\zeta^2}) \qquad (3.3.10)$$

It then follows that

$$P.O.= 100\exp(-\zeta\pi/\sqrt{1-\zeta^2})\% \qquad (3.3.11)$$

The decay ratio ($D.R.$) and period of oscillation (P) are derived in similar fashion:

$$D.R.= \exp(-2\pi\zeta/\sqrt{1-\zeta^2}) \qquad (3.3.12)$$

$$P = 2\pi/(\omega_n\sqrt{1-\zeta^2}) \qquad (3.3.13)$$

Unfortunately, the rise time and settling time are not so easily computed; approximate relationships [Kuo 1991] have been included in Table 3.4 along with the exact expressions (3.3.9)–(3.3.13) derived above.

The performance criteria listed in the table may also be applied to higher order systems having a step response similar to that of Fig. 3.3.1.

Transient response of a system with zeros

When the transfer function of a system has one or more zeros, i.e., the numerator of the transfer function is not a constant, overshoot may be observed in the response even when $\zeta > 1$. This point is illustrated by the following example.

Example 3.3.1

Consider a second order system with transfer function:

$$G(s) = \frac{K(T_a s + 1)}{(T_b s + 1)(T_c s + 1)} \qquad (3.3.14)$$

The response of the system to a step input of size R may be obtained as

$$y(t) = KR(1 + \frac{T_a - T_b}{T_b - T_c} e^{-t/T_b} + \frac{T_a - T_c}{T_c - T_b} e^{-t/T_c}) \qquad (3.3.15)$$

using the procedure of partial fraction expansion and inverse Laplace transformation. From this equation it is clear that the steady state value of $y(t)$ is KR. If we assign specific values to the constants in (3.3.15), e.g.,

$$T_a = 11, \quad T_b = 3, \quad T_c = 1, \quad K = 1 \quad \text{and} \quad R = 1$$

then

$$y(t) = 1 + 4e^{-t/3} - 5e^{-t} \qquad (3.3.16)$$

This function is sketched in Fig. 3.3.3. Notice that the response exhibits overshoot despite the fact that $\zeta = 2/\sqrt{3} > 1$.

Another interesting consequence of zeros in the system transfer function is the potential for inverse or wrong-way response. The initial reaction of a process exhibiting inverse response to a step change in input is to move further away from its eventual value. This phenomenon arises due to the presence of

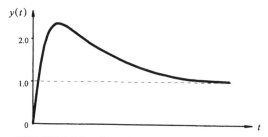

FIGURE 3.3.3. Overshoot in the step response.

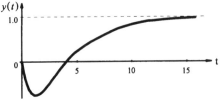

FIGURE 3.3.4. Inverse response.

right half s-plane zeros in the transfer function. Systems with 'positive' zeros are particularly difficult to control and require special attention from the control engineer.

Example 3.3.2

Substituting the alternative values

$$T_a = -5, \quad T_b = 3, \quad T_c = 1, \quad K = 1 \quad \text{and} \quad R = 1$$

into Eq. (3.3.14) results in the transfer function $G(s)$ having a positive zero at $s = 0.2$. (This is a non-minimum phase system, and we will discuss it in Chapter 5.) The system output now follows the expression

$$y(t) = 1 - 4e^{-t/3} + 3e^{-t} \tag{3.3.17}$$

and displays wrong-way response (see Fig. 3.3.4.).

On a physical level, inverse response may be viewed as the net effect of competing forces acting on the process. An important chemical engineering example is level control in the base of a distillation column as shown in Fig. 3.3.5 [Seborg et al. 1989]. Inverse response of liquid level is often encountered when sudden changes are made to steam pressure in the reboiler. This is

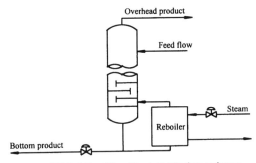

FIGURE 3.3.5. Simplified distillation column.

because an increase in steam pressure results in both increased frothing and decreased liquid flow in the downcomer. The former effect predominates initially, causing the level to increase, but the reduction in liquid flow eventually leads to the expected decrease in level.

Transient response to an impulse input
The response of a system with transfer function $G(s)$ to an impulse input of unit area is

$$\mathcal{L}^{-1}[Y(s)] = \mathcal{L}^{-1}[G(s) \cdot \mathcal{L}[\delta(t)]] = \mathcal{L}^{-1}[G(s)] \tag{3.3.18}$$

because

$$\mathcal{L}[\delta(t)] = 1$$

From the convolution theorem,

$$y(t) = \int_0^t g(t - \tau)\delta(\tau)d\tau \tag{3.3.19}$$

The integrand is only nonzero at $\tau = 0$, and therefore

$$y(t) = g(t) \tag{3.3.20}$$

For this reason, the function $g(t) = \mathcal{L}^{-1}[G(s)]$ is known as the impulse response of the system $G(s)$.

The Laplace transformation of the transient response of a system to a unit impulse input is the transfer function of the system. In other words, the inverse Laplace transform of the system transfer function is the response of the system to a unit impulse. This fact may in principle be used to find the transfer function of an unknown system. However, as discussed earlier, exact impulse functions are not physically realizable. In practice, the width of the 'impulse' is chosen to be less than half the smallest time constant of interest, and the height is designed for unity area.

General information for using subprogram TDA
The PCET software package included with this textbook will enable the reader to examine the dynamics of up to 5th order systems using the subprogram TDA. The process/system models can be in three types as follows:
   Model 1. ($n \le 5$, $m \le 4$) A polynomial model:

$$G(s) = \frac{a_m s^m + a_{m-1}s^{m-1} + \cdots + a_1 s + a_0}{b_n s^n + b_{n-1}s^{n-1} + \cdots + b_1 s + b_0} e^{-\tau s} \tag{3.3.21}$$

   Model 2. ($n \le 5$, $m \le 4$) A zero-pole model:

$$G(s) = K_0 \frac{(s-z_m)(s-z_{m-1})\cdots(s-z_0)}{(s-p_n)(s-p_{n-1})\cdots(s-p_0)} e^{-\tau s} \qquad (3.3.22)$$

Model 3. A typical process model:

$$G(s) = \frac{K_0}{(T_1 s + 1)(T^2 s^2 + 2T\zeta s + 1)} e^{-\tau s} \qquad (3.3.23)$$

The following limitations exist:

1. The order of the process denominator should be less than or equal to 5, and that of numerator should be less than or equal to 4.

2. For the process model, the order of the numerator has to be less than that of the denominator, if the closed loop analysis is used and the controller has a derivative action.

3. Long time scale may bring errors. Usually, keep the abscissa less than 100 time units.

Plotting system open loop responses with *PCET*

The procedure for plotting the open loop system response to a unit step input is as follows:

1. Type *PCET* to begin, and select *TDA*. (Alternatively, type *TDA* to begin.)
2. Select the type of open loop models.
3. Press 1 for "Analysis of the process (open loop)".
4. Input the order of the system and the system parameters.
5. Verify the data you have entered.
6. Select autoscaling or input the scaling dimensions.
7. The result is displayed on the screen.

Example 3.3.3

Plot the step response of a third order system

$$G(s) = \frac{4}{(5s+3)(18s^2+5s+2)}$$

This transfer function can be rewritten in standard form as

$$G(s) = \frac{0.67}{(1.67s+1)(3^2 s^2 + 2 \times 3 \times 0.417s + 1)}$$

Comparing this transfer function with (3.3.23) (Model 3 used in *PCET*), we have the following parameters:

$$K_0 = 0.67, \qquad T_1 = 1.67, \qquad T = 3, \qquad x = 0.417$$

where x is used to represent the damping factor ζ in *PCET*. After selecting model 3 and entering these data on the *TDA* screen (cf. Fig. 3.3.6a), we obtain the step response curve shown in Fig. 3.3.6b.

Models obtained from the transient response
Mathematical models derived from first principles are often too complex to be of practical value to control engineers. In these situations, low order linear models may be obtained empirically as follows:

1. Select a test input (e.g., step or impulse) and perturb the system.

2. Propose a model structure consistent with the shape of the transient response.

3. Fit the model parameters to the response data.

Model structures selected in Step 2 are normally required to be of degree two

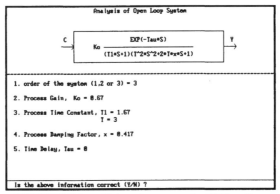

FIGURE 3.3.6a.   Data entry screen for TDA option.

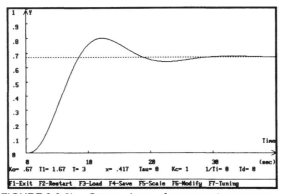

FIGURE 3.3.6b.   Screen dump of process step response.

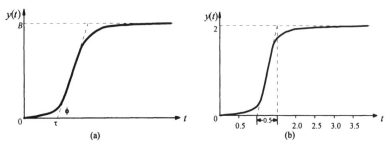

FIGURE 3.3.7. Unit step response for estimating model parameters.

or lower, because of difficulties associated with estimating the parameters of higher order models. For many industrial processes, the plant dynamics are well approximated as first-order-plus-time-delay, i.e.,

$$G(s) = \frac{Ke^{-\tau s}}{Ts + 1} \qquad (3.3.24)$$

where K, T and τ are as defined previously in Section 2.3. In estimating these constants from the step response, K is taken as the ratio of the change in steady state value of the response to the size of the step input, T as the ratio of the change in steady state to the maximum slope of the response, and τ as the time elapsed before the system begins to respond. Time delay τ can be measured on the response curve, while the parameters K and T may be calculated from

$$K = B/A \qquad (3.3.25)$$

$$T = B/S \qquad (3.3.26)$$

where A is the amplitude of the step input, B is the steady state value of the response, and $S = \tan\phi$ is the slope of the response as shown in Fig. 3.3.7a.

Example 3.3.4
    Fit a first-order-plus-time-delay model to the unit step response shown in Fig. 3.3.7b, where $B = 2$, $\tau = 1$ and $\tan\phi = 4$.
    The change in steady state of the process output is 2 units, and thus $K = 2/1 = 2$. The maximum slope of the response is 4:1, giving a time constant of $2/4 = 0.5$. With the time delay to be unity, the transfer function is

$$G(s) = \frac{2e^{-s}}{0.5s + 1}$$

3.4. *PID* CONTROL

The vast majority of industrial feedback systems employ either on-off or proportional-integral-derivative (*PID*) control laws. A *PID* controller includes three modes: proportional, integral and derivative actions. The purpose of this section is to examine each of these strategies in detail.

On-off control

The output of an on-off controller may assume one of only two possible values:

$$c(t) = \begin{cases} P_{on} & if\ e \geq 0 \\ P_{off} & if\ e < 0 \end{cases} \qquad (3.4.1)$$

where e is the input of the controller, i.e., the actuating error. The on-off controller is also referred to as a 'two-position' or 'bang-bang' controller.

If the actuating error signal $e(t)$ is noisy, an on-off controller will try to 'control the noise'. This effect can be reduced by modifying the original design to include a dead band, i.e.,

$$c(t) = \begin{cases} P_{on} & if\ e \geq e_0 \\ no\ change & if\ -e_0 < e \leq e_0 \\ P_{off} & if\ e < -e_0 \end{cases} \qquad (3.4.2)$$

The input-output relationships of the on-off controllers (3.4.1) and (3.4.2) are illustrated in Fig. 3.4.1. Obviously, with an on-off controller, the control strategy is different from that used with other controllers. The output of the system with an on-off controller cannot follow the setpoint.

On-off controllers are inexpensive and easy to implement. They are used industrially in many level control loops, and have found wide commercial application as thermostats in domestic heating systems and refrigerators. The main problem with on-off control is the cycling of the controlled variable, which is unacceptable in a quality control problem, and usually leads to excessive wear and tear on the final control element.

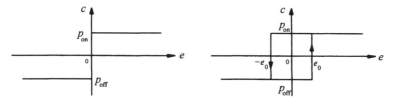

a. On-off controller　　　　b. Modified on-off controller

FIGURE 3.4.1.　Relationship between output and input of on-off controller.

Proportional control

The output $c(t)$ of a proportional controller (or P controller) is proportional to the actuating error signal $e(t)$:

$$c(t) = K_c e(t) + c_0 \qquad (3.4.3)$$

K_c is a user-specified constant called the proportional gain, and c_0 is a bias term corresponding to the value of controller output when $e = 0$.

Expressing the output of the P controller as the deviation variable (cf. Section 2.1)

$$\Delta c(t) = c(t) - c_0(t) \qquad (3.4.4)$$

permits (3.4.3) to be rewritten as

$$\Delta c(t) = K_c e(t) \qquad (3.4.5)$$

From this it is clear that the transfer function of a proportional controller is a constant that is equal to the proportional gain K_c. (It is of course unnecessary to transform $e(t)$ to deviation variable form, as $e(0) = 0$ is implied in the assumption $y(0) = r(0)$).

Alternative expressions of the proportional control law replace K_c with the proportional band (PB), defined as

$$PB = 100/K_c$$

for dimensionless K_c. The larger the proportional gain (or equivalently, the smaller the proportional band) is, the more the controller output $c(t)$ changes for a given error, and vice versa.

As illustrated later in this section, the principal disadvantage of proportional control is the presence of offset in response to step changes in disturbance and/or setpoint.

Integral control

The relationship between the output and input of an integral controller is given by

$$c(t) = c_0 + \frac{1}{\tau_I} \int_0^t e(t) dt \qquad (3.4.6)$$

where τ_I is referred to as the integral (or reset) time, and $1/\tau_I$ is called reset rate. In terms of deviation variables, (3.4.6) becomes

$$\Delta c(t) = \frac{1}{\tau_I} \int_0^t e(t) dt \qquad (3.4.7)$$

It is clear from these expressions that the controller output changes continuously until the actuating error is zero; it is this feature which allows the integral controller to eliminate steady state errors caused by step changes in input.

Integral control normally appears in conjunction with proportional control, since integral-only controllers require a long period of time to generate a strong response. The proportional-integral (*PI*) control equation is

$$\Delta c(t) = K_c[e(t) + \frac{1}{\tau_I}\int_0^t e(t)dt] \tag{3.4.8}$$

Taking Laplace transforms, we obtain the transfer function

$$\frac{\Delta c(s)}{E(s)} = K_c(1 + \frac{1}{\tau_I s}) = K_c(\frac{\tau_I s + 1}{\tau_I s}) \tag{3.4.9}$$

The response of a *PI* controller to a unit step actuating error is shown in Fig. 3.4.2. At time zero, the controller output $c(t) = K_c$, since the integral term is zero. As time passes, the integral mode contributes K_c additional units to the output signal every τ_I time units. For this reason, the integral time is usually specified in units of minutes/repeat.

The introduction of integral action has an adverse effect on system stability because it usually leads to oscillations in the closed loop response. The integral time τ_I (reset rate $1/\tau_I$) must be chosen small (large) enough to eliminate offset within a reasonable length of time, yet large (small) enough to ensure that oscillations damp out quickly.

A further difficulty with integral control arises when the actuating error is sustained over a relatively long period of time, e.g., after an abrupt setpoint change or during start-up of a batch process. In these instances, the integral term increases continuously (cf. Eq. (3.4.8)), leading to eventual saturation of the manipulated variable. If at this point the error remains nonzero, then reset windup occurs, i.e., the integral increases still further, demanding control actions

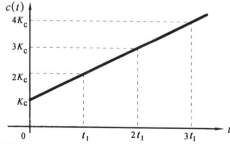

FIGURE 3.4.2. Response of *PI* controller to a step change in error.

that cannot be implemented. To prevent this, all commercial *PID* controllers possess an anti-reset-windup feature designed to restrict growth of the integral term when the system is operating at a physical constraint.

Derivative control
The idea behind derivative control is to anticipate future control errors and compensate for them before they occur. This is accomplished by making the controller output proportional to the rate of change of the actuating error signal, i.e.,

$$c(t) = c_0 + \tau_D \frac{de(t)}{dt} \qquad (3.4.10)$$

or

$$\Delta c(t) = \tau_D \frac{de(t)}{dt} \qquad (3.4.11)$$

where τ_D is a user-specified parameter known as the derivative time. Derivative action is used only in combination with proportional or proportional-integral control. The equation of a *PD* controller is

$$\Delta c(t) = K_c(e(t) + \tau_D \frac{de(t)}{dt}) \qquad (3.4.12)$$

Interpreting de/dt as the slope of the error curve, we see that the derivative mode in (3.4.12) acts in opposition to the current trend of the error. This helps to reduce excessive overshoot, settling time and oscillatory behavior in the closed loop response.

***PID* controller**
The *PID* controller equation is formed by summing the contributions of the three basic control modes:

$$\Delta c(t) = K_c[e(t) + \frac{1}{\tau_I} \int_0^t e(t)dt + \tau_D \frac{de(t)}{dt}] \qquad (3.4.13)$$

which has Laplace domain equivalent:

$$\frac{\Delta C(s)}{E(s)} = K_c(1 + \frac{1}{\tau_I s} + \tau_D s) \qquad (3.4.14)$$

The ideal *PD* and *PID* control laws given here cannot actually be implemented in practice. This is because the order of the numerator of the transfer function (3.4.14) (and of the *PD* transfer function corresponding to

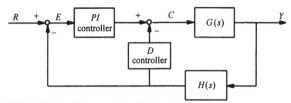

FIGURE 3.4.3.   *PID* control with derivative on measurement.

(3.4.12)) exceeds that of the denominator. Commercial *PID* controllers are based instead upon transfer functions of the form:

$$\frac{\Delta C(s)}{E(s)} = K_c (\frac{\tau_I s + 1}{\tau_I s})(\frac{\tau_D s + 1}{\alpha \tau_D s + 1}) \qquad (3.4.15)$$

with α chosen to be a small number. This relationship can be viewed as the transfer function of a *PI* controller placed in series with an 'approximate' *PD* controller.

An additional modification to the ideal *PID* controller must be introduced to prevent derivative kick on setpoint changes. From (3.4.13) it can be seen that a step change in setpoint (and hence in actuating error $e(t)$) results in a theoretically infinite change or 'kick' in the controller output. As illustrated in Fig. 3.4.3, this problem is alleviated by implementing derivative action on the measurement signal only. However, if the measurement is noisy (as is often the case with flow control), then a similar problem occurs. Derivative action should not be used in such applications, as these large swings in controller output produce excessive wear and tear on the final control element.

The key features of the on-off, proportional, integral and derivative modes are displayed in Table 3.5. Table 3.6 summarizes those of the *P*, *PI* and *PID* control laws.

TABLE 3.5.   Feedback control modes.

Control action	Symbol	Driving force	Response	Effect of tuning parameter
On-off		Existence of error	Switch between limits P_{on} and P_{off}	
Proportional	P	Magnitude of error	Proportionate	$K_c \uparrow$, P action$\uparrow$
Integral	I	Accumulation of error	Resets bias	$\tau_I \uparrow$, I action$\downarrow$
Derivative	D	Rate of error	Anticipatory	$\tau_D \uparrow$, D action$\uparrow$

TABLE 3.6.   Basic properties of *PID*-type controller.

Item	Proportional	Proportional-integral	Proportional-integral-derivative
Acronym	P	PI	PID
Adjustable parameters	K_c	K_c, τ_I	K_c, τ_I, τ_D
Time domain expression	K_c	$K_c[e(t)+\dfrac{1}{\tau_I}\displaystyle\int_0^t e(t)dt]$	$K_c[e(t)+\dfrac{1}{\tau_I}\displaystyle\int_0^t e(t)dt +\tau_D\dfrac{de(t)}{dt}]$
Transfer function	K_c	$K_c(\dfrac{\tau_I s+1}{\tau_I s})$	$K_c(1+\dfrac{1}{\tau_I s}+\tau_D s)$
Potential problems	Offset	Oscillations/ reset windup	Derivative kick
Solution	Increase K_c	Increase τ_I/ Implement antireset-windup	Derivative action on measurement only

Controller analysis with *PCET*

The response of a *PID* controller to a unit step change in actuating error may be simulated with the *PCET* subprogram *TDA*, as described below.

   1. Type *PCET* to begin, and select *TDA*. (Alternatively, type *TDA* to begin.)

   2. Select the type of open loop models.

   3. Press 2 for "Analysis of the controller action".

   4. Input the controller parameters.

   5. Verify the input data.

   6. Select autoscaling or input the scaling dimensions.

   7. The result should now be shown on the screen.

This option provides a convenient means of constructing open loop response curves of the form of Fig. 3.4.2 (cf. Problems 13~16).

3.5. EFFECT OF CONTROLLER PARAMETERS ON CLOSED LOOP RESPONSE

Consider the unity-feedback control system shown in Fig. 3.5.1. $Y(s)$, $R(s)$ and $D(s)$ represent Laplace transforms of the output, setpoint and disturbance signals, while $G(s)$, $G_c(s)$ and $F(s)$ denote the plant, controller and load transfer functions. Routine algebraic manipulation yields:

$$Y(s) = \frac{G_c(s)G(s)}{1+G_c(s)G(s)} R(s) + \frac{F(s)}{1+G_c(s)G(s)} D(s) \qquad (3.5.1)$$

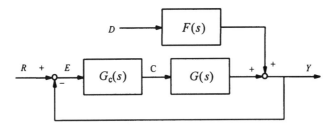

FIGURE 3.5.1 Feedback control system with disturbance at output

When the dynamics of the plant, load and input signals are specified, this expression may be used to describe the closed loop performance of the system in terms of the *PID* parameters K_c, τ_I and τ_D.

Effect of proportional feedback
Assume that $G(s)$ and $F(s)$ are first order transfer functions with a common time constant. Then

$$G(s) = \frac{K_g}{Ts+1}, \qquad F(s) = \frac{K_f}{Ts+1}, \qquad G_c(s) = K_c \qquad (3.5.2)$$

Substituting (3.5.2) into (3.5.1) yields

$$Y(s) = \frac{K_{0g}}{T_0 s+1} R(s) + \frac{K_{0f}}{T_0 s+1} D(s) \qquad (3.5.3)$$

where

$$T_0 = \frac{T}{1+K_c K_g}, \qquad K_{0g} = \frac{K_c K_g}{1+K_c K_g}, \qquad K_{0f} = \frac{K_f}{1+K_c K_g} \qquad (3.5.4)$$

The parameters K_{0g} and K_{0f} are closed loop static gains. It is evident from (3.5.3) and (3.5.4) that the closed loop transfer functions are first order, with time constant and gains smaller than those of the open loop systems. For a servo problem with unit step input, $R(s) = 1/s$, $D(s) = 0$, and thus

$$Y(s) = \frac{K_{0g}}{T_0 s+1} \cdot \frac{1}{s} \qquad (3.5.5)$$

The steady state output of the system is given by

$$Y(\infty) = \lim_{s \to 0} s \frac{K_{0g}}{T_0 s+1} \cdot \frac{1}{s} = K_{0g} \qquad (3.5.6)$$

which results in the steady state error:

$$e(\infty) = r(\infty) - y(\infty) = 1 - K_{0g} = \frac{1}{1 + K_c K_g} \qquad (3.5.7)$$

For the analogous regulator problem (i.e., $R(s) = 0$ and $D(s) = 1/s$),

$$Y(s) = \frac{K_{0f}}{T_0 s + 1} \cdot \frac{1}{s} \qquad (3.5.8)$$

The final value of the output and its corresponding steady state error now become

$$Y(\infty) = \lim_{s \to 0} s \frac{K_{0f}}{T_0 s + 1} \cdot \frac{1}{s} = K_{0f} \qquad (3.5.9)$$

and

$$e(\infty) = r(\infty) - y(\infty) = 0 - K_{0f} = \frac{-K_f}{1 + K_c K_g} \qquad (3.5.10)$$

Remarks

1. For both servo and regulatory problems, a larger proportional gain results in a smaller steady state error. Theoretically speaking, the steady state error tends to zero as K_c approaches infinity. In practice, however, large gains are known to give rise to other problems, such as instability and amplification of measurement noise.

2. If the transfer function $G(s)$ is pure integral element, then the closed loop system remains first order, but exhibits zero steady state error for step changes in setpoint (see Problem 3.9).

3. When the time constants of first order transfer functions $G(s)$ and $F(s)$ are different, the closed loop transfer function for the regulator problem is second order in structure with a single zero.

Consider next the case of a proportional feedback system where the plant and load transfer functions are given by

$$G(s) = \frac{K_g}{(T_1 s + 1)(T_2 s + 1)}, \qquad F(s) = \frac{K_f}{(T_1 s + 1)} \qquad (3.5.11)$$

$G(s)$ may be written in standard form as

$$G(s) = \frac{K_g}{T^2 s^2 + 2T\zeta s + 1} \qquad (3.5.12)$$

where

$$T = T_1 T_2, \qquad \zeta = \frac{T_1 + T_2}{2\sqrt{T_1 T_2}} \qquad (3.5.13)$$

The damping factor is nonnegative because the time constants T_1 and T_2 are nonnegative; that $\zeta \geq 1$ may be proven using the inequality

$$(\sqrt{T_1} - \sqrt{T_2})^2 \geq 0$$

For the regulator problem, $R(s) = 0$ and (3.5.1) yields

$$\frac{Y(s)}{D(s)} = \frac{K_f(T_2 s + 1)}{(T_1 s + 1)(T_2 s + 1) + K_c K_g} \qquad (3.5.14)$$

which is itself a second order system. In standard form,

$$\frac{Y(s)}{D(s)} = \frac{K_{0f}(T_2 s + 1)}{T_0^2 s^2 + 2 T_0 \zeta_0 s + 1} \qquad (3.5.15)$$

where

$$K_{0f} = \frac{K_f}{1 + K_c K_g} \qquad (3.5.16)$$

$$T_0 = \frac{T_1 T_2}{1 + K_c K_g} \qquad (3.5.17)$$

and

$$\zeta_0 = \frac{T_1 + T_2}{2\sqrt{T_1 T_2 (1 + K_c K_g)}} \qquad (3.5.18)$$

The inequalities $|K_{0f}| \leq |K_f|$, $T_0 \leq T$ and $\zeta_0 \leq \zeta$ result from the fact that K_c and K_g are necessarily of the same sign in a stable, negative feedback system. The magnitude of the static gain, the natural period of oscillation and the damping factor decrease as the gain of the controller increases. Because $\zeta \geq 1$ and $\zeta_0 \leq \zeta$, it follows that increasing K_c beyond some critical value leads to an underdamped closed loop response.

Effect of derivative action
We now wish to examine the effect of adding derivative action to the proportional controller (with $G(s)$ and $F(s)$ as in (3.5.11) above). The controller transfer function $G_c(s)$ is given by

$$G_c(s) = K_c(1 + \tau_D s) \qquad (3.5.19)$$

and the closed loop transfer function between the disturbance and the system output becomes

$$\frac{Y(s)}{D(s)} = \frac{K_f(T_2 s + 1)}{T_1 T_2 s^2 + (T_1 + T_2 + \tau_D K_c K_g)s + 1 + K_c K_g} \tag{3.5.20}$$

This is once again a second order system with natural period of oscillation

$$T = \frac{T_1 + T_2}{1 + K_c K_g} \tag{3.5.21}$$

and damping factor

$$\zeta = \frac{T_1 + T_2 + \tau_D K_c K_g}{2\sqrt{T_1 T_2 (1 + K_c K_g)}} \tag{3.5.22}$$

The denominators of (3.5.18) and (3.5.22) are identical, but the numerator of (3.5.22) is made larger than that of (3.5.18) by the addition of the positive term $\tau_D K_c K_g$. It follows that incorporating derivative action into a proportional controller improves system stability by increasing the closed loop damping factor.

Application of the final value theorem to (3.5.20) with $D(s) = 1/s$ gives

$$e(\infty) = r(\infty) - y(\infty) = 0 - y(\infty)$$

$$= -\lim_{s \to 0} s \cdot \frac{1}{s} \cdot \frac{K_f(T_2 s + 1)}{T_1 T_2 s^2 + (T_1 + T_2 + \tau_D K_c K_g)s + 1 + K_c K_g}$$

$$= \frac{-K_f}{1 + K_c K_g} \tag{3.5.23}$$

which is consistent with (3.5.10), i.e., the derivative mode has no effect on the steady state error.

Effect of integral action

Let us consider the effect of adding integral control action into a simple proportional controller. The transfer function of a *PI* controller is

$$G_c(s) = K_c(1 + \frac{1}{\tau_I s}) \tag{3.5.24}$$

For the regulator problem with $G(s)$ and $F(s)$ defined as

$$G(s) = \frac{K_g}{Ts + 1}, \qquad F(s) = \frac{K_f}{Ts + 1} \tag{3.5.25}$$

we obtain

$$\frac{Y(s)}{D(s)} = \frac{K_f \tau_I s}{T\tau_I s^2 + (1 + K_c K_g)\tau_I s + K_c K_g} \qquad (3.5.26)$$

This is a second order system with damping factor

$$\zeta = (1 + K_c K_g) / 2 \sqrt{\frac{T K_c K_g}{\tau_I}} \qquad (3.5.27)$$

It is apparent that ζ may be made greater than one if τ_I is made large enough, and may be less than one if τ_I is small enough. In other words, the smaller the integral time τ_I, the stronger the integral control action, and the more oscillatory the response becomes to changes in load. The nature of the closed loop response depends just as strongly, however, upon the value chosen for the proportional gain K_c.

The steady state error

$$e(\infty) = r(\infty) - y(\infty) = 0 - y(\infty)$$

$$= -\lim_{s \to 0} s \cdot \frac{A}{s} \cdot \frac{K_f \tau_I s}{T\tau_I s^2 + (1 + K_c K_g)\tau_I s + K_c K_g} = 0 \qquad (3.5.28)$$

implies that integral action eliminates offset resulting from step changes in load. This important property justifies the inclusion of integral action in the three-term (*PID*) controller.

The general qualitative effects of proportional, derivative and integral control on the response of a system to a step change in input are summarized in Table 3.7. It should be noticed that the effects in the table are not true if a pure integral element is included in the system.

Closed loop analysis using *PCET* software
The unit step response of a closed loop system under *PID* control may be viewed

Table 3.7.   Effect of controller parameters on system performance.

Item	Increase proportional gain	Increase derivative time	Increase reset rate
Damping factor	Decreases	Increases	Decreases
Overshoot	Increases	Decreases	Increases
Decay ratio	Increases	Decreases	Increases
Steady state error	Decreases	No change	Eliminated
Stability	Decreases	Increases	Decreases
Closed loop gain	Decreases	No change	No change

using the following procedure:
1. Type *PCET* to begin, and select *TDA*. (Alternatively, type *TDA* to begin.)
2. Select the type of open loop models.
3. Press 3 for "Analysis of the closed loop response".
4. Input the process data.
5. Verify the process data you have entered.
6. Input the controller parameters.
7. Verify the controller data.
8. Select autoscaling or input the scaling dimensions.
9. Observe the result.

Summary of *TDA*

In these projects, *PCET* subprogram *TDA* is used to obtain the unit step response of the process, controller and closed loop system. The keystroke sequences required to execute each of these options are summarized for convenience in Table 3.8.

Figure 3.5.2 was obtained in this manner for the plant model

Table 3.8. Plotting the unit step response using *PCET*.

Step	User input	Function
1	Type *PCET*	Start program
2	Select *TDA*	Time domain analysis
3	Press 1, 2 or 3 for model selection	1. Polynomial model 2. Pole-zero model 3. A typical process model
4	Press 1, 2, 3 or 4	1. Analysis of open loop process 2. Analysis of controller action 3. Analysis of closed loop response 4. Change setpoint/disturbance
5	Input system model and controller parameters	Specify system dynamics.
6	Revise input if necessary	Verify input data
7	Select autoscaling, or input figure dimension	Autoscaling computes axis dimension from response data
8		Desired response is displayed on the screen
9	Press a function key to continue	F1-Exit, back to main Menu F2-Restart, restart *TDA* F3-Load, load a saved data file F4-Save, save the data in a file F5-Scale, change the dimension F6-Modify, change the model F7-tuning, change controller

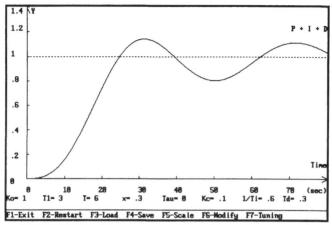

FIGURE 3.5.2.   Screen dump of closed loop step response.

$$G(s) = \frac{K_0}{(T_1 s + 1)(T^2 s^2 + 2T\zeta s + 1)}$$

The model parameters, and those of the *PID* controller, are displayed in the bottom of the figure.

3.6. PERFORMANCE CRITERIA

The concept of a performance criterion was defined earlier as a quantitative measure of system performance. Much of what is known as 'modern control theory' is concerned with finding the controller which minimizes a particular performance criterion, e.g., mean square actuating error. Such a control law is then said to be optimal with respect to that criterion. In conventional (as opposed to optimal) controller synthesis, the performance criteria may be defined on the step or impulse response plots. These criteria may be classified as either steady state (as in the case of offset) or transient; the latter can be further categorized according to whether they measure characteristics of the entire transient, or make use of only a few points of the closed loop response.

Criteria that use only a few points of the response
Each of the performance measures introduced in Section 3.3 falls into this category. When tuning a *PID* controller, the engineer often seeks to minimize two or more of these criteria simultaneously. Because they are to some extent

inversely related (e.g., one cannot achieve both minimum rise time and minimum settling time), the tuning process involves trade-offs among competing performance objectives. In this connection, the decay ratio has been identified as a single parameter with which a compromise can be effected between the various criteria. A quarter decay response (i.e., decay ratio = 1/4) was suggested by Cohen and Coon. For a second order system response, this would imply a damping factor $\zeta \approx 0.2$ and a 50% overshoot to a step change in input. The Cohen and Coon design method will be studied in Chapter 7.

Time integral performance criteria
Performance measures which utilize the entire transient response usually assume the form of a time integral of the actuating error function. Common examples are the integral of the squared error (*ISE*)

$$J_1 = \int_0^\infty e^2(t)dt \tag{3.6.1}$$

and integral of the absolute error (*IAE*)

$$J_2 = \int_0^\infty |e(t)|dt \tag{3.6.2}$$

In general, tuning a *PID* controller for minimum *ISE* and minimum *IAE* will lead to different results. The *ISE* is a better criterion for suppressing large errors because a squared error greater than unity contributes more to the integral than does an absolute error. Conversely, the *IAE* leads to controllers which better suppress small errors.

In an effort to reduce the importance of large initial errors, and place greater emphasis on small errors occurring later in the response, the integral of the time-weighted absolute error (*ITAE*)

$$J_3 = \int_0^\infty t|e(t)|dt \tag{3.6.3}$$

and integral of the time-weighted squared error (*ITSE*)

$$J_4 = \int_0^\infty te^2(t)dt \tag{3.6.4}$$

criteria have been proposed. The *ITAE* index is generally preferred over the *ITSE* because it places greater penalty on small errors occurring at large time. Indeed, the *ITAE* is the most popular among the four criteria, probably because its use results in the most conservative controller design [Lopez et al. 1967]. The four integral performance criteria are summarized in Table 3.9.

Table 3.9. Integral error performance criteria.

Integral error	Acronym	Definition
Integral of squared error	ISE	$J_1 = \int_0^\infty e^2(t)dt$
Integral of absolute error	IAE	$J_2 = \int_0^\infty |e(t)|dt$
Integral of time-weighted absolute error	ITAE	$J_3 = \int_0^\infty t|e(t)|dt$
Integral of time-weighted squared error	ITSE	$J_4 = \int_0^\infty te^2(t)dt$

Remarks

1. For computational purposes, the upper limit of the integrals may be replaced by the settling time t_s.

2. Different controllers will result when tuning for different criteria and/or different test inputs.

3. As shown in Section 3.5, proportional controllers lead to non-zero steady state errors in response to step inputs. Because offset implies infinite *ISE*, *IAE*, *ITAE* and *ITSE*, these tuning methods are restricted for use with controllers whose structure guarantees zero steady state error for the test input in question.

Problems

3.1. For a unity-feedback control system, the open loop transfer function is

$$G(s) = \frac{10}{(0.1s + 1)(0.5s + 1)}$$

What is the step error constant? the ramp error constant? Determine (separately) the steady state errors resulting from a unit step input and a ramp input with unit slope.

3.2. Consider a unity-feedback control system with open loop transfer function

$$G(s) = \frac{2}{s(s + 1)(0.5s + 1)}$$

Compute the step and ramp error constants for this system. Find the steady state errors for a unit step input and for a ramp input with $r(t) = 5t$.

3.3. Calculate the parabolic error constant of a unity-feedback control system with open loop transfer function

$$G(s) = \frac{8(0.5s + 1)}{s^2(0.1s + 1)}$$

Find the steady state error in response to the input $r(t) = 5t^2$.

3.4. Consider the system of Fig. 3.P.1. Let the dynamic elements G_1 and G_2 be defined as

$$G_1(s) = \frac{K(\tau s + 1)}{s}, \qquad G_2(s) = \frac{1}{s(Ts + 1)}$$

with $K = 316$, $\tau = 0.1$ sec and $T = 0.01$ sec.

i) Find the error constants K_p, K_v and K_a.

ii) Calculate the steady state error when the input signals are given by

   a. $r(t) = 2u(t)$ and $f(t) = 0$
   b. $r(t) = (2 + 4t)u(t)$ and $f(t) = 0$
   c. $r(t) = (2 + 4t + 6t^2)u(t)$ and $f(t) = 0$
   d. $f(t) = 3u(t)$ and $r(t) = 0$
   e. $r(t) = (2 + 4t)u(t)$ and $f(t) = 3u(t)$

where $u(t)$ is a unit step input.

3.5. Consider the feedback system illustrated in Fig. 3.P.2. Can you find a suitable coefficient f which guarantees zero steady state error for the ramp input $r(t) = at$?

3.6. Does the feedback system of Figure 3.P.3 exhibit inverse response to a unit step change in $r(t)$? Verify your prediction by computing the time response.

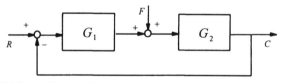

FIGURE 3.P.1.   Feedback control system with disturbance at plant input.

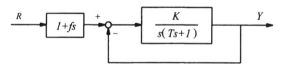

FIGURE 3.P.2.   A control system of Problem 3.5.

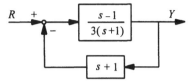

FIGURE 3.P.3. A Feedback control system.

3.7. Calculate the time response of the second order process (3.3.3) with $\zeta = 0.6$ and $\omega_n = 5 \text{ sec}^{-1}$ to a unit step change in input. Evaluate the system overshoot, rise time, settling time and peak time.

3.8. For the system shown in Fig. 3.P.4a, evaluate the rise time, peak time, maximum overshoot and settling time. Determine the value of the velocity feedback coefficient K which guarantees $\zeta = 0.5$ for the system of Fig. 3.P.4b. Reevaluate the performance criteria and compare with the original values.

3.9. For the unity-feedback control system shown in Fig. 3.P.1 with

$$G_1(s) = K_c, \qquad G_2(s) = 1/s$$

prove that

a. The steady state error of the system to a step change in setpoint is zero.

b. The steady state error of the system to a step change in disturbance is nonzero.

3.10. (Open loop response of first order system) The transfer function of a first-order-plus-time-delay system is

$$G(s) = \frac{Ke^{-\tau s}}{T_1 s + 1}$$

Use *PCET* to examine the effect upon the process step response of:

a. Varying K, given $T_1 = 5$ and $\tau = 0$.

b. Varying T_1 , given $K = 1$ and $\tau = 0$.

c. Varying τ, given $T_1 = 5$ and $K = 1$.

3.11. (Open loop response of second order system) The transfer function of a

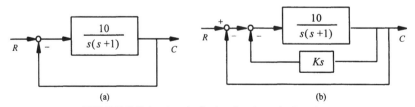

(a)   (b)

FIGURE 3.P.4. A velocity feedback control system.

second-order-plus-time-delay system may be written in standard form as

$$G(s) = \frac{Ke^{-\tau s}}{T^2 s^2 + 2T\zeta s + 1}$$

Using *PCET*, describe the effect upon the process step response of:
 a. Varying K, given $T = 1$, $\zeta = 1$ and $\tau = 0$.
 b. Varying T, given $K = 1$, $\zeta = 1$ and $\tau = 0$.
 c. Selecting $\zeta = 1.5, 0.45, 0.1, 0, -0.5$, given $T = 1$, $K = 1$ and $\tau = 0$.
 d. Varying τ, given $T = 1$, $\zeta = 0.5$ and $K = 1$.
 3.12. Compare the unit step responses obtained in *PCET* for the following systems:

 a) $G(s) = \dfrac{1}{s+1}$,   b) $G(s) = \dfrac{1}{s^2 + 2s + 1}$

 c) $G(s) = \dfrac{1}{(s+1)(s^2 + 2s + 1)}$ d) $G(s) = \dfrac{e^{-2s}}{(s+1)(s^2 + 2s + 1)}$

 3.13. (Open loop behavior of proportional controller)   Set the controller parameters $1/T_i = 0$ and $T_d = 0$ in *PCET*, and obtain the unit step response for a proportional controller when $K_c = 1$. Use *PCET* to examine the effect upon the step response of the controller when K_c increases or decreases.
 3.14. (Open loop behavior of proportional-integral controller)   Set the derivative time $T_d = 0$ and use *PCET* to examine the unit step responses of a *PI* controller and compare their integral actions when
 a. $K_c = 1$ and varies, given $1/T_i = 1$;
 b. $1/T_i = 1$ and varies, given $K_c = 1$.
 c. With the initial value theorem, prove the initial values in the figures you obtained are correct.
 3.15. (Open loop behavior of proportional-derivative controller)   Set the inverse integral time $1/T_i = 0$ and use *PCET* to examine the unit step response of a *PD* controller when
 a. $K_c = 1$ and varies, given $T_d = 1$;
 b. $T_d = 1$ and varies, given $K_c = 1$;
 c. With the final value theorem prove that the final values you determined in the figures are correct.
 3.16. (Open loop behavior of proportional-integral-derivative controller)   Use *PCET* to examine the unit step response of a *PID* controller when
 a. $K_c = 1$ and varies, given $1/T_i = 1$ and $T_d = 1$;
 b. $1/T_i = 1$ and varies, given $K_c = 1$ and $T_d = 1$;
 c. $T_d = 1$ and varies, given $K_c = 1$ and $1/T_i = 1$.

3.17. (Controller design for a first order system) Consider a first order system with open loop transfer function

$$G(s) = \frac{2}{5.2s + 1}$$

a. Find the open loop response to a unit step input.

b. Examine the closed loop response obtained using a proportional controller with $K_c = 1.5$. Does offset appear?

c. Design a proportional controller by trial and error so that the closed loop response satisfies the following performance requirements:

Offset ≤ 0.1

Settling time ≤ 4 sec

d. Design a proportional-integral (PI) controller by trial and error so that the closed loop system meets the performance specifications:

Overshoot ≤ 0.3

Settling time ≤ 5 sec

e. Compare the closed loop step responses obtained using the controllers developed in parts c) and d). What is the effect of integral action on the process response?

3.18. (Controller design for a second order system) Two tanks are connected in series, i.e., the outlet flow of Tank 1 is the inlet flow of Tank 2. The dynamics of this system may be described in deviation variable form by the linearized differential equations

$$10\frac{dh_1}{dt} + 2h_1 = f_1$$

$$15\frac{dh_2}{dt} + 1.5h_2 = 2h_1$$

where f_1 is the inlet flow rate of Tank 1; h_1 and h_2 are the liquid levels in Tanks 1 and 2, respectively.

a. Find the overall transfer function of the process between the inlet flow rate of Tank 1 and the level of Tank 2.

b. Use PCET to design a proportional controller by trial and error such that the closed loop step response exhibits less than 30% overshoot, and offset is as small as possible.

c. Using PCET, design a PI controller by trial and error such that the response exhibits less than 60% overshoot, and the settling time is as small as possible.

3.19. (Controller design for a third order system) Consider the third order system with open loop transfer function

$$G(s) = \frac{45}{(3s+1)(6.76s^2 + 1.56s + 1)}$$

Use *PCET* to design a *PID* controller by trial and error so that the closed loop system satisfies the time domain performance criteria:

Overshoot ≤ 0.3

Settling time ≤ 35 seconds

CHAPTER 4

STABILITY AND ROOT LOCUS TECHNIQUE

4.1. POLES AND STABILITY

A very important characteristic of a system is its stability. An unstable system can cause a disaster. The damage of the first bridge across the Tacoma Narrows at Puget Sound, Washington, is an example. It was put into use on July 1, 1940, and was found to oscillate under windy conditions, i.e., the system was close to being unstable. Only four months later, on November 7, 1940, the bridge collapsed after severe oscillation caused by wind [Farquharson 1950].

A dynamic system is stable if the system output response is bounded for all bounded inputs, regardless of the initial conditions of the system. Otherwise, the system is unstable.

Characteristic polynomials and characteristic equations
For a system with the transfer function $KZ(s)/Q(s)$, the response of the system is primarily characterized by $Q(s)$, the denominator of the transfer function. Therefore, the denominator of the system transfer function is referred to as the characteristic polynomial of the system. Setting the characteristic polynomial equal to zero, we obtain the characteristic equation, i.e.,

$$Q(s) = 0$$

The roots of the characteristic equation are referred to as poles of the system. The characteristic equation plays a decisive role in determining system stability, as discussed later.

Usually, the transfer functions of realizable systems are rational functions. One exception is a time delay system with a transfer function $e^{-\tau s}$. For a rational system transfer function, all the coefficients of the characteristic equation are real. The poles may be real or complex. If they are complex, the poles appear in complex conjugate pairs. Any real coefficient polynomial can be factored as the product of first and second order polynomials, where every first order

93

polynomial has a real root and every second order polynomial has a pair of conjugate roots. (If a second order polynomial has two real roots, it can be factored into two first order polynomials.)

Stability and location of poles

As an example, let us consider a system with a transfer function

$$F(s) = \frac{KZ(s)}{Q(s)} = \frac{KZ(s)}{(s - p_1)(s - p_2)\cdots(s - p_n)} \tag{4.1.1}$$

where p_i (i = 1, 2, $\cdots$, n) are distinct poles. $Q(s)$ of order n and $Z(s)$ of order m ($m \leq n$) are monic polynomials. (A polynomial is monic if the coefficient of the highest order term is 1.) The response of the system to a unit step input will be

$$y(t) = \sum_{i=1}^{n} \frac{KZ(p_i)}{[sQ(s)]_{s=p_i}} e^{p_i t} + \frac{KZ(0)}{[sQ(s)]_{s=0}} \tag{4.1.2}$$

Obviously, the bound of output depends on the values of all poles. If there is a real positive pole $p_i > 0$, then it is clear from (4.1.2) that the system output $y(t)$ is unbounded, i.e., the system is unstable. If the real part of a pair of conjugate complex poles is positive, i.e., the poles are on the right half s-plane, then the output of the system is oscillatory divergent, as we studied in the last chapter, i.e., the system is unstable.

It should be noted that if there are a pair of poles on the imaginary axis, the system will yield oscillation, the amplitude of which is neither decaying nor growing with time. By definition, the system is stable because the oscillatory response is bounded. However in practice, sustained oscillation cannot be tolerated. Thus, conventionally we still consider such a system to be unstable in the following chapters except in Chapter 9. If a system has a pole at $s = 0$ as an integrating process, the system response to a step input will go to infinity, which we already studied in Chapter 2.

In general, if the system poles are represented by

$$s = \alpha \pm j\omega$$

the real part α affects system stability, while imaginary part ω determines whether the system response to a step input is oscillatory or not. It is summarized in Table 4.1.

TABLE 4.1. Effect of system poles on the system behavior.

$s = \alpha \pm j\omega$	Real part α		Imaginary part ω	
Effect on	$\alpha \geq 0$	System is unstable.	$\omega = 0$	System is not oscillatory.
system	$\alpha < 0$	System is stable.	$\omega \neq 0$	System is oscillatory.

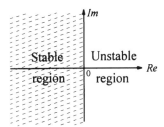

FIGURE 4.1.1. Stable and unstable regions in s-plane.

With respect to stability then, the s-plane can be divided into two regions: stable region (left half s-plane) and unstable region (right half s-plane, including the imaginary axis) as shown in Fig. 4.1.1. The relationship between the stability and the location of the system poles can be found in Table 4.2. The reader may compare this table to Table 2.7 for a second order system.

Poles and dynamics of a system
The location of poles affects not only the stability, but also the dynamics of the system. Let us review the effect of pole location on the dynamics. The transfer function of a second order underdamped system is assumed to be

$$\frac{Y(s)}{R(s)} = \frac{K}{T^2 s^2 + 2T\zeta s + 1} = \frac{K\omega_n^2}{s^2 + 2\zeta\omega_n s + \omega_n^2} \tag{4.1.3}$$

Two conjugate poles of the system may be written as

$$s_1, s_2 = -\alpha \pm j\omega = -\zeta\omega_n \pm j\omega_n\sqrt{1-\zeta^2} \tag{4.1.4}$$

It is easy to see that $\|s_1\| = \|s_2\| = \omega_n$. Thus Fig. 4.1.2 can be obtained, where θ is the angle between the negative real axis and the line on which the pole located.

TABLE 4.2. System stability versus the location of system poles.

Location of poles	Stability of system	Step response
All on the real negative axis	Stable	Monotonic convergent
All on the left half s-plane, some are complex	Stable	Oscillatory convergent
At least one pair of conjugate poles on imaginary axis, others on left half s-plane	Unstable	Sustained oscillation
Some complex poles on right half s-plane	Unstable	Oscillatory divergent
At least one pole at origin, others on left half s-plane	Unstable	Monotonic divergent

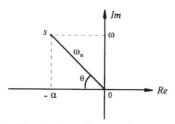

FIGURE 4.1.2. Magnitude and angle of a pole ($s = -\alpha + j\omega$).

Clearly,

$$\zeta = \cos\theta \qquad (4.1.5)$$

The smaller angle θ ($\theta \geq 0$) gives larger damping factor ζ. If we restrict the poles in the region between two constant ζ-lines with the angle θ as shown in Fig. 4.1.3a, then the decay ratio (3.3.12) will be limited as

$$D.R. \leq \exp(-2\pi\zeta / \sqrt{1-\zeta^2}) = \exp(-2\pi\cot\theta) \qquad (4.1.6)$$

As we studied in Chapter 2, the response of this underdamped system to a unit step input can be obtained from (2.3.22) as

$$y(t) = 1 - e^{-\zeta\omega_n t}(\sqrt{1-\zeta^2})^{-1}\sin(\omega_n\sqrt{1-\zeta^2}\, t + \phi)$$
$$= 1 - e^{-\alpha t}(\sqrt{1-\zeta^2})^{-1}\sin(\omega t + \phi) \qquad (4.1.7)$$

where

$$\alpha = \zeta\omega_n, \quad \omega = \omega_n\sqrt{1-\zeta^2}, \quad \text{and} \quad \phi = \arctan(\omega/\alpha) \qquad (4.1.8)$$

From Fig. 4.1.2, $\phi = \theta$. It may be seen from (4.1.7) that $-\alpha$, the real part of the poles, determines the rise and decay rate of the response. In other words, α determines the 'damping' of the system, and is called the damping constant. The settling time may be found in Table 3.4 as

$$t_s \approx 3/\zeta\omega_n = 3/\alpha \qquad (4.1.9)$$

It follows that if the poles of a second order system are located to the left of the constant α-line as shown in Fig. 4.1.3b, then the settling time of the system will be less than $3/\alpha$.

Eq. (4.1.7) also tells us that ω, the imaginary part of the poles, determines the frequency of the decaying oscillation. When $\zeta = 0$, the response is a pure sinusoidal wave with frequency ω. If the system poles are placed in the area shown in Fig. 4.1.3c, then the decaying oscillation frequency will be less than ω. In other words, the period of oscillation P in (3.3.13) will be

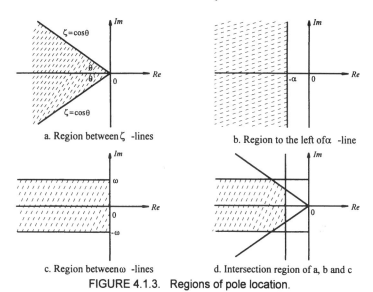

FIGURE 4.1.3. Regions of pole location.

$$P \geq 2\pi/\omega \qquad (4.1.10)$$

Summarizing the cases discussed, in order to keep the decay ratio, settling time and decaying oscillation frequency in specified ranges, the poles of a second order system should be placed in the area as shown in Fig. 4.1.3d. The effect of the location of poles of a second order system on the response to a step input is summarized in Table 4.3.

From Fig. 4.1.2 and the performance criterion formulas of a second order system learned in the last chapter, the relationship between pole location and the system performance criteria can be listed in Table 4.4.

TABLE 4.3. Effect of pole location of a second order system on the response.

Boundary loci	Constant ζ-loci	Constant α-loci	Constant ω-loci	Conjunction of ζ, α, ω-loci
Formula	$\zeta = \cos\theta$	$\alpha = \zeta\omega_n$	$\omega = \omega_n\sqrt{1-\zeta^2}$	
Covered region	Section between two -lines	Half-plane to the left of -line	Band between two ω-lines	Conjunction part
Figures	Fig. 4.1.3a	Fig. 4.1.3b	Fig. 4.1.3c	Fig. 4.1.3d
Effect	$D.R. \leq \exp(-2\pi\cot\theta)$	$t_s \leq 3/\zeta\omega_n = 3/\alpha$	$P \geq 2\pi/\omega$	Joint effect of ζ, α, ω

TABLE 4.4. Pole location and performance criteria of a second order system.

Name	Acronym	Polar coordinate expression	Rectangular coordinate expression
Maximum overshoot	$M.O.$	$\exp(-\pi\cot\theta)$	$\exp(-\pi\alpha/\omega)$
Percent overshoot	$P.O.$	$100\exp(-\pi\cot\theta)$	$100\exp(-\pi\alpha/\omega)$
Rise time	t_r	$\approx (0.8+2.5\cos\theta)/\omega_n$	$0.8/\sqrt{\Delta}+2.5\alpha/\Delta$ where $\Delta = \alpha^2 + \omega^2$
Settling time	t_s	$\approx 3/\omega_n\cos\theta$	$\approx 3/\alpha$
Peak time	t_p	$\pi/\omega_n\sin\theta$	π/ω
Period of oscillation	P	$2\pi/\omega_n\sin\theta$	$2\pi/\omega$
Decay ratio	$D.R.$	$\exp(-2\pi\cot\theta)$	$\exp(-2\pi\alpha/\omega)$

Dominant poles

Strictly speaking, all the formulas we obtained in Table 4.3 and 4.4 are valid only for a second order system without zeros. Fortunately, many high order systems possess a dominant pair of roots of the characteristic equations. Thus, the step response of the system may be estimated by a second order system.

Dominant poles (dominant roots) are the roots of the characteristic equation, which primarily determine the transient response of the system.

According to the analyses on first order and second order systems in the last two chapters, it can be observed that

1) The farther out from the origin the poles lie on the negative real axis, the faster the system response will be;

2) The closer the poles lie to the imaginary axis, the slower the response converges to its steady state;

Therefore, the closest poles to the imaginary axis would be the dominant poles. If the absolute values of the real parts of all non-dominant poles are more than 5 times larger than that of the two dominant poles, and there are no zeros nearby, then the system can be approximated by a second order system with the dominant poles. The farther the other poles are away from the dominant poles, the better the approximation will be.

Remarks

1. The response of a system also depends on the system zeros, especially when the zeros are relatively close to the dominant poles [Clark 1962]. The effect of zeros on the response is much more complicated to discuss than that of

the poles. However, the stability of a system does only depend on the poles of the system.

2. The dominant poles may be one real pole, two complex conjugate poles or more. For example, if a system has 5 poles at $-1 \pm j1$, -12, $-11 \pm j1$, then there are two dominant poles at $-1 \pm j1$.

3. We should distinguish open loop poles and closed loop poles. The stability of an open loop system is determined by open loop poles, while the stability of a closed loop system is determined by closed loop poles.

Example 4.1.1

For a unity feedback system with an open loop transfer function

$$G(s) = \frac{1}{s(0.5s+1)(0.2s+1)}$$

the closed loop dominant poles may be determined as follows: The closed loop transfer function is

$$W(s) = \frac{G(s)}{1+G(s)} = \frac{1}{s(0.5s+1)(0.2s+1)+1}$$

$$= \frac{10}{s^3 + 7s^2 + 10s + 10}$$

The closed loop transfer function has no zeros. The closed loop characteristic equation is

$$s^3 + 7s^2 + 10s + 10 = (s + 5.52)(s^2 + 1.48s + 1.8304) = 0$$

By solving the closed loop characteristic equation, the closed loop poles can be found as

$$s_{1,2} = -0.74 \pm j1.1326, \qquad s_3 = -5.52$$

The ratio of the real parts is

$$\text{Re}(s_{1,2})/\text{Re}(s_3) = 0.74/5.52 = 0.135 < 1/5$$

Therefore, $s_{1,2}$ can be considered as a pair of dominant poles.

The response of the system may be estimated by a second order system with the transfer function

$$G(s) = \frac{1.8304}{s^2 + 1.48s + 1.8304}$$

where we figure out static gain as 1.8304 for matching the steady state error with that of the original system.

The reader can compare the responses of two systems to a unit step input using the software package *PCET* supplied with this book.

4.2. ROUTH CRITERION

If the poles of a system are known, the stability of the system can be determined easily. However, it is often difficult to find the poles by solving a high order algebraic equation.

With the Routh criterion, also called the Routh-Hurwitz criterion, it is not necessary to calculate actual values of the characteristic roots to determine system stability. The criterion provides a way to investigate how many poles of the system are located in the right half s-plane without solving algebraic equations.

The characteristic equation of a system is assumed to be

$$a_n s^n + a_{n-1} s^{n-1} + \cdots + a_1 s + a_0 = 0 \qquad (4.2.1)$$

where $a_n > 0$ is specified without loss of generality. It may be factored as

$$a_n (s + r_1)(s + r_2) \cdots (s + r_n) = 0 \qquad (4.2.2)$$

where $-r_1, -r_2, \cdots, -r_n$ are roots of (4.2.1), i.e., poles of the system. Comparing (4.2.1) with (4.2.2), we can conclude that all the coefficients of the system characteristic equation must be positive (have the same sign as a_n), if all the poles are in the left half s-plane. It follows that if any coefficients of the characteristic equation are negative or zero, then at least one root of the characteristic equation lies to the right of, or on, the imaginary axis, and thus the system is unstable. However, it must be noted that it is a necessary but not sufficient condition. In other words, if all the coefficients of the characteristic equation are positive, stability is not guaranteed.

If all the coefficients of characteristic equation (4.2.1) are positive, we may construct a Routh array as follows:

$$
\begin{array}{c|cccc}
s^n & a_n & a_{n-2} & a_{n-4} & \cdots \\
s^{n-1} & a_{n-1} & a_{n-3} & a_{n-5} & \cdots \\
s^{n-2} & b_1 & b_2 & b_3 & \cdots \\
s^{n-3} & c_1 & c_2 & \cdots & \\
\vdots & \vdots & & & \\
s^0 & h_1 & & &
\end{array}
$$

where

$$b_1 = \frac{a_{n-1}a_{n-2} - a_n a_{n-3}}{a_{n-1}} = \frac{-1}{a_{n-1}} \begin{vmatrix} a_n & a_{n-2} \\ a_{n-1} & a_{n-3} \end{vmatrix}$$

$$b_2 = \frac{a_{n-1}a_{n-4} - a_n a_{n-5}}{a_{n-1}} = \frac{-1}{a_{n-1}} \begin{vmatrix} a_n & a_{n-4} \\ a_{n-1} & a_{n-5} \end{vmatrix} \qquad (4.2.3)$$

$$c_1 = \frac{b_1 a_{n-3} - a_{n-1}b_2}{b_1} = \frac{-1}{b_1} \begin{vmatrix} a_{n-1} & a_{n-3} \\ b_1 & b_2 \end{vmatrix}$$

$$c_2 = \frac{b_1 a_{n-5} - a_{n-1}b_3}{b_1} = \frac{-1}{b_1} \begin{vmatrix} a_{n-1} & a_{n-5} \\ b_1 & b_3 \end{vmatrix}$$

and so on. A Routh array has $n + 1$ rows. From here on, Routh arrays will be written without the column identifying the powers of s from s^n to s^0.

The Routh stability criterion says that the number of roots of a polynomial on the right half s-plane is equal to the number of changes in sign of the first column of the Routh array, i.e., the signs of a_n, a_{n-1}, b_1, c_1, ..., h_1. It follows that a necessary and sufficient condition for a stable system is that all the elements in the first column of the Routh array, constructed by the coefficients of its characteristic equation, are positive.

The mathematical proof of the Routh criterion can be found elsewhere [Routh 1877].

Example 4.2.1

The characteristic equation of a third order system is

$$s^3 + 3s^2 + 6s + 6K = 0$$

What values of K make the system stable?

The Routh array is

$$\begin{array}{cc} 1 & 6 \\ 3 & 6K \\ \dfrac{18 - 6K}{3} & 0 \\ 6K & \end{array}$$

In order to ensure a stable system, it is required that

$$(18 - 6K)/3 > 0 \qquad \text{and} \qquad 6K > 0$$

or

$$0 < K < 3$$

TABLE 4.5. Direct criteria for the stability of 1st-3rd order system.

System	First order	Second order	Third order
Characteristic equation	$a_1 s + a_0 = 0$ $(a_1 > 0)$	$a_2 s^2 + a_1 s + a_0 = 0$ $(a_2 > 0)$	$a_3 s^3 + a_2 s^2 + a_1 s + a_0 = 0$ $(a_3 > 0)$
Routh array	a_1 a_0	$\begin{matrix} a_2 & a_0 \\ a_1 & 0 \\ a_0 \end{matrix}$	$\begin{matrix} a_3 & a_1 \\ a_2 & a_0 \\ \dfrac{a_2 a_1 - a_3 a_0}{a_2} & 0 \\ a_0 \end{matrix}$
Stability criterion	$a_1, a_0 > 0$	$a_2, a_1, a_0 > 0$	$a_3, a_2, a_1, a_0 > 0$ $\begin{vmatrix} a_3 & a_1 \\ a_2 & a_0 \end{vmatrix} < 0$

Direct criteria for 1st~3rd order systems
As in the example above, it is easy to check the stability of first order, second order and third order systems. The direct criteria can be determined by reader. The results are listed in Table 4.5.

Case of a zero appearing in the first column
If a zero appears in the first column of a Routh array, rearrangement may be needed to use the Routh criterion. The following three methods are introduced:

i) *Reciprocal root checking*
It is easy to see that $1/s$ is located on the same half s-plane as s is. One way to avoid the appearance of a zero element in the first column of the Routh array is to change the characteristic equation with the transformation $s = 1/p$.

Example 4.2.2
    If the characteristic equation of a system is

$$s^4 + s^3 + 2s^2 + 2s + 5 = 0 \qquad (4.2.4)$$

The Routh array will be

$$\begin{matrix} 1 & 2 & 5 \\ 1 & 2 & 0 \\ 0 & 5 \end{matrix}$$

There is a zero in the first column of the Routh array. Let $s = 1/p$, and substitute it into (4.2.4). The new equation multiplied by p^4 becomes

$$5p^4 + 2p^3 + 2p^2 + p + 1 = 0 \qquad (4.2.5)$$

If p_1 is a root of (4.2.5), $s_1 = 1/p_1$ is a root of (4.2.4). The roots s_1 and p_1 are in the same half s-plane. The new Routh array is

$$
\begin{array}{ccc}
5 & 2 & 1 \\
2 & 1 & 0 \\
-0.5 & 1 & \\
5 & 0 & \\
1 & &
\end{array}
$$

The signs in the first column of the Routh array change twice, and there are two poles of the system (4.2.5) on the right half s-plane. Therefore the system (4.2.4) also has two poles on the right half s-plane, and the system is unstable.

ii) ε *replacement*
We can replace zero in the first column of the Routh array by a small positive number ε to continue the Routh array as in the following example.

Example 4.2.3
Recalling Example 4.2.2, the Routh array can be rewritten by the replacement of zero with ε in the first column as

$$
\begin{array}{ccc}
1 & 2 & 5 \\
1 & 2 & \\
\varepsilon & 5 & \\
2 - 5/\varepsilon & 0 & \\
5 & &
\end{array}
$$

Since ε is a small positive number, $2 - 5/\varepsilon$ is negative, and the number of changes in sign of the first column of the array is 2. Thus, there are two poles with positive real parts, and the system is unstable.

iii) *Multiplication by a stable polynomial*
If the characteristic polynomial of the system is multiplied by a stable polynomial, the resulting polynomial is stable if the characteristic polynomial is stable, or vice versa. This property may be used to change the Routh array to avoid the appearance of zero in the first column.

Example 4.2.4
Recalling Example 4.2.2 again, the characteristic equation multiplied by a stable polynomial $(s + 1)$ yields

$$(s^4 + s^3 + 2s^2 + 2s + 5)(s + 1)$$

$$= s^5 + 2s^4 + 3s^3 + 4s^2 + 7s + 5 = 0$$

The Routh array will be

$$
\begin{array}{ccc}
1 & 3 & 7 \\
2 & 4 & 5 \\
1 & 9/2 & 0 \\
-5 & 5 & 0 \\
11/2 & 0 & \\
5 & & \\
\end{array}
$$

The signs of the first column elements change twice. The same conclusion can be obtained as in the last two examples.

These three methods of dealing with the appearance of a zero in the first column of a Routh array are summarized in Table 4.6.

Remarks

1. The Routh criterion is valid only for linear systems. The characteristic equation must be a linear algebraic one, and all the coefficients must be real.

2. The Routh criterion is generally qualitatively used to determine only the

TABLE 4.6.   Method for handling a zero in the first column of Routh array.

Method	Principle	Procedure
Reciprocal root checking	$1/s$ is located on the same half s-plane as s is.	Reverse the order of the coefficients of the characteristic equation to form a new Routh array.
ε-replacement	Approximate zero in the first column with a small positive number.	Replace 0 with ε in the first column of Routh array and check the sign in the first column.
Multiplying a stable polynomial	Stability does not change by multiplying the characteristic polynomial with a stable polynomial.	Obtain the new characteristic equation and form a new Routh array.

TABLE 4.7.   Steps for checking stability with Routh criterion.

Step	Condition	Content to check	Conclusion
1	All poles available	Any pole in right half s-plane or on imaginary axis	If yes, unstable.
2	Linear algebraic characteristic equation	Any coefficient is zero or negative	If yes, unstable.
3	Routh array completed without 0 in the first column	The number of sign changes in the first column	If not zero, unstable.
4	Routh array completed with 0 in the first column and 0 replaced by small ε > 0	The number of sign changes in the first column	If not zero, unstable.

stability of a system. For two stable systems, the criterion does not indicate which one is more stable, or which one has better dynamic characteristics.

The steps for checking the stability of a system are tabulated in Table 4.7, where step 4 can be replaced with other methods in Table 4.6.

Routh criterion in *PCET* software

The Routh criterion is also programmed into the software *PCET*. The sub-program *RSC* (Routh Stability Criterion) determines Routh array, the roots of the system characteristic equation and then the system stability. The procedure for running this program is shown in Table 4.8.

For Step 6 in the procedure of Table 4.8, selection of F6, or F7, will display the Routh array, or the roots of the characteristic equation.

When zero appears in the first column of the Routh array, ε-replacement method is used in *PCET* package. ε is a small positive number. The value used by the package is $\varepsilon = 10^{-16}$, but this value can be changed by selecting F8.

4.3. ROOT LOCUS TECHNIQUE

The stability of a system is of prime consideration for control engineers. The degree of stability, called relative stability, provides information about not only

TABLE 4.8.   Procedure for determining the system stability with *PCET*.

Step	User input		Function
1	Type *PCET*		Start program
2	Select *RSC*		Routh stability criterion
3	Select F3	"F1-Main menu"	Return to main menu
		"F2-Restart"	Return to very beginning of *RSC*
		"F3-Data"	Ready to input system data
		"F4-Load"	Load a saved file to get data
4	Input system characteristic equation		Specify system dynamics
5	Confirm the correctness of the data		Verify input data
6	Select	"F1-Main menu"	Return to main menu
		"F2-Restart"	Return to very beginning of *RSC*
		"F3-Data"	Change the system data
		"F4-Load"	Load a saved file to get data
		"F5-Save"	Save the update data in a file
		"F6-Routh"	Give the Routh array
		"F7-Root"	Obtain the equation roots
		"F8-Epsilon"	Change the small number ε

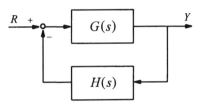

FIGURE 4.3.1. A single loop system.

how stable the system is, but also to some extent what dynamics the system exhibits.   The relative stability and transient performance of a closed loop control system are directly related to the location of its poles in the s-plane, i.e., the roots of the closed loop characteristic equation. It is useful to determine the locus of the characteristic equation roots in the s-plane as a parameter varies. Root locus is the locus or path of the closed loop roots traced out in the s-plane as a system parameter is changed.

The root locus technique is a graphical method for drawing the root locus. The root locus method provides an approximate sketch of closed loop poles that can be used to obtain qualitative information concerning the stability and performance of the system.

Magnitude and angle requirements
For a single loop system as shown in Fig. 4.3.1, the closed loop characteristic equation is

$$1 + G(s)H(s) = 0 \qquad (4.3.1)$$

Since s is a complex variable, Eq. (4.3.1) may be rewritten in a polar form as

$$|G(s)H(s)|\underline{/G(s)H(s)} = -1 \qquad (4.3.2)$$

This requires that

$$|G(s)H(s)| = 1 \qquad (4.3.3)$$

and

$$\underline{/G(s)H(s)} = \arg[G(s)H(s)] = (2k + 1)\pi \qquad (4.3.4)$$

where $k = 0, \pm1, \pm2, \pm3, \cdots$.   Eq. (4.3.3) is referred to as the magnitude requirement of root locus, and (4.3.4) the angle requirement.

In general, the open loop transfer function $G(s)H(s)$ can be written as

$$G(s)H(s) = \frac{K(s+z_1)(s+z_2)\cdots(s+z_m)}{(s+p_1)(s+p_2)\cdots(s+p_n)} \qquad (4.3.5)$$

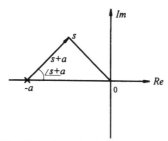

FIGURE 4.3.2. Magnitude and angle of $(s + a)$.

where $m < n$. Then the magnitude and angle requirements for the root locus are

$$|G(s)H(s)| = \frac{K|(s+z_1)\|(s+z_2)|\cdots|(s+z_m)|}{|(s+p_1)\|(s+p_2)|\cdots|(s+p_n)|} = \frac{K\prod_{i=1}^{m}|s+z_i|}{\prod_{j=1}^{n}|s+p_j|} = 1 \qquad (4.3.6)$$

and

$$\arg[G(s)H(s)] = \sum_{i=1}^{m}\arg(s+z_i) - \sum_{j=1}^{n}\arg(s+p_j) = (2k+1)\pi \qquad (4.3.7)$$

where $k = 0, \pm 1, \pm 2, \pm 3, \cdots$. The magnitude and angle of $s + a$ are shown in Fig. 4.3.2, where $-a$ is the root of $s + a = 0$, and $s + a$ is a vector from $-a$ to s.

The angle requirement (4.3.7) enables us to determine the root locus. For a given s on the root locus, the value of K corresponding to this point can be found by using the magnitude requirement (4.3.6).

The magnitude and angle requirements are summarized in Table 4.9.

Starting and end points of the root locus

We usually determine the root locus as the open loop gain K in (4.3.5) varies

TABLE 4.9. Magnitude requirement and angle requirement.

Requirement	General Formula	Pole-zero form formula	Usefulness
Angle requirement	$\underline{/G(s)H(s)}$ $= (2k+1)\pi$	$\arg[G(s)H(s)] = \arg(s+z_1) +$ $\arg(s+z_2) + \cdots + \arg(s+z_m)$ $-\arg(s+p_1) - \arg(s+p_2) - \cdots$ $- \arg(s+p_n) = (2k+1)\pi$	To determine the root locus
Magnitude requirement	$|G(s)H(s)| = 1$	$\dfrac{K|(s+z_1)\|(s+z_2)|\cdots|(s+z_m)|}{|(s+p_1)\|(s+p_2)|\cdots|(s+p_n)|} = 1$	To dimension K for points on the root locus

from zero to infinity. Substituting (4.3.5) into (4.3.1) and rewriting, we obtain

$$\prod_{j=1}^{n}(s+p_j) + K\prod_{i=1}^{m}(s+z_i) = 0 \qquad (4.3.8)$$

Therefore, when $K = 0$, the roots of the characteristic equation are simply the poles of $G(s)H(s)$. When K approaches infinity, the roots of the characteristic equation tend to the zeros of $G(s)H(s)$. This means that the root locus starts at the poles of the open loop transfer function ($K = 0$) and moves to its zeros as K approaches ∞. For most common open loop transfer functions, there may be several zeros at infinity in the s-plane.

Usually we mark each pole of an open loop transfer function in a root locus as $\times$, and each zero as O.

Root locus on the real axis

The angle requirement shows that the sum of all angles contributed by the zeros minus that contributed by the poles equals $180° \pm n360°$, where n is any integer. Obviously, the complex roots must appear as pairs of complex conjugate roots, and a pair of conjugate complex zeros or poles contribute nothing to the angle requirement when the locus is on the real axis, because one's contribution is just equal and opposite to that of the other. Therefore, we may conclude that the locus on the real axis lies in a section to the left of an odd number of poles and zeros.

Asymptotes of root loci

The loci begin at the poles and tend to the zeros, including zeros at infinity, and every distinct pole or zero can only start or end one locus. Therefore, the number of separate loci is equal to the larger of m or n which are the numbers of zeros and poles respectively. Usually, the number of poles n is greater than the number of zeros m. Thus, the number of separate loci is equal to the number of poles. Therefore, $n - m$ sections of loci must tend to zeros at infinity. These sections of loci go to the zeros at infinity along asymptotes as K approaches infinity.

An asymptote is characterized by its angle with respect to the real axis, and by its point of intersection with the real axis. The intersection is called the asymptote centroid, and may be obtained as follows:

The open loop transfer function (4.3.5) may be rewritten as

$$G(s)H(s) = K\frac{s^m + a_1 s^{m-1} + \cdots}{s^n + b_1 s^{n-1} + \cdots} \qquad (4.3.9)$$

where

$$a_1 = z_1 + z_2 + \cdots + z_m = -\sum \text{zeros of } G(s)H(s) \qquad (4.3.10)$$

$$b_1 = p_1 + p_2 + \cdots + p_m = - \sum \text{poles of } G(s)H(s) \qquad (4.3.11)$$

Dividing the numerator and denominator of (4.3.9) by the numerator yields

$$G(s)H(s) = \frac{K}{s^{n-m} + (b_1 - a_1)s^{n-m-1} + \cdots} \qquad (4.3.12)$$

Using (4.3.1), substitute $G(s)H(s) = -1$ into (4.3.12), which gives

$$s^{n-m} + (b_1 - a_1)s^{n-m-1} + \cdots = -K \qquad (4.3.13)$$

The asymptote will coincide with the root locus as $|s|$ tends to infinity. When $|s|$ is very large, Eq. (4.3.13) is equivalent to a binomial equation as

$$(s + \frac{b_1 - a_1}{n-m})^{n-m} = -K = Ke^{j(2k+1)\pi} \qquad (4.3.14)$$

where $Ke^{j(2k+1)\pi}$ is a polar coordinate expression of $-K$. The first two terms of (4.3.14) are identical to those in (4.3.13). Eq. (4.3.14) may be rewritten as

$$s + \frac{b_1 - a_1}{n-m} = K^{\frac{1}{n-m}} e^{j\frac{(2k+1)\pi}{n-m}} \qquad (4.3.15)$$

It may then be viewed as an asymptote equation. Since K is small for the root locus near the real axis, the asymptote centroid may be obtained from (4.3.15), (4.3.10) and (4.3.11) as follows

$$\sigma_A = -\frac{b_1 - a_1}{n-m} = \frac{\sum p - \sum z}{n-m}$$

$$= \frac{\sum \text{poles of } G(s)H(s) - \sum \text{zeros of } G(s)H(s)}{n-m} \qquad (4.3.16)$$

From (4.3.15), the angles of the asymptotes with respect to the real axis are

$$\phi_A = \frac{(2k+1)\pi}{n-m} \qquad (4.3.17)$$

where k is an integer, and $n - m$ angles can be selected with $k = 0, \pm 1, \pm 2, \pm 3, \cdots$.

Breakaway point on the real axis

The root locus leaves the real axis at a breakaway point. Usually, it occurs when there are two open loop poles next to each other on the real axis, and between them is a part of locus.

First express the characteristic equation $1 + G(s)H(s) = 0$ as

$$1 + KQ(s) = 0 \qquad (4.3.18)$$

where the denominator and numerator of $Q(s)$ are monic. Then, K may be expressed as

$$K = -1/Q(s) \qquad (4.3.19)$$

Since the root locus begins at poles corresponding to $K = 0$, the K value corresponding to the breakaway point must be larger than the K values corresponding to all the other points in the same segment of root locus on the real axis. In other words, the K value of the breakaway point is the largest one in the segment of the locus on the real axis. The complex variable s reduces to a real variable while s changes along the real axis. Therefore, the K corresponding to the breakaway point may be found by taking the derivatives of (4.3.19), i.e.,

$$\frac{dK}{ds} = -\frac{d}{ds}(1/Q(s)) = 0 \qquad (4.3.20)$$

This equation may be difficult to solve if its order is high, and it may be necessary to draw a graph to find the maximum K. A trial and error method may also be used to evaluate the K value of the breakaway point.

In solving Eq. (4.3.20) to find the breakaway point, several solution points may be found. Only the solution point on the real axis and on the root locus is the breakaway point.

The breakaway point can also be calculated by solving the following equation:

$$(\frac{1}{z_1 + a} + \cdots + \frac{1}{z_m + a}) - (\frac{1}{p_1 + a} + \cdots + \frac{1}{p_n + a}) = 0 \qquad (4.3.21)$$

where a is the breakaway point, z_i ($i = 1, 2, \cdots, m$) are zeros of $G(s)H(s)$, and p_i ($i = 1, 2, \cdots, n$) are its poles.

Angles of departure and arrival
The angle of departure of the locus from a pole and the angle of arrival of the locus at a zero can be determined from the angle requirement (4.3.7). The procedure is demonstrated in the following example.

Example 4.3.1
The open loop transfer function of a given system of third order has a pair of conjugate complex poles p_1, p_2, and a real pole p_3 as shown in Fig. 4.3.3, and the system has no finite open loop zeros. Taking a point s_1 near p_1, which is assumed to be on the root locus, we may denote the vectors from all poles to this

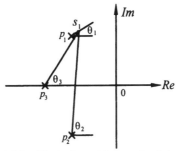

FIGURE 4.3.3. Illustration of the angle of departure.

point and corresponding angles θ_1, θ_2 and θ_3 as shown in the figure. From the angle requirement, we have

$$\theta_1 + \theta_2 + \theta_3 = 180°$$

where $\theta_2 \approx 90°$ since p_1 and p_2 are a conjugate pair and s_1 is close to p_1. It follows that

$$\theta_1 = 90° - \theta_3$$

θ_3 may be measured in the figure, then θ_1 can be obtained. The angle of departure at p_1 is θ_1, and then the angle of departure at p_2 is $-\theta_1$.

The points at which the loci cross the imaginary axis
The root loci can be drawn with the method above. It is useful to calculate the cross points of the root loci on the imaginary axis. These cross points can tell us when the system will be unstable as the parameter K changes from 0 to ∞, and how the system behaves as the system becomes critical.

On the imaginary axis, we have $s = j\omega$. The cross points of root loci to the imaginary axis must satisfy the closed loop characteristic equation since they are on the loci, and satisfy $s = j\omega$ since they are on the imaginary axis. By substituting $s = j\omega$ into the closed loop characteristic equation, and separating the real and imaginary parts, we can obtain two equations. The cross points $(0, \pm j\omega)$ and the corresponding parameter K can be obtained by solving these two equations.

If only the parameter K of the cross points is needed, the Routh criterion is an easy method. The characteristic equation of the system includes the open loop parameter K. As explained in Example 4.2.1, the range of values of K that make the system stable can be found by the Routh criterion. The critical value corresponds to the cross points.

Dimensioning the root locus

Sometimes, we need to dimension the root locus by the parameter K. For a point s on the locus, the corresponding value of parameter K may be found from the magnitude requirement. From (4.3.6), we have

$$K = \frac{|(s+p_1)||(s+p_2)|\cdots|(s+p_n)|}{|(s+z_1)||(s+z_2)|\cdots|(s+z_m)|} \qquad (4.3.22)$$

where $|s + p_i|$ and $|s + z_i|$ may be measured in the s-plane.

Remarks

1. The root locus can be calculated directly from the closed loop characteristic equation by computers without using the technique introduced here. However, the root locus technique may give us a quick overview of the closed loop poles.

2. The root locus gives us the sufficient information about the stability of the closed loop system, but insufficient information except for some very special cases about the system dynamics, which also depends on the location of the

TABLE 4.10. General steps for drawing root locus.

Step	Contents	Formulas or rules
1	Write the closed loop characteristic equation with a explicit K.	$1 + KG(s)H(s) = 0$
2	Locate open loop poles and zeros in s-plane.	× for poles O for zeros
3	Plot root loci on the real axis.	The segments to the left of an odd number of poles and zeros
4	Determine the number of separate loci.	Equal to the number of poles
5	Plot asymptotes.	Angle Centroid $\phi_A = \dfrac{(2k+1)\pi}{n-m}$ $\sigma_A = \dfrac{\sum p - \sum z}{n-m}$
6	Determine the breakaway point.	$\dfrac{dK}{ds} = -\dfrac{d}{ds}[1/(G(s)H(s))] = 0$
7	Determine the cross point on imaginary axis.	$1 + G(j\omega)H(j\omega) = 0$ or, Routh Criterion
8	Estimate departure or arrival angle of complex poles or zeros	Use angle requirement directly.
9	Draw the root locus and dimension it if needed.	Measure distance of p_i and z_i from origin, and measure K using magnitude requirement.

zeros.

The procedure for drawing the root locus is summarized in Table 4.10.

Example 4.3.2
Consider a system with an open loop transfer function

$$G(s)H(s) = \frac{K(s+2)}{s(s+1)(s+3)}$$

The open loop zero and poles are marked in Fig. 4.3.4. The locus on the real axis is between 0 and −1, and between −2 and −3.

There must be a breakaway point between 0 and −1. The closed loop characteristic equation can be written as

$$s(s+1)(s+3) + K(s+2) = 0$$

or

$$K = \frac{-s(s+1)(s+3)}{(s+2)} \tag{4.3.23}$$

Differentiating, then equating the result to zero yields

$$(3s^2 + 8s + 3)(s+2) - (s^3 + 4s^2 + 3s) = 0$$

or

$$s^3 + 5s^2 + 8s + 3 = 0$$

The breakaway point is given by the real solution $s = -0.534$. We discard the complex solutions $s = -2.233 \pm j0.7926$, since they are not on the real axis.

The angles of asymptotes with respect to the real axis are

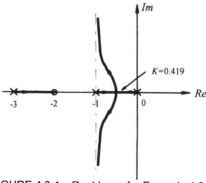

FIGURE 4.3.4.   Root locus for Example 4.3.2.

$$\phi_A = -\frac{(2k+1)\pi}{n-m} = 90°, 270°$$

The asymptote centroid may be obtained by

$$\sigma_A = \frac{-1-3+2}{(3-1)} = -1$$

The root locus is sketched in Fig. 4.3.4. The value of K corresponding to the breakaway point may be obtained by substituting $s = -0.534$ into (4.3.23), which yields $K = 0.419$. If $K > 0.419$, the closed loop system has a pair of complex conjugate poles.

Example 4.3.3

Consider the system shown in Fig. 4.3.5. By the block diagram reduction technique, the open loop transfer function can be found as

$$G(s)H(s) = \frac{\dfrac{1}{s(s+2)}}{1+\dfrac{2}{s(s+2)}} \cdot \frac{K}{s} = \frac{K}{s(s^2+2s+2)}$$

Thus, $n = 3$ and $m = 0$. The closed loop characteristic equation is

$$s^3 + 2s^2 + 2s + K = 0 \qquad (4.3.24)$$

The poles of the open loop transfer function are $s = 0, -1 \pm j1$.

A root locus is on the negative real axis from $s = 0$ to $s = -\infty$, and there is no breakaway point on it. There are 3 separate loci, and the two besides the one on the negative real axis begin at two complex conjugate poles. The angles of the asymptotes are

$$\phi_A = -\frac{(2k+1)\pi}{n-m} = \frac{\pi}{3}, \pi, \frac{5\pi}{3}$$

And the centroid is

$$\sigma_A = \frac{0-1+j1-1-j1}{3} = -\frac{2}{3}$$

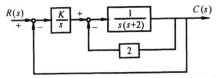

FIGURE 4.3.5. Block diagram for Example 4.3.3.

Let the departure angle of the pole $-1 + j1$ be θ; then

$$(\frac{3}{4}\pi + \theta + \frac{\pi}{2}) = \pm(2k+1)\pi$$

It follows that

$$\theta = -\pi/4$$

Similarly, the departure angle of the pole $(-1, -j1)$ is $\pi/4$.

The cross points may be found as follows: Let $s = j\omega$. Substituting this into the closed loop characteristic equation yields

$$(K - 2\omega^2) + j\omega(2 - \omega^2) = 0$$

From the imaginary part of the equation, we have

$$\omega = 0 \qquad \text{and} \qquad \omega = \pm\sqrt{2}$$

Substituting this into the real part of the equation yields

$$K = 0 \quad \text{when} \quad \omega = 0; \quad K = 4 \quad \text{when} \quad \omega = \pm\sqrt{2}$$

The Routh criterion can be used to determine K at cross points, since the closed loop characteristic equation (4.3.24) is known. Write the Routh array as

$$
\begin{array}{cc}
1 & 2 \\
2 & K \\
\dfrac{4-K}{2} & 0 \\
K &
\end{array}
$$

From this we find that $0 < K < 4$ makes the system stable. In other words, the roots of the characteristic equation are on the imaginary axis in the s-plane while $K = 4$ or $K = 0$. From the closed loop characteristic equation, $\omega = \pm\sqrt{2}$ is obtained if $K = 4$, and $\omega = 0$ if $K = 0$.

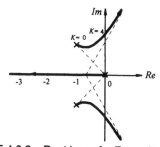

FIGURE 4.3.6. Root locus for Example 4.3.3.

The root locus is drawn in the Fig. 4.3.6.

Plot root locus with teachware package *PCET*

The root locus of a system may be plotted with the software package *PCET* provided with this book. The open loop transfer function may be written in the

TABLE 4.11. Plotting the root locus using *PCET*.

Step	User input	Function
1	Type *PCET*	Start program
2	Select *RLT*	Root locus technique
3	Press 1,2 or 3 for model selection	1. Polynomial model 2. Pole-zero model 3. A typical process model
4	Input maximum gain K_{max}	The root locus will be drawn as the open loop gain K varies from 0 to the maximum gain K_{max}.
5	Input system data	The order and parameters of the system model are determined.
6	Revise input if necessary	Verify input data.
7		Root locus is displayed on the screen.
8	Select function keys	More information can be obtained. Different functions are shown in Table 4.12.

TABLE 4.12. Function keys and their functions in *RLT*.

Function keys	Title on screen	Functions
F1	Decrease Maximum K	The maximum open loop gain K on the drawn root locus decreases to half of the previous one, and a new root locus is plotted.
F2	Increase Maximum K	The maximum open loop gain K on the drawn root locus increases to twice the previous one, and a new root locus is plotted.
F3	Gain at Imag. Axis	Gain K at the cross point of the root locus to the imaginary axis is shown on screen.
F4	Frequency at Imag. Axis	The value of the cross point on the imaginary axis is shown on screen.
F5	Closed Loop Roots	A specified open loop gain is required, for which the screen will show the closed loop roots both on the root locus and in numerical form.
F6	Open Loop Zeros	The open loop zeros will be shown on screen in numerical form.
F7	End	Back to *RLT* and ready for another system.

polynomial form or zero-pole form, both of which are acceptable for the software. The procedure for plotting the root locus on the screen with *PCET* is shown in Table 4.11. After the root locus is obtained, there are several function keys that can be selected. These are listed in Table 4.12.

4.4. SYSTEM ANALYSIS AND DESIGN WITH ROOT LOCUS

For a single loop feedback control system as shown in Fig. 4.3.1, the selection of an open loop gain is essentially a P controller design. Thus, the root locus may be used for the analysis and design of a system with a P controller. In fact, the root locus can also be used in the analysis and design of the system with other controllers, such as a PD controller, PI controller and so on.

Root locus with respect to a parameter of characteristic equation
The system root locus may be drawn with respect to a system parameter other than the open loop gain K. For example, if the characteristic equation of a system is

$$a_n s^n + a_{n-1} s^{n-1} + \cdots + a_1 s + a_0 = 0 \qquad (4.4.1)$$

then the root locus of the system may be drawn with respect to any coefficient of the equation. As for the effect of a_{n-1} on the system poles, rearranging (4.4.1) by dividing both sides by the whole polynomial minus the term including a_{n-1} yields

$$1 + \frac{a_{n-1} s^{n-1}}{a_n s^n + a_{n-2} s^{n-2} + \cdots + a_1 s + a_0} = 0 \qquad (4.4.2)$$

Using the root locus technique, the effect of the coefficient a_{n-1} on the system poles may be ascertained. The effects of other coefficients in the equation on the system poles may be studied in the same way.

Example 4.4.1
    Consider a system with the characteristic equation

$$s^3 + 3s^2 + 4s + 2 = 0$$

Describe the variation of the system poles when the coefficient of s^2 varies from 3 to 10.
    Rewrite the characteristic equation as

$$s^3 + (3 + K)s^2 + 4s + 2 = 0$$

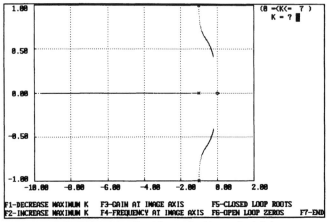

FIGURE 4.4.1.   Root locus for Example 4.4.1.

or

$$1 + \frac{Ks^2}{s^3 + 3s^2 + 4s + 2} = 0$$

Take the second term of the equation above as an open loop transfer function that has poles at $-1 \pm j1$, -1 and double zeros at 0. The system root locus can be obtained with the software $PCET$. The maximum open loop gain is set at 7. The result, shown in Fig. 4.4.1, illustrates the variation of system poles as K varies from 0 to 7. By selecting F5, the system poles at $K = 7$ can be found to be $-0.20 \pm j0.41$ and -9.61.

Root locus of a system with a *PD* controller

The transfer function for an ideal *PD* controller is

$$G_c(s) = K_c(1 + \tau_D s) \tag{4.4.3}$$

Incorporating a *PD* controller into a system is equivalent to adding a zero into the open loop transfer function of the system. If the open loop transfer function is $G(s)H(s)$, after the change, the open loop transfer function becomes $K_c(1 + \tau_D s)G(s)H(s)$, and the closed loop characteristic equation of the system is

$$1 + K_c(1 + \tau_D s)G(s)H(s) = 0 \tag{4.4.4}$$

In order to choose a suitable derivative time τ_D, it is convenient to draw a root locus with respect to the derivative time τ_D rather than proportional gain K_c. Consider another system with an open loop transfer function

$$\hat{G}(s) = \frac{K_c \tau_D s G(s) H(s)}{1 + K_c G(s) H(s)} \qquad (4.4.5)$$

Simple calculation shows that the closed loop characteristic equation of this system will be the same as (4.4.4). For this reason, we call (4.4.5) the equivalent transfer function to the open loop transfer function $K_c(1 + \tau_D s) G(s) H(s)$. The root locus of the system with the open loop transfer function (4.4.5) is the same as that of the system with the open loop transfer function $K_c(1 + \tau_D s) G(s) H(s)$. After K_c has been chosen, (4.4.5) is used to draw the root locus with respect to the parameter τ_D.

Even if the zeros and poles of $K_c G(s) H(s)$ are known, the poles of (4.4.5) may be difficult to calculate. Eq. (4.4.5) shows that the poles of the equivalent transfer function are the closed loop poles of a system with an open loop transfer function $K_c G(s) H(s)$. Therefore, the poles of the equivalent transfer function, which are the open loop poles for the root locus drawn with respect to the parameter τ_D, may be obtained from another root locus drawn with an open loop transfer function $K_c G(s) H(s)$. These poles can also be obtained by solving the equation $1 + K_c G(s) H(s) = 0$, if it is not too complicated.

Example 4.4.2

An open loop transfer function of a system is given as

$$G(s) H(s) = \frac{1}{(s+4)(s+1)(s-1)}$$

With the *PD* controller in (4.4.3), the open loop transfer function becomes

$$G_c(s) G(s) H(s) = \frac{K_c(1 + \tau_D s)}{(s+4)(s+1)(s-1)}$$

The equivalent transfer function may be found to be

$$\hat{G}(s) = \frac{K_c \tau_D s}{(s+4)(s+1)(s-1) + K_c}$$

In order to find the roots of its denominator, consider the system with open loop transfer function

$$K_c G(s) H(s) = \frac{K_c}{(s+4)(s+1)(s-1)}$$

The closed loop root locus of this system will give the poles of the equivalent transfer function.

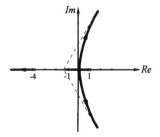
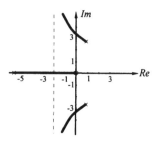

a. Root locus of equivalent transfer function   b. Root locus with respect to derivative time
FIGURE 4.4.2.   Root locus for Example 4.4.2.

For drawing the root locus, we can find through calculation that the angles of the asymptotes are 60°, 180° and 300°, the asymptote centroid is -1.33, and the breakaway point is at 0.12. The root locus may be drawn as shown in Fig. 4.4.2a. If the proportional gain is assumed to be 50, three poles may be obtained on the root locus as -5.64, $0.82 \pm j2.7$, which are the poles of the equivalent transfer function.

The root locus with respect to derivative time τ_D can now be developed. The locus on the real axis is between $s = 0$ and $s = -5.6$. The angles of the asymptotes are 90° and 270°. The asymptote centroid is at -2. The departure angle at the pole $s = 0.82 \pm j2.7$ is $\pm 138°$. The root locus is shown in Fig. 4.4.2b. It can be found that if we want to locate the closed loop poles close to the line $s = -1$, then $\tau_D = 0.56$ may be designated. The reader can check this result with the software package *PCET*.

Root locus of a system with *PI* controller
The transfer function of a *PI* controller is

$$G_c(s) = K_c(1 + \frac{1}{\tau_I s}) \tag{4.4.6}$$

or

$$G_c(s) = K_c \frac{1}{s}(s + \frac{1}{\tau_I}) \tag{4.4.7}$$

This means that incorporating a *PI* controller into a system is equivalent to adding a pole and a zero into the open loop transfer function. The pole is always at origin, whereas the zero is changed with the integral time.

If the open loop transfer function of a system is $G(s)H(s)$, the closed loop characteristic equation of the system with a *PI* controller (4.4.7) will be

$$s + K_c s G(s)H(s) + \frac{K_c}{\tau_I} G(s)H(s) = 0 \qquad (4.4.8)$$

As with the case of a *PD* controller, an open loop equivalent transfer function may be proposed as

$$\hat{G}(s) = \frac{K_c}{\tau_I} \cdot \frac{G(s)H(s)}{s + K_c s G(s)H(s)} \qquad (4.4.9)$$

Therefore, with the equivalent open loop transfer function, the root locus of the system may be drawn with respect to the integral time τ_I, or its inverse $1/\tau_I$.

Pole $s = 0$ is one pole of the equivalent transfer function $\hat{G}(s)$, and other poles of $\hat{G}(s)$ may be found by solving the algebraic equation, or by plotting another root locus that corresponds to an open loop transfer function $K_c G(s)H(s)$.

Example 4.4.3

The open loop transfer function of a system is given by

$$G(s)H(s) = \frac{\hat{K}}{(s+0.2)(s+0.4)} \qquad (4.4.10)$$

With a *PI* controller (4.4.7), the equivalent transfer function may be found to be

$$\hat{G}(s) = \frac{K/\tau_I}{s[(s+0.2)(s+0.4) + K]}$$

where $K = K_c \hat{K}$. We will discuss the effect of the integral time on the root locus assuming that the decay ratio is 0.25.

When the controller has only a proportional action, the open loop transfer function is (4.4.10) with the replacement of $\hat{K}$ by K. The root locus of the

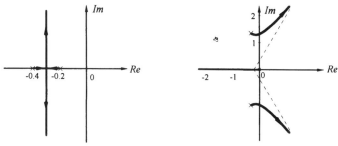

a. Root locus with respect to proportional gain   b. Root locus with respect to integral time

FIGURE 4.4.3.   Root locus for Example 4.4.3.

TABLE 4.13.   Effect on root locus of adding PD or PI controller.

Controller	PD	PI
Action	Derivative action	Integral action
Adding pole or zero	A zero at s $= -1/\tau_D$	A zero at s $= -1/\tau_I$ A pole at s=0
Closed loop characteristic equation	$1 + K_c(1 + \tau_D s)G(s)H(s)$ $= 0$	$s + K_c sG(s)H(s) + \dfrac{K_c}{\tau_I}G(s)H(s) = 0$
Equivalent transfer function	$\hat{G}(s) = \dfrac{K_c \tau_D sG(s)H(s)}{1 + K_c G(s)H(s)}$	$\hat{G}(s) = \dfrac{K_c}{\tau_I}\dfrac{G(s)H(s)}{s + K_c sG(s)H(s)}$

system with only a proportional action is shown in Fig. 4.4.3a.   According to Table 4.4, a decay ratio of 0.25 yields

$$\exp(-2\pi\alpha/\omega) = 0.25$$

It follows that $\omega/\alpha = 4.532$, and $\omega = 1.36$ because $\alpha = 0.3$.   For decay ratio 0.25, K may be obtained from (4.3.22) as

$$K = |-0.3 + j1.36 + 0.2||-0.3 - j1.36 + 0.4| = 1.86$$

Then, the equivalent transfer function becomes

$$\hat{G}(s) = \frac{1.86/\tau_I}{s[(s+0.2)(s+0.4)+1.86]}$$

$$= \frac{1.86/\tau_I}{s[(s+0.3+j1.36)(s+0.3-j1.36)]}$$

The root locus with respect to the integral time is shown in Fig. 4.4.3b, where it should be noted that the root locus traces to infinity as the integral time decreases to zero rather than going to infinity.

The effects on the root locus of adding derivative action and integral action to the controller are summarized in Table 4.13.

Problems

4.1. Determine the system stability with the Routh criterion, if the characteristic equation of the system is given as

a. $s^4 + 2s^3 + s^2 + 4s + 2 = 0$

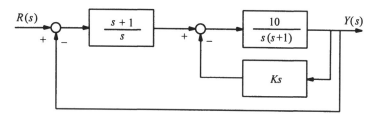

FIGURE 4.P.1.   A system with two feedback loops.

b. $s^4 + s^3 + 3s^2 + 5s + 9 = 0$

c. $s^6 + 24s^5 + 10s^4 + 100s^3 + 20s^2 + 40s + 2 = 0$

d. $s^6 + 13s^5 + 64s^4 + 154s^3 + 193s^2 + 121s + 30 = 0$

4.2. Find K to stabilize a system with the characteristic equation

$$s^3 + 30s^2 + 650s + 700K = 0$$

4.3. For a system with the characteristic equation

$$s^4 + s^3 + 3s^2 + 3s + 6 = 0$$

how many poles of the system are located in the right half s-plane? Use three different methods with the Routh criterion to solve the problem.

4.4. For a unity feedback system, the open loop transfer function is given as follows:

$$G(s) = \frac{K(s+1)}{s(s-1)(s+5)}$$

Find the values of K with Routh criterion to keep the closed loop system stable.

4.5. Consider a system as shown in Fig. 4.P.1, which has two feedback loops.

a. Find the values of inner loop feedback coefficient K with the Routh criterion to keep the system stable.

b. Can you adjust the inner loop feedback coefficient K to locate all the system poles to the left of $s = -1$ in the s-plane? If yes, find the K values with the Routh criterion; if no, give proof with the Routh criterion technique.

4.6. Draw the root locus of the system in Problem 4.4 and calculate the value of K at the intersection with the imaginary axis.

4.7. Draw the root locus for a system with an open loop transfer function

$$G(s)H(s) = \frac{K}{s(s+1)(s+2)}$$

and find the values of K to stabilize the closed loop system.

4.8. Sketch the root locus for a system with an open loop transfer function

a. $G(s) = \dfrac{K}{s^2(s+2)}$

b. $G(s) = \dfrac{K(s+1)}{s^2(s+2)}$

c. $G(s) = \dfrac{K(s^2+2s+2)}{s^2(s+2)}$

4.9. With the *RSC* in *PCET*, determine the stability of the systems in Problem 4.1 and Problem 4.3.

4.10. With the software *PCET*, draw the root locus of the system in Problem 4.2 with respect to the coefficient K.

4.11. With the *RLT* in *PCET*, draw the root loci for the systems in Problems 4.6, 4.7 and 4.8.

4.12. With the *RLT* in *PCET*, draw the root locus of the system below with respect to the coefficient of s^3, and determine the stability of the system. The system characteristic equation is

$$s^4 + 2s^3 + 3s^2 + 4s + 2 = 0$$

4.13. Consider a system with open loop transfer function

$$G(s) = \dfrac{1}{(s+3)(s-2)(s+5)}$$

With the *RLT* in *PCET* design a *PD* controller to locate the closed loop poles at a suitable place in the s-plane.

CHAPTER 5

FREQUENCY DOMAIN ANALYSIS AND DESIGN

5.1. FREQUENCY RESPONSES AND NYQUIST PLOT

Until now we have studied system behaviors in the time domain, such as system response to a unit step input. In reality, a system rarely incurs a single disturbance but rather a series of disturbances. To simulate those disturbances, we can apply sinusoidal signals with specified frequencies. In other words, we can analyze them in the frequency domain.

The behavior of control systems in the frequency domain will be studied in this chapter. The system frequency response to a sinusoidal input, rather than the transient response to a step input, will be investigated. The Nyquist plot and Bode diagram are two basic frequency responses of a system, and both are means of analyzing the stability and dynamics of a system in the frequency domain.

Sinusoidal signals

For a sinusoidal signal, there are three elements: frequency, amplitude and phase angle.

For a linear system with a sinusoidal input, is the ultimate output of the system still a sinusoidal signal? If yes, how are the three elements of the sinusoidal output affected by the input and the system transfer function?

Let us consider a stable system with the transfer function

$$G(s) = \frac{(s+z_1)(s+z_2)\cdots(s+z_m)}{(s+p_1)(s+p_2)\cdots(s+p_n)} \tag{5.1.1}$$

where z_i ($i = 1, 2, \cdots, m$) and p_j ($j = 1, 2, \cdots, n$) are distinct. A sinusoidal input is chosen as

125

$$r(t) = A_m \sin \omega t \tag{5.1.2}$$

where A_m is the amplitude, and ω is the angular frequency. The phase angle is 0 at $t = 0$. From Example 2.2.2, the Laplace transform of (5.1.2) is

$$R(s) = \frac{A_m \omega}{s^2 + \omega^2} = \frac{A_m \omega}{(s + j\omega)(s - j\omega)} \tag{5.1.3}$$

Then, the Laplace transform of the system output is

$$Y(s) = G(s)R(s) = \frac{(s + z_1)(s + z_2)\cdots(s + z_m)}{(s + p_1)(s + p_2)\cdots(s + p_n)} \cdot \frac{A_m \omega}{(s + j\omega)(s - j\omega)} \tag{5.1.4}$$

Employing in partial fraction expansion gives

$$Y(s) = \frac{A}{(s + j\omega)} + \frac{B}{(s - j\omega)} + \sum_{i=1}^{n} \frac{C_i}{(s + p_i)} \tag{5.1.5}$$

where

$$A = \lim_{s \to -j\omega} [(s + j\omega)Y(s)] = \lim_{s \to -j\omega} \frac{A_m \omega G(s)}{s - j\omega} = -\frac{A_m}{2j} G(-j\omega)$$

$$B = \lim_{s \to j\omega} [(s - j\omega)Y(s)] = \lim_{s \to j\omega} \frac{A_m \omega G(s)}{s + j\omega} = \frac{A_m}{2j} G(j\omega)$$

$$C_i = \lim_{s \to -p_i} [(s + p_i)Y(s)]$$

Therefore, the output of the system is

$$y(t) = [-\frac{A_m}{2j} G(-j\omega)]e^{-j\omega t} + [\frac{A_m}{2j} G(j\omega)]e^{j\omega t} + \sum_{i=1}^{n} C_i e^{-p_i t} \tag{5.1.6}$$

For the ultimate state ($t \to \infty$), the last term of (5.1.6) vanishes, because the system is stable, and thus all the poles of the system must have negative real parts, i.e., $-p_j < 0$ or $p_j > 0$. Hence, the ultimate output $y_u(t)$ to a sinusoidal input will be

$$y_u(t) = \frac{A_m}{2j} [G(j\omega)e^{j\omega t} - G(-j\omega)e^{-j\omega t}] \tag{5.1.7}$$

Because

$$G(j\omega) = |G(j\omega)|e^{j \arg G(j\omega)}$$

$$G(-j\omega) = |G(-j\omega)|e^{j \arg G(-j\omega)} = |G(j\omega)|e^{-j \arg G(j\omega)}$$

we obtain

$$y_u(t) = \frac{A_m}{2j}|G(j\omega)|[e^{j[\omega t + \arg G(j\omega)]} - e^{-j[\omega t + \arg G(j\omega)]}]$$

$$= A_m|G(j\omega)|\sin(\omega t + \arg G(j\omega)) \qquad (5.1.8)$$

where $|G(j\omega)|$ is referred to as the amplitude ratio, denoted AR, since it is the ratio of the amplitude output to the input. $\text{Arg}G(j\omega)$, or ϕ, is referred to as the phase shift, because it represents the phase angle difference between sinusoidal output and input. Eq. (5.1.8) shows that the ultimate output of a system with a sinusoidal input $A_m\sin\omega t$ is also a sustained sinusoidal signal. The frequency of the ultimate output is identical to that of the input. Its amplitude is $A_m|G(j\omega)|$, where $G(j\omega)$ is the transfer function of the system with s replaced by $j\omega$. Since the ultimate output wave lags behind the input wave, the phase shift $\arg G(j\omega)$ is also called the phase lag.

All the ultimate signals in a linear system are sinusoidal signals, with the frequency identical to that of the sinusoidal input, because any signal in the system may be chosen as an output.

The reader may have noticed that we use 'ultimate' instead of 'steady state' here to represent the state as $t \to \infty$. This is done to avoid confusion. It no longer refers to a final value, but a state as time goes to infinity, which may change regularly.

Frequency response
The frequency response of a system is defined as the ultimate response of the system to a sinusoidal input signal.

Frequency response is easy to determine. For a linear system, as we discussed above, the ultimate state of any signal in the system is sinusoidal with the identical frequency as the sinusoidal input. As a matter of fact, if the response is determined by experiment, only the amplitude and phase shift of the output need to be measured for every designated frequency. If the transfer function $G(s)$ of the system is known, the frequency response can be determined simply by calculating the amplitude ratio and phase shift of $G(j\omega)$. That is why we also call $G(j\omega)$ the frequency response of the system. Table 5.1 shows the procedure for finding the frequency response.

Both frequency response and transient response reflect the performance of a system, but except for some special cases the correlation between them is not straightforward. However, a system that has a satisfactory frequency response will normally have a satisfactory transient response, or vice versa.

The differences between frequency response and time response are summarized in Table 5.2.

TABLE 5.1.   Method of finding frequency response.

Step	Item	Content
1	Replacement	Replace s in $G(s)$ with $j\omega$
2	Rationalization	Rearrange $G(j\omega)$ as $Re(\omega) + jIm(\omega)$ by multiplying the denominator and numerator by the complex conjugate term of the denominator.
3	Polarization	$AR = \sqrt{Re(\omega)^2 + Im(\omega)^2}$, $\phi = arctg(Im(\omega)/Re(\omega))$

TABLE 5.2.   Difference between time and frequency responses.

Response	Time response	Frequency response
Input signal	Step (or impulse) input	Sinusoidal input
Concerned state	Transient state	Ultimate state
Independent variable	Time t	Angular frequency ω
Advantage	Easy to understand	Easy to determine

Nyquist plots

A Nyquist plot is the polar plot of frequency response $G(j\omega)$ as the angular frequency ω varies from 0 to infinity. This plot is used to represent the system characteristics in the frequency domain. Nyquist plots use the imaginary part of $G(j\omega)$, denoted $Im[G(j\omega)]$, as the ordinate, and the real part of $G(j\omega)$, denoted $Re[G(j\omega)]$, as the abscissa. The shape and location of a Nyquist plot reflect the characteristics of the system.

Nyquist plots for first, second and high order systems

Consider a first order system with a transfer function

$$G(s) = \frac{K}{Ts+1} \qquad (5.1.9)$$

Replacing s with $j\omega$ yields

$$G(j\omega) = \frac{K}{1+jT\omega} = \frac{K(1-jT\omega)}{(1+jT\omega)(1-jT\omega)}$$

$$= \frac{K}{1+T^2\omega^2}(1-jT\omega) \qquad (5.1.10)$$

Thus, AR and phase lag are

$$|G(j\omega)| = \frac{K}{\sqrt{1+\omega^2T^2}} \qquad (5.1.11)$$

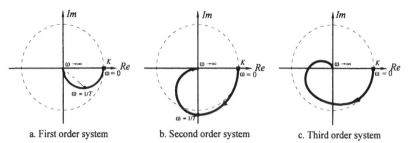

a. First order system   b. Second order system   c. Third order system

FIGURE 5.1.1. Nyquist plots for first to third order systems.

$$\phi = \arg G(j\omega) = \text{arctg}(-T\omega) = -\text{arctg}(T\omega) \tag{5.1.12}$$

When $\omega = 0$, we have $AR = |G(j\omega)| = K$ and $\phi = 0°$. Therefore, the Nyquist plot starts at the point $(K, j0)$ on the real axis. As $\omega \to \infty$, we find $AR = |G(j\omega)| \to 0$ and $\phi \to -90°$. Hence, the Nyquist plot ends at the origin. In fact, the Nyquist plot of a first order system is a half circle as shown in Fig 5.1.1a, with a special point at $\omega = 1/T$, which is located half way across the arc with $\phi = -45°$ and $AR = K\sqrt{2}/2$.

For a second order system with a transfer function

$$G(s) = \frac{K}{T^2 s^2 + 2\zeta Ts + 1} \tag{5.1.13}$$

we have

$$AR = \frac{K}{\sqrt{(1 - T^2\omega^2)^2 + (2\zeta T\omega)^2}} \tag{5.1.14}$$

$$\phi = \text{arctg}(-\frac{2\zeta T\omega}{1 - T^2\omega^2}) = -\text{arctg}(\frac{2\zeta T\omega}{1 - T^2\omega^2}) \tag{5.1.15}$$

The Nyquist plot begins at $s = K$ on the real axis, where $\omega = 0$, $AR = K$ and $\phi = 0°$. When $\omega \to \infty$, then $AR \to 0$ and $\phi \to -180°$, i.e., the Nyquist plot ends at

TABLE 5.3. Starting and ending points of Nyquist plot.

Order of system	Starting point	Ending point	Approaching angle at ending point
First	K on the real axis	Origin	$-\pi/2$
Second	K on the real axis	Origin	$-\pi$
Third	K on the real axis	Origin	$-3\pi/2$
nth	K on the real axis	Origin	$-n\pi/2$

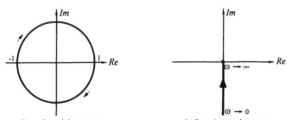

a. Pure time delay system   b. Pure integral system
FIGURE 5.1.2 Nyquist plots for pure time delay and pure integral systems.

the origin, where the plot approaches with an angle $-180°$. The Nyquist plot of a
second order system is shown in Fig. 5.1.1b. If $\omega = 1/T$, we have $AR = K/2\zeta$ and
$\phi = -90°$. This is the point on the negative imaginary axis.

Similarly, the Nyquist plot of a third order system starts at $s = K$ and ends at
the origin with an approaching angle $-270°$ as shown in Fig. 5.1.1c.

We summarize the Nyquist plots for up to an nth-order system in Table 5.3.

Nyquist plots of other systems

A system of pure time delay τ has a transfer function $G(s) = e^{-\tau s}$. Following the
steps in Table 5.1 yields

$$AR = 1 \quad \text{and} \quad \phi = -\omega\tau$$

The Nyquist plot of a pure time delay system is shown in Fig. 5.1.2a.

For a pure integral system with transfer function $G(s) = 1/s$, the quantities AR
and ϕ are found to be $1/\omega$ and $-90°$, respectively. Therefore, the Nyquist plot of
a pure integral system coincides with the negative part of the imaginary axis, as
shown in Fig. 5.1.2b.

Similarly, the Nyquist plots of P, PI, PD and PID controllers can be
constructed as shown in Figs. 5.1.3a, b, c and d, respectively. The reader can
draw these Nyquist plots with the software $PCET$.

The type of system determines the starting point of the Nyquist plot. As
shown in Fig. 5.1.3, when ω goes to 0, the Nyquist plot of a PI or PID controller
traces a line parallel to the negative imaginary axis. In fact all the systems of

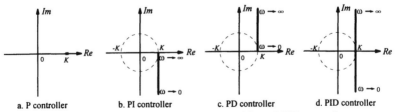

a. P controller   b. PI controller   c. PD controller   d. PID controller
FIGURE 5.1.3. Nyquist plots for P, PI, PD, and PID controllers.

TABLE 5.4. The effect of system type on the Nyquist plot.

System type	Starting point of the Nyquist plot
0	On positive real axis
1	Approaching a line parallel to negative imaginary axis as $\omega \to 0$
2	Approaching a line parallel to negative real axis as $\omega \to 0$
3	Approaching a line parallel to positive imaginary axis as $\omega \to 0$
N	A line parallel to $N \times (-90°)$ axis as $\omega \to 0$

type 1 have a Nyquist plot that approaches a line parallel to the negative imaginary axis as $\omega \to 0$. For a system of type 2, the Nyquist plot traces a line parallel to the negative real axis as ω goes to 0. Other cases can be found in Table 5.4.

Nyquist plots with *PCET* software

Nyquist plots can be drawn on the screen or printed out with the sub-program *FDA* (Frequency Domain Analysis) of the software package *PCET*. Three different input transfer function models can be selected: the polynomial model, the factored polynomial model and the zero-pole model. After typing the order and parameters of the model as well as the maximum and minimum frequency of the plot, function key *F*6 gives the Nyquist plot, assuming that a *P* controller is

TABLE 5.5. Procedure for plotting Nyquist plot with *PCET*.

Step	User input		Function
1	Type *PCET*		Start program
2	Select *FDA*		Frequency domain analysis
3	Select models	1	Polynomial model
		2	Factored polynomial model
		3	Zero-pole model
4	Input: 1. Order & parameters of the model 2. Maximum & minimum frequency		Define the system and determine the frequency range.
5	Select F6	"F1-Exit"	Return to main menu
		"F2-Restart"	Return to very beginning of *FDA*
		"F3-Load"	Load a saved file to get data
		"F4-Save"	Save the updated data in a file
		"F5-Modify"	Change system parameters
		"F6-Nyquist"	Display the Nyquist plot
		"F7-Bode"	Display the Bode diagram
6	F1 - F7 Same as step 5		
	"F8-Scale"		Change the dimensions
	"F9-*PID*"		Tune the *PID* controller

used with $K_c = 1$. Function key $F8$ is used to change the dimension, and $F9$ to tune the controller.

The procedure for plotting Nyquist plots with $PCET$ is described in Table 5.5.

5.2. BODE DIAGRAMS

The Bode diagram or Bode plot is a plot of amplitude ratio (AR) and phase shift (ϕ) of $G(j\omega)$ with respect to the angular frequency ω.

The Phase shift is usually plotted against the logarithm of frequency on semilog coordinates, i.e., the abscissa is dimensioned by $\log \omega$. With such a dimension, the distance between 0.1 and 1 is equal to that between 1 and 10, or between 10 and 100, and so on. The distance between 1 and 10, or an equivalence, is called a decade.

The amplitude ratio is also plotted against the logarithm of angular frequency in semilog coordinates. For convenience, the amplitude ratio is usually converted to log modulus (logarithmic modulus) defined by the equation

$$L = \log \text{modulus} = 20 \log|G(j\omega)| \qquad (5.2.1)$$

The units of log modulus are decibels (dB), a term originally used in communication engineering to indicate the ratio of two values of power.

However, some control engineers plot the amplitude ratio in log-log coordinates, with dimensions in real values instead of log modulus. In fact, it is just the same plot with a different scale.

Composition of a transfer function

Since the numerator and denominator polynomials of a transfer function can be factored into the combination of first and second order polynomials, there are only five different kinds of factors that may occur in a transfer function. They are

1. constant gain K;
2. poles (or zeros) at origin;
3. poles (or zeros) on real axis;
4. complex conjugate poles (or zeros);
5. pure delay time.

The transfer function of a general system

$$G(s) = \frac{K \prod\limits_{i=1}^{m}(1 + T_i s)}{s^p \prod\limits_{j=1}^{n}(1 + \overline{T}_j s) \prod\limits_{k=1}^{l}(1 + 2\zeta_k \hat{T}_k s + \hat{T}_k^2 s^2)} e^{-j\omega\tau} \qquad (5.2.2)$$

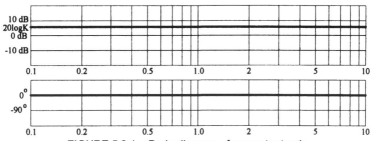

FIGURE 5.2.1. Bode diagram of a constant gain.

has a constant gain K, m zeros on the real axis, p poles at the origin, n poles on the real axis, l pairs of complex conjugate poles and a delay time τ.

After the individual Bode diagram for these five factors are studied, it can be seen that the Bode diagram of a transfer function can be obtained by simply adding graphically the separate contributions.

This procedure can be simplified further by using the asymptotic approximations of the Bode diagrams of these five factors.

Constant gain

The logarithmic modulus of a constant gain K is $20\log K$ in units of dB, and the phase shift is zero. Both the log modulus curve and phase shift curve are horizontal lines on a Bode diagram, as shown in Fig. 5.2.1.

A pole (or a zero) at the origin

For a pure integral transfer function $1/s$, only a pole at the origin exists. The log modulus of a pole will be

$$L = 20 \log|1/j\omega| = -20\log\omega \text{ (dB)} \tag{5.2.3}$$

and the phase shift is a constant, $\phi(\omega) = -90°$. Since the abscissa is dimensioned with $\log\omega$, the log modulus curve is a straight line with a slope of -20 dB/decade. When $\omega = 1$, the log modulus is equal to 0 dB.

Similarly, for a pure derivative transfer function s, only a zero at the origin appears. Its logarithmic modulus will be $20\log\omega$ (dB), and the phase shift is $90°$.

The Bode diagrams for a pole and a zero are shown in Fig. 5.2.2a and b, respectively.

A pole (or a zero) on the real axis

The transfer function (5.1.9) of a first order system has a pole on the real axis. The amplitude ratio and phase shift are shown in (5.1.11) and (5.1.12). Thus, when $K = 1$, the log modulus is

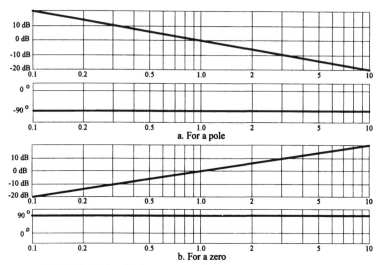

FIGURE 5.2.2. Bode diagram of a pole and a zero at origin.

$$L = 20\log|G(j\omega)| = 20\log\frac{1}{\sqrt{1+\omega^2 T^2}} = -10\log(1+\omega^2 T^2) \qquad (5.2.4)$$

One of the most convenient features of a Bode diagram is that the log modulus curve can be easily sketched by considering the low and high frequency asymptotes. As ω goes to zero, L goes to $-20\log 1 = 0$ dB. As ω becomes very large, (5.2.4) reduces to

$$L = -10\log(\omega^2 T^2) = -20\log\omega - 20\log T \qquad (5.2.5)$$

This is the equation of a straight line of L versus $\log\omega$, which has a slope of -20dB/decade. The high frequency asymptote intersects the low frequency asymptote $L = 0$ at $\omega = 1/T$, which is called a breakpoint frequency, or break frequency. From (5.2.4), the value of log modulus at the break frequency is

$$L(\omega = 1/T) = -10\log 2 = -3\,\text{dB}$$

This indicates that the actual log modulus at break frequency is 3 dB lower than the intersection of asymptotes. The differences between asymptotes and actual log moduli at various frequencies for a first order system are given in Table 5.6.

TABLE 5.6. Difference between asymptotes and actual log modulus.

ωT	0.1	0.25	0.4	0.5	1.0	2.0	2.5	4.0	10.0
Difference	−0.04	−0.32	−0.65	−1.0	−3.01	−1.0	−0.65	−0.32	−0.04

TABLE 5.7. Phase shift data of a first order system.

ωT	0.01	0.1	0.25	0.5	1.0	2.0	4.0	10.0	100
$\phi(°)$	−0.67	−5.7	−14.1	−26.6	−45	−63.4	−75.9	−84.3	−89.4

It can be found from (5.1.12) that the phase shift varies from 0° to −90° as ω changes from 0 to infinity. When $\omega = 1/T$, $\arg G(\omega) = -\arctan 1 = -45°$. The data for drawing the phase shift of a first order system are listed in Table 5.7.

The Bode diagram of a first order system is shown in Fig. 5.2.3, where the abscissa is normalized with $\log \omega T$.

The Bode diagram of the transfer function $(1 + Ts)$ can be obtained, as shown in Fig. 5.2.4, in the same manner as that of a pole. However, the slope of the log modulus is positive (20 dB/decade), and the phase shift is $\phi(\omega) = \arctan \omega T$. The data in Table 5.6 and Table 5.7 can still be used by changing their signs.

The procedure for drawing a Bode diagram of a first order system is given in Table 5.8.

Complex conjugate poles (zeros)
A second order system may have a pair of complex conjugate poles. The amplitude ratio and phase shift of a second order system (5.1.13) were given in (5.1.14) and (5.1.15).

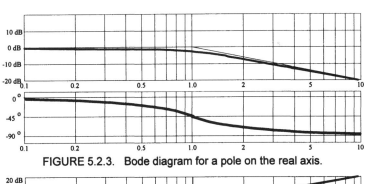

FIGURE 5.2.3. Bode diagram for a pole on the real axis.

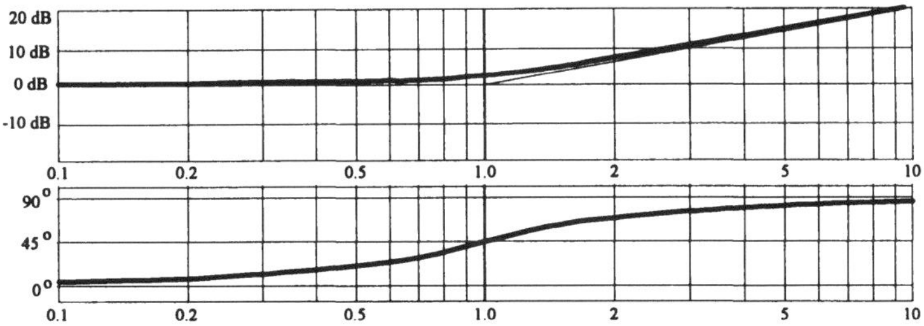

FIGURE 5.2.4. Bode diagram for a zero on the real axis.

TABLE 5.8.   Procedure for drawing Bode diagram of a first order system.

Step	Item	Content of formula
1	Break frequency	Locate the break frequency at $\omega = 1/T$.
2	Low frequency asymptote	Draw horizontal asymptote $L = 20 \log K$ for $\omega < 1/T$.
3	High frequency asymptote	Draw asymptote with slope = -20dB/decade intersecting $L = 20 \log K$ at $\omega = 1/T$.
4	Actual log modulus	-3 dB lower at $\omega = 1/T$
5	Phase shift at break frequency	$-45°$
6	Phase shift asymptote	Two horizontal asymptotes at $\phi = 0°$ and $\phi = -90°$
7	Phase shift curve	Use Table 5.7 for other points.

The log modulus of system (5.1.13) with $K = 1$ is

$$L = 20 \log \frac{1}{\sqrt{(1 - T^2\omega^2)^2 + (2\zeta T\omega)^2}}$$

$$= -10 \log[(1 - T^2\omega^2)^2 + (2\zeta T\omega)^2] \qquad (5.2.6)$$

When $\omega \to 0$, we can find the low frequency asymptote

$$\lim_{\omega \to 0} L = -10 \log 1 = 0$$

As ω becomes large, we have

$$L = -10 \log(T^4\omega^4) = -40 \log \omega - 40 \log T$$

Thus the high frequency asymptote has a slope of -40dB/decade, and intersects the low frequency asymptote at $\omega = 1/T$.

The difference between the actual magnitude curve and the asymptotic approximation is a function of the damping factor ζ. When $\zeta \geq 1$, the system is composed of two first order systems, for which the Bode diagram has been studied, and the amplitude ratio is attenuated for all ω. For an underdamped system, there are a pair of complex conjugate poles, and the log modulus curve exhibits a maximum for the values of ζ between 0 and $\sqrt{2}/2$. The reader may prove that the resonant frequency satisfies

$$\omega_r = (\sqrt{1 - 2\zeta^2})/T \qquad (5.2.7)$$

From this the maximum value of the amplitude at the resonant frequency is

$$M_p = |G(j\omega_r)| = (2\zeta\sqrt{1 - \zeta^2})^{-1} \qquad (5.2.8)$$

TABLE 5.9. Difference at break frequency between asymptotic and actual log modulus of a second order system.

Damping factor ζ	0.05	0.1	0.15	0.2	0.25	0.3	0.4	0.5	0.6	0.7	0.8	1.0
Difference (dB)	20	14	10.4	8	6	4.4	2.0	0	1.6	3.0	4.0	6.0

TABLE 5.10. Phase shift data (in degrees) of a second order system.

ζ \\ ωT	0.01	0.1	0.5	1.0	2.0	5.0	10	30	100
0.1	−0.1	−1.2	−7.6	−90	−173.4	−177.6	−178.8	−179.6	−179.9
0.2	−0.2	−2.3	−14.9	−90	−165.1	−175.2	−177.6	−179.2	−179.8
0.3	−0.3	−3.5	−21.8	−90	−158.2	−172.9	−176.6	−178.9	−179.7
0.4	−0.5	−4.6	−28.1	−90	−151.9	−170.6	−175.4	−178.5	−179.5
0.5	−0.6	−5.8	−33.7	−90	−146.3	−168.2	−174.2	−178.2	−179.4
0.6	−0.7	−6.9	−38.7	−90	−141.3	−166.0	−173.1	−177.7	−179.3
0.7	−0.8	−8.1	−43.0	−90	−137.0	−163.7	−171.9	−177.3	−179.2
0.8	−0.9	−9.2	−46.9	−90	−133.1	−161.6	−170.8	−176.9	−179.1
0.9	−1.0	−10.3	−50.2	−90	−129.8	−159.4	−169.7	−176.6	−179.0
1.0	−1.1	−11.4	−53.1	−90	−126.9	−157.4	−168.6	−176.2	−178.9

With respect to damping factor, the differences between asymptotic and actual log modulus at break frequency of a second order system are shown in Table 5.9.

The phase shift of the second order system (5.1.13), given in (5.1.15), varies from 0° to −180° as ω changes from 0 to infinity, and equals −90° when ω = 1/T. Obviously, phase shift is also a function of damping factor ζ. The phase shift data are shown in Table 5.10.

The Bode diagram of a second order system is shown in Fig. 5.2.5, where the abscissa is normalized with $\log \omega T$.

For a pair of complex conjugate zeros, the Bode diagram may be obtained in the same way. The only differences are that the slope of the high frequency asymptote of log modulus is 40 dB/decade, and the phase shift varies from 0° to 180° as ω changes from 0 to infinity.

The procedure for drawing a Bode diagram of a second order system is summarized in Table 5.11.

Pure time delay

A pure time delay system possesses a transfer function $e^{-s\tau}$, where τ is the pure delay time. The log modulus and the phase shift of the system are

$$L = 20\log\left|e^{-j\omega\tau}\right| = 20\log 1 = 0$$

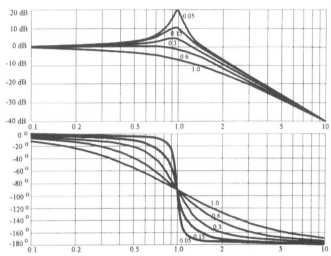

FIGURE 5.2.5. Bode diagram of a second order system.

$$\arg G(j\omega) = -\omega\tau$$

As shown in Fig. 5.2.6, the log modulus curve is a straight line at 0 dB, and the phase shift drops off to minus infinity. The phase shift is $-180°$ at the angular frequency $\omega = \pi/\tau$, or $-90°$ at $\omega = \pi/2\tau$. Thus, the bigger delay time implies that the phase shift drops off more rapidly.

Bode diagram of a general system

Recall the transfer function of a general system

TABLE 5.11. Procedure for drawing Bode diagram of a second order system.

Step	Item	Content or formula
1	Break frequency	Locate the break frequency, $\omega = 1/T$.
2	Low frequency asymptote	Draw horizontal asymptote $L = 20 \log K$ for $\omega < 1/T$.
3	High frequency asymptote	Draw asymptote with slope $= -40$dB/decade intersecting $L = 20\log K$ at $\omega = 1/T$.
4	Actual log modulus	Calculate several points on the actual curve near $\omega = 1/T$.
5	Phase shift at break frequency	$-90°$
6	Phase shift asymptote	Two horizontal asymptotes at $\phi = 0°$ and $\phi = -180°$.
7	Phase shift curve	Calculate several points on the phase shift curve and connect them.

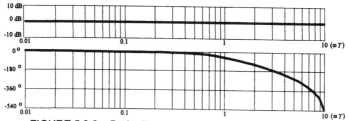

FIGURE 5.2.6. Bode diagram of a pure time delay system.

$$G(s) = \frac{K \prod_{i=1}^{m}(1+T_i s)}{s^p \prod_{j=1}^{n}(1+\overline{T}_j s) \prod_{k=1}^{l}(1+2\zeta_k \hat{T}_k s + \hat{T}_k^2 s^2)} e^{-j\omega\tau} \qquad (5.2.2)$$

The system has a constant gain K, m zeros on the real axis, p poles at origin, n poles on the real axis, l pairs of complex conjugate poles and a delay time τ. The advantage of taking the log modulus appears in the following equations:

$$20\log|G(j\omega)| = 20\log K + 20\sum_{i=1}^{m}\log|1+j\omega T_i| - 20p\log|j\omega|$$

$$-20\sum_{j=1}^{n}\log|1+j\omega\overline{T}_j| - 20\sum_{k=1}^{l}\log|1+2j\omega\zeta_k\hat{T}_k + (j\omega\hat{T}_k)^2| \quad (5.2.9)$$

and

$$\phi(\omega) = -\omega\tau + \sum_{i=1}^{m}\text{arctg}(\omega T_i) - p90°$$

$$-\sum_{j=1}^{n}\text{arctg}(\omega\overline{T}_j) - \sum_{k=1}^{l}\text{arctg}(\frac{2\zeta_k\hat{T}_k}{1-\hat{T}_k^2\omega^2}) \qquad (5.2.10)$$

Thus the Bode diagram of a general system can be obtained by adding the five kinds of individual Bode diagrams that we have just studied.

Sometimes, for a primary approximate analysis or design, only the asymptotes of log modulus, rather than the actual data, are plotted. By doing this we simplify the problem at the expense of accuracy.

Example 5.2.1

Assume a system with an open loop transfer function of

$$G(s) = \frac{6(1+0.2s)}{s(1+0.05s+0.0025s^2)}$$

which has a constant gain of 6, a zero $s = -5$ on the real axis, a pole at the origin and a pair of complex conjugate poles at $-10 \pm j17.32$. The asymptotes of the log modulus curves are shown in Fig. 5.2.7a. Combining them, we obtain Fig. 5.2.7b. Through correction, the actual log modulus curve may be obtained. The separate phase shift curves are shown in Fig. 5.2.8a, and the combined one is shown in Fig 5.2.8b.

Plot Bode diagram with *PCET* software
Bode diagrams can also be drawn on the screen or printed out with the sub-

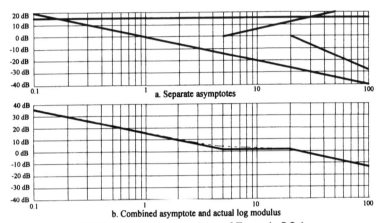

FIGURE 5.2.7.   Log modulus of Example 5.2.1.

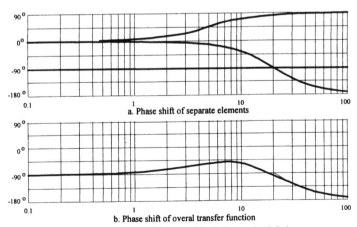

FIGURE 5.2.8.   Phase shift of Example 5.2.1.

program *FDA* in the software package *PCET*. Since the first several steps for drawing a Bode diagram on screen are the same as those for drawing a Nyquist diagram, the procedure for plotting Bode diagrams can also be found in Table 5.5. Since the computer screen is limited, the log modulus and the phase shift are overlapped on the screen. The left side dimension is for log modulus, and the right side for phase shift. The stability margin (See next section) can be displayed on the screen after the Bode diagram is drawn.

5.3. STABILITY ANALYSIS IN FREQUENCY DOMAIN

The stability of a system is of prime consideration for control engineers. Also important is the relative stability, which gives the degree of stability and to some extent, system dynamics.

As we studied in Chapter 4, the root locus method can be used to investigate the relative stability of a system. Frequency response can also be used for the same purpose. A frequency domain stability criterion, the Nyquist stability criterion, has been developed, which still plays an important role in conventional control theory.

Principle of the argument

Some complex variable theorems will be reviewed before introducing the Nyquist stability criterion.

A complex function $F(s)$ of a complex variable s has its own real part and imaginary part. For each specific complex value s, the complex value of $F(s)$ can be calculated. Therefore, $F(s)$ is also a mapping function. A point in the s-plane is mapped to a point in the $F(s)$-plane, and a line in the s-plane will trace out a line in the $F(s)$-plane.

A closed contour in the s-plane may be mapped by $F(s)$ into another closed contour in the $F(s)$-plane as shown in Fig. 5.3.1. For a closed contour in a plane, we assume that if the direction of transversal is clockwise, the area inside the

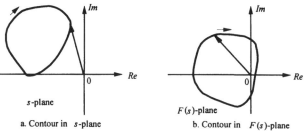

a. Contour in s-plane   b. Contour in $F(s)$-plane

FIGURE 5.3.1. Mapping from s-plane to $F(s)$-plane.

contour is enclosed by the contour; if the direction is counter-clockwise, the area outside the contour is enclosed. In other words, as you go along the contour, the area on the right side is enclosed. This assumption goes against the one employed in complex variable theory, but is equally applicable and generally used in control system theory.

The principle of the argument, known as the Cauchy theorem, states that if a contour C in the s-plane, which does not pass through any poles of $F(s)$, encircles Z zeros and P poles of $F(s)$ traversing clockwise, the corresponding mapping contour C' in the $F(s)$-plane encircles the origin of the $F(s)$-plane N times clockwise, where $N = Z - P$.

Consequently, if contour C encircles a zero (a pole) of $F(s)$ traversing clockwise, the mapping C' will encircle the origin once traversing clockwise (counter-clockwise). Some examples are listed in Table 5.12 and shown in Fig. 5.3.2.

Nyquist stability criterion

Consider a single loop feedback control system as shown in Fig. 2.4.3a. With an open loop transfer function $G(s)H(s)$, the closed loop characteristic equation may be written as

$$F(s) = 1 + G(s)H(s) \qquad (5.3.1)$$

It has been shown that the necessary and sufficient condition for a system to be stable is that all the roots of system characteristic equation $F(s) = 0$ lie to the left of the imaginary axis in the s-plane. Therefore, as shown in Fig. 5.3.3, a

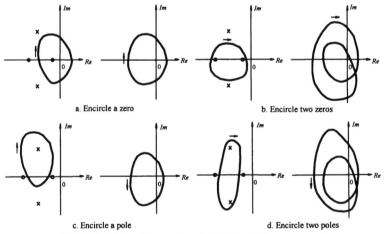

a. Encircle a zero   b. Encircle two zeros

c. Encircle a pole   d. Encircle two poles

FIGURE 5.3.2. Examples of principle of the argument.

TABLE 5.12. Mapping based on Principle of the Argument.

s-plane		F(s)-plane		
Encirclement	Direction	Encirclement	Times	Direction
A pole	Clockwise	Origin	1	Counter clockwise
Two poles	Clockwise	Origin	2	Counter clockwise
A zero	Clockwise	Origin	1	Clockwise
Two zeros	Clockwise	Origin	2	Clockwise
A pole and a zero	Clockwise	Origin	0	
P poles and Z zeros	Clockwise	Origin	$Z - P$	Clockwise

contour can be chosen in such a way that it will enclose the entire right half s-plane as the radius $R \to \infty$. This contour, known as a Nyquist contour, will be used for determining whether any roots of $F(s)$ lie in the right half s-plane.

If the open loop transfer function is assumed to be

$$G(s)H(s) = \frac{b_m s^m + b_{m-1} s^{m-1} + \cdots + b_0}{a_n s^n + a_{n-1} s^{n-1} + \cdots + a_0} \qquad (5.3.2)$$

where $n \geq m$, then

$$\lim_{s \to \infty} F(s) = \lim_{s \to \infty} (1 + G(s)H(s)) = \begin{cases} 1 & (n > m) \\ 1 + b_m / a_n & (n = m) \end{cases} \qquad (5.3.3)$$

Thus $F(s)$ is a real constant as $s \to \infty$, i.e., the arc of the Nyquist contour is mapped into a point on the real axis in the $F(s)$-plane. The part of the contour on imaginary axis, i.e., $s = j\omega$, is mapped into the polar plot $F(j\omega)$ in the $F(s)$-plane, which traces out a closed contour as ω changes from $-\infty$ to ∞. A mapping contour of the Nyquist contour is shown in Fig. 5.3.4.

From the principle of the argument, the mapping contour of the Nyquist contour will encircle the origin in the $F(s)$-plane $Z - P$ times, where Z and P are the numbers of zeros and poles of $F(s)$ on the right half s-plane, respectively.

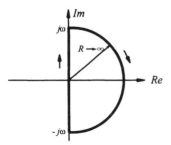

FIGURE 5.3.3. Nyquist contour.

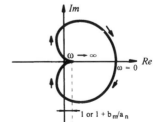

FIGURE 5.3.4. A mapping contour of a Nyquist contour.

According to (5.3.1), $F(j\omega)$ may be obtained by moving $G(j\omega)H(j\omega)$ a unit distance to the right. In other words, $F(j\omega)$ encircling the origin is equivalent to $G(j\omega)H(j\omega)$ encircling the point $(-1, j0)$. The relationship between $F(j\omega)$ and $G(j\omega)H(j\omega)$ is shown in Fig. 5.3.5.

Therefore, the principle of the argument may be translated as follows: when the Nyquist contour encircles Z zeros and P poles of $F(s)$ in the s-plane, the mapping contour $G(j\omega)H(j\omega)$ in the $G(s)H(s)$-plane must encircle point $(-1, j0)$ N times, where $N = Z - P$.

For a stable system, the characteristic polynomial of the closed loop system, $F(s) = 1 + G(s)H(s)$, has no roots on the right half s-plane. $F(s)$ and $G(s)H(s)$ have identical poles (the open loop poles of the system). Therefore, the Nyquist stability criterion can be stated as follows: when the open loop polar plot $G(j\omega)H(j\omega)$ of a system does not pass through the point $(-1, j0)$, the system is stable if and only if $G(j\omega)H(j\omega)$ encircles the point $(-1, j0)$ counter clockwise N times, where N is equal to P, the number of poles of $G(s)H(s)$ located on the right half s-plane.

Remarks

1. As we studied in Section 5.1, the polar plot $G(j\omega)H(j\omega)$ is the Nyquist plot of the transfer function $G(s)H(s)$. Therefore, the Nyquist stability criterion uses an open loop Nyquist plot to determine the stability of the closed loop system.

2. For an unstable open loop system ($P \neq 0$), the closed loop system may be stable. The only condition for having a stable closed loop system is $N = -P$, where N is the number of encirclements of $G(j\omega)H(j\omega)$ about the point $(-1, j0)$.

3. The polar plot of $G(j\omega)H(j\omega)$ from $\omega = -\infty$ to $\omega = 0^-$ is symmetrical to the polar plot from $\omega = \infty$ to $\omega = 0^+$ about the real axis, as shown in Fig. 5.3.6. Thus, drawing $G(j\omega)H(j\omega)$ from $\omega = \infty$ to $\omega = 0$, which is the Nyquist plot we studied in section 5.1, is sufficient for ascertaining the stability of systems. In fact, the negative frequency is physically meaningless.

4. Assume that an open loop transfer function includes a pole at the origin, i.e.,

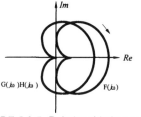

FIGURE 5.3.5 Relationship between $G(j\omega)$ and $F(j\omega)$.

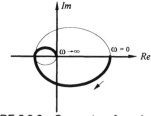

FIGURE 5.3.6. Symmetry of a polar plot.

$$G(s)H(s) = \frac{b_m s^m + b_{m-1} s^{m-1} + \cdots + b_0}{s(a_n s^n + a_{n-1} s^{n-1} + \cdots + a_0)} \tag{5.3.4}$$

The Nyquist contour should be revised as shown in Fig. 5.3.7a, where a small semicircular detour around the pole at the origin is introduced to avoid the contour passing through a pole of $F(s)$. The semicircle detour can be represented by setting $s = \varepsilon e^{j\phi}$ and allowing ϕ to vary from $-90°$ at $\omega = 0^-$ to $90°$ at $\omega = 0^+$. The corresponding part of the mapping contour will be

$$\lim_{\varepsilon \to 0} G(j\omega)H(j\omega) = \lim_{\varepsilon \to 0} \frac{b_0}{a_0 \varepsilon e^{j\phi}} = \lim_{\varepsilon \to 0} \frac{b_0 e^{-j\phi}}{a_0 \varepsilon} \tag{5.3.5}$$

Therefore, the angle of the contour in the $G(j\omega)H(j\omega)$-plane changes from $90°$ at $\omega = 0^-$ to $-90°$ at $\omega = 0^+$ with a very large radius (tending to infinity). The mapping point corresponding to $\omega = \pm\infty$ is at the origin in the $G(j\omega)H(j\omega)$-plane. A mapping contour corresponding to this contour is given in Fig. 5.3.7b.

Minimum phase systems and non-minimum phase systems

If a system has a transfer function with all poles and zeros in the left half s-plane, the system is referred to as a minimum phase system, or the system has minimum phase. If there are one or more poles or zeros in the right half s-plane, the system is called a non-minimum phase system. Sometimes, only the location of zeros is used to distinguish a minimum phase system from a non-minimum phase system, since most open loop industrial control systems are stable, and the zeros are more difficult to arrange than poles through feedback.

It is not difficult to find that if the zeros and poles of two transfer functions are identical, except for some zeros or poles at the symmetrical locations about the imaginary axis, there is no difference between the two amplitude ratio curves or log modulus curves of the two systems, while the two phase shift curves are different. The net phase shift of a system with all zeros located in the left half s-plane over the frequency range from zero to infinity is less than that of the

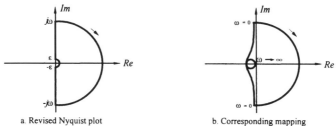

a. Revised Nyquist plot   b. Corresponding mapping

FIGURE 5.3.7.   Revised Nyquist plot and its mapping.

system with some zeros moved to symmetrical locations about the imaginary axis. This is how the minimum phase system came to be named. Obviously, a system with time delay is a non-minimum phase system, since its phase shift is greater than the same system without time delay.

Most industrial processes are minimum phase processes, that is, the corresponding open loop systems are stable. In other words, their open loop transfer functions have no poles on the right half s-plane, i.e., $P = 0$. Therefore, the Nyquist stability criterion may be simplified as follows: if the open loop polar plot does not encircle or pass through point $(-1, j0)$, the closed loop system is stable. The stability may be determined by finding in which side of the Nyquist plot, the point $(-1, j0)$ is located as the plot traverses from $\omega = 0$ to $\omega = \infty$. If point $(-1, j0)$ is to the left of $G(j\omega)H(j\omega)$, as shown in Fig. 5.3.8a, the system is stable. Otherwise, as demonstrated in Fig 5.3.8b, the system is unstable. From now on in this chapter, we assume that the open loop systems are stable.

Measurement of relative stability
The Nyquist stability criterion is defined in terms of point $(-1, j0)$ in the $G(j\omega)H(j\omega)$-plane. The relative stability, or the performance of a system, may be measured by the proximity of the Nyquist plot to the point $(-1, j0)$. Usually, we use two specifications to describe the relative stability: gain margin and phase margin. Together, we call them stability margin for convenience.

The gain margin is defined as the reciprocal of the intersection value of polar plot $G(j\omega)H(j\omega)$ on the negative real axis, i.e.,

$$GM = 1/|G(j\omega)H(j\omega)| \quad \text{when} \quad \arg(G(j\omega)H(j\omega)) = -180°$$

where ω satisfying the condition $\arg(G(j\omega)H(j\omega)) = -180°$ is referred to as the critical frequency, denoted ω_c. Gain margin may be described in terms of a logarithmic measure as

$$20\log GM = -20\log|G(j\omega_c)H(j\omega_c)| \qquad (5.3.6)$$

It is clear from the definition that if polar plot $G(j\omega)H(j\omega)$ passes through point

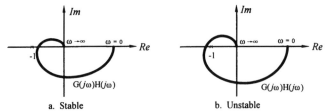

a. Stable   b. Unstable
FIGURE 5.3.8.   Closed loop stability of an open loop stable system.

$(-1, j0)$, we have $|G(j\omega)H(j\omega)| = 1$, i.e., $GM = 1$ and $20\log GM = 0$. For an open loop stable system, if polar plot $G(j\omega)H(j\omega)$ does not encircle or pass through point $(-1, j0)$, i.e., the closed loop system is stable, we have $|G(j\omega)H(j\omega)| < 1$, i.e., $GM > 1$ or $20\log GM > 0$. On the other hand, if $|G(j\omega)H(j\omega)| > 1$, i.e., $GM < 1$ or $20\log GM < 0$, the closed loop system is unstable. The gain margin is a measure of how far the system is from instability. It may be viewed as a safety factor, which is usually designed between 6 and 10 dB. In other words, GM is between 2 and 3, or $|G(j\omega)H(j\omega)|$ between 0.3 and 0.5.

An alternative measure of relative stability can be specified in terms of the phase margin, which is defined as the phase angle between the negative real axis and a radial line drawn from the origin to the intersection point of $G(j\omega)H(j\omega)$ on the unit circle. Since $G(j\omega)H(j\omega)$ usually has negative phase shift, the phase margin may be written

$$PM = 180 + \arg(G(j\omega)H(j\omega)) \quad \text{when} \quad |G(j\omega)H(j\omega)| = 1$$

where ω satisfying the condition $|G(j\omega)H(j\omega)| = 1$ is referred to as the gain crossover frequency, and is denoted ω_g.

If the $G(j\omega)H(j\omega)$ polar plot passes through point $(-1, j0)$, the phase margin is zero. The system is stable if the phase margin is positive. A bigger phase margin indicates a more stable closed loop system. The phase margin is equal to the additional phase lag required before the system becomes unstable. Phase margin is another safety factor, which is designed between 30° and 45°.

The gain margin and phase margin in a Nyquist plot are shown in Fig. 5.3.9, and a comparison between them is drawn in Table 5.13.

Stability analysis in Bode diagrams

Bode diagrams are more often used, and easier to obtain, than Nyquist plots. The gain and phase margins of a system can be easily evaluated from its Bode diagram. The unit circle in a polar plot corresponds to the 0 dB line in the log modulus plot of a Bode diagram, and the negative real axis corresponds to the −180° line in a phase shift plot. Since the gain margin defined in (5.3.6) can be rewritten as

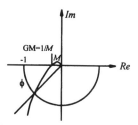

FIGURE 5.3.9. Gain margin and phase margin in a Nyquist.

TABLE 5.13 Gain margin and phase margin.

Item	Gain margin	Phase margin
Acronym	*GM*	*PM*
Definition	$GM = 1/\|G(j\omega)H(j\omega)\|$ when $\arg(G(j\omega)H(j\omega)) = -180°$	$PM = 180 + \arg(G(j\omega)H(j\omega))$ when $\|G(j\omega)H(j\omega)\| = 1$
Measuring in Nyquist plot	On negative real axis	On unit circle
Measuring in Bode diagram	On log modulus at critical frequency	On phase shift at gain crossover frequency
Logarithmic measure	$20\log GM = -20\log\|G(j\omega_c)H(j\omega_c)\|$	
Requirement for stability	$GM > 1$ or $20\log GM > 0$	$PM > 0$
Suggested value	6 dB ~ 10 dB (2 ~ 3)	30° ~ 45°

$$20\log GM = 0 \text{ dB} -20\log|G(j\omega_c)H(j\omega_c)| \text{ dB} \qquad (5.3.7)$$

the gain margin on a Bode diagram is the distance in dB between the 0 dB line and the point on the log modulus curve at critical frequency. If the 0 dB line is above the point on the log modulus at critical frequency, the gain margin is positive, and the closed loop system is stable. However, for a stable system, the log modulus at critical frequency is negative, while the gain margin is positive as mentioned above. The phase margin on a Bode diagram is the distance in degrees between the $-180°$ line and the point on the phase shift curve at gain crossover frequency ω_g. If the $-180°$ line is above the point on the phase shift curve, the phase margin is negative, and the closed loop system is unstable. Otherwise, the phase margin is positive, and the system is stable. The Bode diagram with marked gain margin and phase margin is displayed in Fig. 5.3.10.

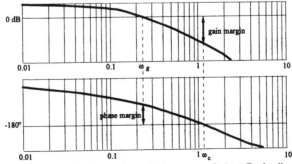

FIGURE 5.3.10. Gain margin and phase margin in a Bode diagram.

TABLE 5.14. Stability analysis from a Bode diagram.

Step	At gain crossover frequency ω_g	At critical frequency ω_c
Drawing	Bode diagram	
Finding	Gain crossover frequency ω_g	Critical frequency ω_c
Measuring	Phase shift ϕ at ω_g	Log modulus at ω_c
Stability	If $(\phi - 180°) > 0$, stable.	If $20\log|G(j\omega_c)| < 0$, stable.
criterion	If $(\phi - 180°) \leq 0$, unstable.	If $20\log|G(j\omega_c)| \geq 0$, unstable.

TABLE 5.15. Stability analysis in time domain and frequency domain.

Domain	Time domain	Frequency domain
Goal	Determine the stability of closed loop system.	
Investigated object	Closed loop characteristic equation	Open loop frequency response
Stability criterion	All poles on the left half s-plane	Gain margin > 0 Phase margin > 0

The procedure for analyzing the stability of a system from its Bode diagram can be found in Table 5.14. The table gives two methods. However, it should be noted that the critical frequency may not exist, e.g., for a first order system.

Comparison of stability analysis between time domain and frequency domain

For a single loop feedback control system as shown in Fig. 2.4.3a, the open loop transfer function is $G(s)H(s)$, whereas the closed loop transfer function is $G(s)/[1 + H(s)G(s)]$. In the time domain, we investigate the system stability by checking the closed loop characteristic equation to find whether or not all the poles are located on the left half s-plane. However, in the frequency domain, we determine the closed loop system stability based on the open loop frequency response $G(j\omega)H(j\omega)$ rather than the closed loop one. The gain margin and phase margin are checked on the open loop frequency response to determine the stability of the closed loop system.

The differences between the stability determination in the time domain and frequency domain are summarized in Table 5.15.

5.4. FREQUENCY DOMAIN ANALYSIS AND DESIGN FOR *PID* CONTROLLERS

Gain margin and phase margin are two important measures of relative stability to be obtained from the frequency response of a system. They can also be taken as performance criteria of a system for analysis and design, since they affect not only the relative stability but also the dynamics of the closed loop system. In this section, we will discuss the frequency domain analysis and controller design based on the Bode diagrams of the minimum phase systems.

Requirements on log modulus for minimum phase systems

For a minimum phase system, the log modulus and phase shift curves of a Bode diagram are one-to-one related. The approximate relationship has been found as

$$\phi(\omega) = 4.5m \text{ (in degrees)} \tag{5.4.1}$$

or

$$\phi(\omega) = m\pi/40 \text{ (in radians)}$$

where ϕ is the phase shift at angular frequency ω, and m is the slope of the log modulus at ω in dB/decade.

Usually, the log modulus of a minimum phase system with satisfactory characteristics must satisfy some requirements in low, medium and high frequency regions, respectively.

At the low frequency region, the log modulus should have enough gain to make the steady state error small. If a pure integral action is included in the controller or the process of a system, the slope of the log modulus in the low frequency region is −20 dB/decade, and the gain will tend to infinity as $\omega \to 0$.

According to the relationship between log modulus and phase shift for a minimum phase system as shown in (5.4.1), if the phase shift at the gain crossover frequency is $\phi = -180°$, i.e., the phase margin is equal to zero, the slope of log modulus at gain crossover frequency will be −40 dB/decade. In other words, the stability of a minimum phase system requires that the slope of its log modulus at gain crossover frequency be greater (less negative) than −40 dB/decade. In most practical cases, a slope of −20 dB/decade for an asymptote of the log modulus is desirable at the gain crossover frequency to ensure the system stability. It is also required that the frequency covering range of this asymptote be long enough. Empirically, it is required that

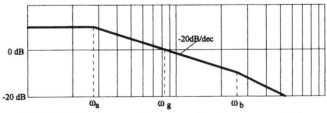

FIGURE 5.4.1. Log modulus at medium frequency region.

$$(\omega_g - \omega_a)/(\omega_b - \omega_g) = 1/3 \sim 1/2 \tag{5.4.2}$$

where ω_a, ω_g and ω_b are shown in Fig. 5.4.1. A slope of -20 dB/decade for an asymptote of the log modulus at the gain crossover frequency and a sufficiently long covering range are the requirements for the medium frequency region.

At the high frequency region, the log modulus should drop off quickly to ensure that the noise acting on the system can be greatly attenuated. Usually, this requirement is inherently satisfied. If not, a low-pass filter needs to be incorporated into the system.

Conventional controller compensation

In order to meet the required performance of a system, a suitable controller may be incorporated in the system, and the parameters of the controller adjusted. This is called compensation. More precisely, compensation is the alteration or adjustment of a system in order to make up for deficiencies or inadequacies.

In many industrial process control cases, it is not necessary to design a controller from scratch. *P*, *PI* and *PID* controllers are the three most commonly used and commercially available conventional controllers. The designing of a controller for the compensation of a system may be reduced to choosing one of three controllers and adjusting its parameters.

For a proportional controller, the transfer function is

$$G_c(s) = K \tag{5.4.3}$$

with a Bode diagram of Fig. 5.2.1. As the proportional gain K of the controller changes, the log modulus curve of the system moves accordingly up or down, but the phase shift curve remains unchanged. Therefore, the phase margin and

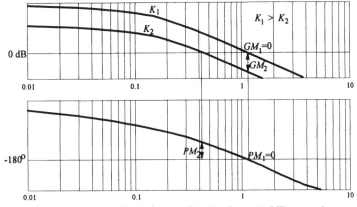

FIGURE 5.4.2.   Effect of proportional gain on stability margin.

gain margin of a system may be adjusted by incorporating a *P* controller and altering the gain. The effect of proportional gain *K* on the stability margin of a system is shown in Fig. 5.4.2, where K_1 is an open loop gain of the system in a critical stable state, and thus the phase margin PM_1 and gain margin GM_1 are both zero. After a *P* controller $K_c < 1$ is incorporated, the system gain becomes $K_2 = K_1 K_c$, and phase margin PM_2 and gain margin GM_2 are obtained. Usually, a bigger gain K_c results in smaller gain margin and phase margin.

In order to eliminate steady state error, a *PI* controller may be used, which has the transfer function

$$G_c(s) = K(1 + \frac{1}{\tau_I s})$$ (5.4.4)

where the two parameters K and τ_I may be adjusted. The controller has a pole at 0 and a zero at $-1/\tau_I$. The Bode diagram of the controller is shown in Fig. 5.4.3, where we find that the *PI* controller amplifies log modulus and contributes $-90°$ of phase shift at low frequencies. The *PI* controller satisfies the requirement at the low frequency region, but may not in the medium frequency region.

In order to minimize the effect on the log modulus in the medium frequency region, the zero, or the break frequency of the *PI* controller should be located to the left of the critical frequency of the uncompensated system Bode diagram. Empirically, $\tau_I = (3 \sim 6)/\omega_c$ is chosen, where ω_c is the critical frequency of the system Bode diagram. The break frequency of a *PI* controller is at $\omega = 1/\tau_I$. With this empirical data, only parameter K of a *PI* controller needs to be specified. K may be chosen for obtaining a suitable phase margin, or gain margin.

For an ideal *PID* controller, the transfer function is

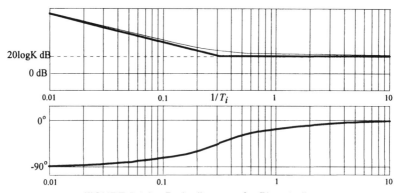

FIGURE 5.4.3.   Bode diagram of a *PI* controller.

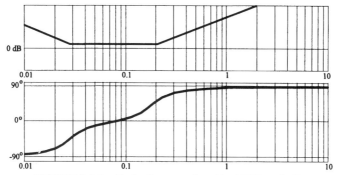

FIGURE 5.4.4. Bode diagram of an ideal *PID* controller.

$$G_c(s) = K(1 + \frac{1}{\tau_I s} + \tau_D s)$$ (5.4.5)

The corresponding Bode diagram is shown in Fig. 5.4.4.

As discussed in Chapter 3, an ideal *PID* transfer function is physically unrealizable since the order of the numerator is greater than that of the denominator after combining the three terms with a common denominator. As we can see in Fig. 5.4.4, the log modulus of an ideal *PID* controller at the high frequency region has a slope of 20 dB/dec, and the log modulus will tend to infinity as $\omega \to \infty$. It is unrealizable. Commercial controllers approximate *PID* controllers with transfer functions of the form

$$G_c(s) = K_c(\frac{\tau_I s + 1}{\tau_I s})(\frac{\tau_D s + 1}{\alpha \tau_D s + 1})$$ (5.4.6)

where α is a small number. The Bode diagram of a commercial *PID* controller is shown in Fig. 5.4.5. The *PID* controller contributes positive phase shift at the

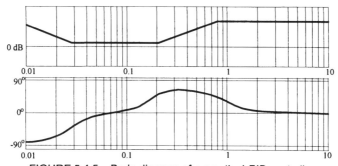

FIGURE 5.4.5. Bode diagram of a practical *PID* controller.

TABLE 5.15.   Design requirement for a minimum phase system.

Frequency	Low frequency region	Medium frequency region	High frequency region
Requirement	High gain	−20 dB/dec slope	Blocking off
Objectives	Eliminating steady-state error	Keeping suitable stability margin	Attenuating noise
Controller selection	P or PI	P or PID	Low-pass filter

medium and high frequency regions.

For convenience empirical data are sometimes used to approximate engineering design. $\tau_I = (3 \sim 6)/\omega_c$, $\tau_D = (0.25 \sim 0.35)\tau_I$ and $\alpha = 0.1 \sim 0.2$ may be considered, where ω_c is the critical frequency of the system Bode diagram. The requirement and measurement for minimum phase systems at the different frequency regions are summarized in Table 5.15.

Example 5.4.1

An uncompensated system has a Bode diagram as shown in Fig. 5.4.6, where the log modulus is denoted L_u. From the Bode diagram it may be found that the critical frequency is 5 sec^{-1}, the gain margin is about 22 dB, and the phase margin is about 70°. Big phase margin or gain margin will bring about slow dynamic characteristics of the system.

With a proportional controller, we can adjust the gain margin to 10 dB as shown in Fig. 5.4.6, where the log modulus, moved up 12 dB, is denoted L_k in the figure. The proportional gain may be found as $20\log K = 12$. It follows that $K = 10^{0.6} \approx 4$. It should be noted that the phase margin of the compensated system is less than 10°, which is too small to get satisfactory dynamic characteristics.

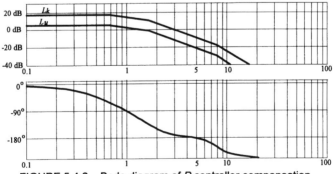

FIGURE 5.4.6.   Bode diagram of P controller compensation.

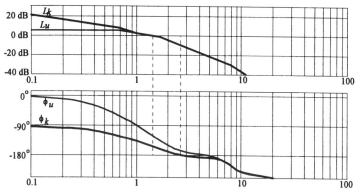

FIGURE 5.4.7. Bode diagram of *PI* controller compensation.

In order to eliminate steady state error, a *PI* controller is incorporated. With the approximation formula $\tau_I = (3 \sim 6)/\omega_c$, we choose $\omega_I = 1/\tau_I = 1$ sec^{-1} for $\omega_c = 5$. Assuming $K_c = 1$, the Bode diagram of the system with this *PI* controller may be drawn as shown in Fig. 5.4.7, where L_u is the log modulus of the uncompensated system. The gain margin may be found in the figure as 10 dB, and the phase margin 30°. Since the stability margin is suitable, the proportional gain $K = 1$ can be determined. Thus, the *PI* controller may be designed as

$$G_c(s) = (s+1)/s$$

For the compensation of a *PID* controller, τ_I may still be chosen as 1 sec, and τ_D may be selected as 0.25 with the formula $\tau_D = (0.25 \quad 0.35)\tau_I$, i.e., $\omega_D = 4$ sec^{-1}. If $\alpha = 0.1$ is taken, the Bode diagram with $K = 1$ can be plotted in Fig. 5.4.8, where the log modulus with $K = 1$ is denoted as L_{k1}. Moving the log modulus up 28 dB yields the curve denoted L_{k2}, from which 10 dB gain margin and 35°

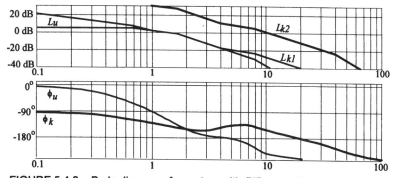

FIGURE 5.4.8. Bode diagram of a system with *PID* controller compensation.

phase margin may be read. Therefore, the proportional gain of the *PID* controller is designed to be 25, which may be found from $20\log K = 28$. The designed *PID* controller will be

$$G_c(s) = 25(\frac{s+1}{s})(\frac{0.25s+1}{0.025s+1})$$

Design a *PID* controller with *PCET*

As we have done, several Bode diagrams should be drawn when designing a controller. Furthermore, the proposed controller may not satisfy the design requirement, and may need to be adjusted to achieve the best performance. The subprogram *FDA* of *PCET* can easily draw the frequency response and simulate the controller tuning.

A Bode diagram of the open loop transfer function can be obtained on the screen by taking the steps in Table 5.5. The procedure for controller tuning can be found in Table 5.16. Through trial and error, a satisfactory *PID* controller can be found by repeating Steps 3 and 4 in Table 5.16.

Example 5.4.2

Consider a single loop feedback control system with process transfer function

TABLE 5.16.   Procedure for tuning controller on the screen with *PCET*.

Step	User input		Function
1	Follow the steps in Table 5.5		Draw the Bode diagram
2	Select F9	"F1-Exit"	Return to main menu
		"F2-Restart"	Return to very beginning of *FDA*
		"F3-Load"	Load a saved file to get data
		"F4-Save"	Save the updated data in a file
		"F5-Modify"	Change system parameters
		"F6-Nyquist"	Give the Nyquist plot
		"F7-Bode"	Obtain the Bode diagram
		"F8-Margin"	Obtain gain and phase margins
		"F9-PID"	Select *PID* controller parameters
3	F2, F3 or F4 then F1	"F1-Exit"	Draw the new Bode diagram
		"F2-Proportional action"	Change *P* parameter
		"F3-Integral action"	Change *I* parameter
		"F4-Derivative action"	Change *D* parameter
4	F1 - F6 Same as step 2		
	"F7-Bode"		Clear the old Bode diagram
	"F8-Margin"		Get the new stability margins
	"F9-PID"		Continue to tune *PID* controller

$$G_p(s) = \frac{2}{(0.2s+1)(0.4s+1)(2s+1)(20s+1)}$$

Design an ideal *PID* controller to satisfy the regular phase margin and gain margin requirement ($GM = 6 \sim 10$ dB and $PM = 30° \sim 45°$).

Following the steps in Table 5.5, the Bode diagram of the open loop transfer function with controller $P = 1$ can be drawn on the screen. By pressing function key $F8$, gain margin 26.71dB and phase margin 107.95° are displayed as shown in Fig. 5.4.9. In the Bode diagram drawn by *PCET*, the log modulus and the phase shift are overlapped. They appear in different colors on the screen. To distinguish them on a black and white print, labels are added in the figure. The left side is the dimension of log modulus, with phase shift on the right side. Four color lines that appear after pressing function key $F8$ help us find the gain crossover frequency, critical frequency, gain margin and phase margin.

From Fig. 5.4.9, the critical frequency can be determined as $\omega_c = 0.9$ sec⁻¹. Based on the formula $\tau_I = (3 \sim 6)/\omega_c$ and $\tau_D = (0.25 \sim 0.35)\tau_I$, we choose $\tau_I = 4$ sec or $1/\tau_I = 0.25$ sec⁻¹ and $\tau_D = 1.2$ sec. By pressing function key $F9$ and inputting the controller parameters, the new Bode diagram gives a 40.32 dB gain margin and 30.31° phase margin. The gain margin does not satisfy the design requirement. With trial and error, the following *PID* controller parameters may be found:

$$K = 30, \qquad 1/\tau_I = 0.25 \text{ sec}^{-1} \qquad \text{and} \qquad \tau_D = 1.5 \text{ sec}.$$

The gain margin of 9.86 dB and the phase margin of 32.09° as shown in Fig. 5.4.10 satisfy the design requirement.

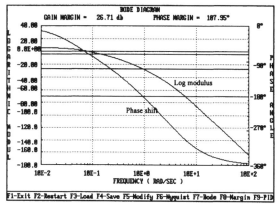

FIGURE 5.4.9.   Bode diagram of Example 5.4.2 before compensation.

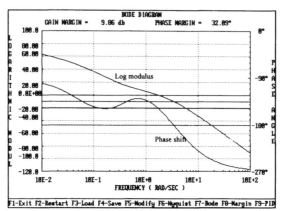

FIGURE 5.4.10.   Bode diagram of Example 5.4.2 after compensation.

5.5. COMPENSATION WITH PHASE LEAD AND PHASE LAG
NETWORKS

Phase lead and phase lag networks are commonly used as compensators to improve system performance. The technique of compensation with phase lead or phase lag networks has been developed using a Bode diagram.

Phase lead and phase lag networks
Phase lead and phase lag networks can be composed of electrical elements as shown in Figs. 5.5.1a and b. The voltages on either side of a network are taken as the input and output of the network.

With simple calculation, the transfer function of a phase lead compensator can be found as

$$G_c(s) = \frac{U_2}{U_1} = \frac{\alpha Ts + 1}{\alpha(Ts + 1)} = \frac{s + 1/\alpha T}{s + 1/T} \qquad (5.5.1)$$

where

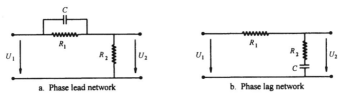

FIGURE 5.5.1.   Phase lead and phase lag networks.

$$T = \frac{R_1 R_2 C}{R_1 + R_2} \qquad \text{and} \qquad \alpha = \frac{R_1 + R_2}{R_2} > 1 \qquad (5.5.2)$$

The phase lag compensation transfer function can be obtained as

$$G_c(s) = \frac{U_2}{U_1} = \frac{Ts+1}{\alpha Ts + 1} = \frac{s + 1/T}{\alpha(s + 1/\alpha T)} \qquad (5.5.3)$$

where

$$T = R_2 C \qquad \alpha = \frac{R_1 + R_2}{R_2} > 1 \qquad (5.5.4)$$

The Bode diagrams of phase lead and phase lag compensators are shown in Figs. 5.5.2a and b, respectively.

Both compensator transfer functions are composed of a zero and a pole on the negative real axis in the s-plane. For a phase lead compensator, the zero $-1/\alpha T$ $(\alpha > 1)$ is to the right of the pole $-1/T$ on the negative real axis in the s-plane, whereas for a phase lag compensator, the zero $-1/T$ is to the left of the pole $-1/\alpha$ T $(\alpha > 1)$.

Remarks

1. The transfer functions of the phase lead and phase lag networks are obtained under the condition that the output impedance of the device before the network is zero, and that the input impedance of the device behind the network is infinity. Practically, the loading problem, as we studied in Chapter 2, deserves much attention.

2. Usually the compensators are used with an amplifier, i.e., a gain K must be

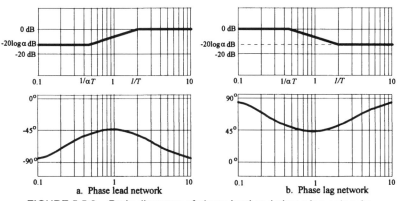

FIGURE 5.5.2.   Bode diagrams of phase lead and phase lag networks.

added in the transfer function of the compensators.

Compensation with a phase lead network

If the phase margin of a system is too small or is negative, a phase lead compensator can be used to increase the phase margin.

As shown in Fig. 5.5.2a, the maximum phase lead of a phase lead network occurs at frequency ω_m that is the geometric mean of the zero and pole. In a logarithmic abscissa, we have

$$\log \omega_m = \frac{1}{2}\left(\log\frac{1}{T} + \log\frac{1}{\alpha T} \right) = \frac{1}{2}\log\left(\frac{1}{T}\frac{1}{\alpha T} \right) = \log\frac{1}{T\sqrt{\alpha}}$$

or

$$\omega_m = 1/T\sqrt{\alpha} \qquad (5.5.5)$$

The maximum phase shift $\phi_m = \phi(\omega_m)$ of the phase lead network can be found from (5.5.1) as

$$\phi_m = \arg[G(j\omega_m)] = \arctan(\omega_m\alpha T) - \arctan(\omega_m T)$$

It follows from the addition theorem of trigonometric function that

$$tg\phi_m = \left.\frac{\omega_m\alpha T - \omega_m T}{1 + \omega_m^2\alpha T^2}\right|_{\omega=\omega_m} = \frac{\alpha - 1}{2\sqrt{\alpha}}$$

or

$$\sin\phi_m = \frac{\alpha - 1}{\alpha + 1} \qquad (5.5.6)$$

This equation can be used for calculating α when ϕ_m is known. It should be noted that ϕ_m is less than 90°.

In order to obtain maximum additional phase lead, ω_m should be located at the new gain crossover frequency at which the compensated log modulus crosses the 0 dB line. The contribution of a phase lead network to log modulus at ω_m is $-10\log\alpha$ dB (Problem 5.13).

The procedure of phase lead compensation is given in Table 5.17.

Example 5.5.1

Consider a system with an open loop transfer function

$$G(s) = \frac{0.08}{s(s+0.5)} = \frac{0.16}{s(2s+1)}$$

Design a phase lead compensator (controller) to satisfy the condition that the ramp error constant $K_v \geq 4$ and phase margin $M \approx 45°$.

TABLE 5.17. Procedure for phase lead compensation.

Step	Item	Content or formula
1	Selecting K	According to the requirement of steady state error, specify gain K.
2	Measuring phase margin	Draw uncompensated Bode diagram of $KG(j\omega)$, measuring the phase margin.
3	Selecting ϕ_m	ϕ_m = specified phase margin − uncompensated phase margin
4	Selecting α	$\sin\phi_m = \dfrac{\alpha-1}{\alpha+1} \qquad \alpha = \dfrac{1+\sin\phi_m}{1-\sin\phi_m}$
5	Finding cross over frequency	Determine the new crossover frequency at which uncompensated log modulus is $-10\log\alpha$.
6	Selecting zero and pole	Draw −20 dB/decade asymptote to determine zero and pole of compensator.
7	Drawing Bode diagram	Draw the compensated Bode diagram and check the phase margin and gain margin.

From the definition of a ramp error constant, we have

$$K_v = \lim_{s\to 0} sG(s) = \lim_{s\to 0} \frac{0.16K}{(2s+1)} = 0.16K \geq 4$$

where K is the gain of a phase lead controller (compensator). Thus $K \geq 25$ is required. We choose $K = 25$, i.e., $K_v = 4$. The Bode diagram of $KG(j\omega)$ is drawn as solid lines in Fig. 5.5.3, where the phase margin may be estimated to be about $10°$. The maximum phase lead can be chosen as

$$\phi_m = \text{specified } PM - \text{uncompensated } PM + 5°$$

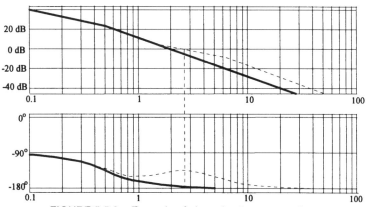

FIGURE 5.5.3. Example of phase lead compensation.

$$= 45° - 10° + 5° = 40°$$

where a 5° phase shift is added, because the uncompensated phase shift at the new crossover frequency is less than that at the uncompensated crossover frequency. The constant α may be calculated from (5.5.6), i.e.,

$$\frac{\alpha - 1}{\alpha + 1} = \sin 40° = 0.643$$

Solving this equation yields $\alpha = 4.6$. The contribution of the phase lead network to the log modulus curve at ω_m is $10\log\alpha = 6.6$ (dB), with which the new log modulus crossover frequency may be found on the Bode diagram as $\omega_m = 2.5$ sec^{-1}. An asymptote of slope −20 dB/dec may be drawn through the 0 dB line at ω_m, which has an intersection on the uncompensated log modulus at $\omega = 1.4$, i.e., $1/\alpha T$. It follows that another break frequency is located at $\omega = 1.4 \times \alpha = 6.4$ sec^{-1}, i.e., $1/T$.

The compensated Bode diagram is drawn with a dotted line in the figure. The phase lead compensator has the transfer function

$$G_c(s) = 25\frac{s + 1.4}{s + 6.4}$$

Compensation with a phase lag network

The phase lag network plays the role of lowering the gain crossover frequency of an uncompensated log modulus curve so that the phase margin of the system is increased. Thus, the pole and zero of the phase lag network should be designed in such a way that their magnitudes are much smaller than that of any pole or zero (not at the origin) of the uncompensated system. In other words, the phase shift of the uncompensated system will not be affected by the phase lag compensator at the medium and high frequency regions. The phase lag at the low frequency region is not important, while the attenuation of −20logα affects the log modulus for compensation at the medium and high frequency regions.

To compensate a system with a phase lag network, we may find a suitable phase shift at the frequency corresponding to the new crossover frequency after the implementation of the phase lag network. The parameter may be found by evaluating the attenuation of the log modulus at this frequency.

The procedure for phase lag compensation is listed in Table 5.18, and illustrated by Example 5.5.2.

Example 5.5.2

For the system in Example 5.5.1, we compensate the system with a phase lag compensator with the transfer function (5.5.3) for the same requirement that the ramp error constant $K_v \geq 4$ and phase margin $M \approx 45°$. The gain $K = 25$

TABLE 5.18. Procedure for phase lag compensation.

Step	Item	Content or formula
1	Selection of K	According to the requirement of steady state error, specify gain K.
2	Location of the gain crossover frequency	Draw uncompensated Bode diagram of $KG(j\omega)$, and determine the new crossover frequency at which the requirement of phase margin is satisfied.
3	Location of zero	Place zero one decade below the crossover frequency.
4	Measurement	Measure log modulus at the crossover frequency.
5	Selection of α	Measured value = $20\log\alpha$
6	Drawing Bode diagram	Draw the compensated Bode diagram and check the phase margin and gain margin.

corresponding to $K_v = 4$ has already been determined in Example 5.5.1. The uncompensated Bode diagram of $KG(j\omega)$ is drawn as solid lines in Fig. 5.5.4.

Considering the effect of phase lag on the new gain crossover frequency, 5° more is added to the phase margin requirement of 45°, i.e., 50° is used for design. The new gain crossover frequency can be determined in the plot at $\omega = 0.45$, where the phase shift is 50°. The zero of the compensator may be chosen at $\omega = 0.045$ (i.e., $1/T$), a decade below the new crossover frequency. It may be found that −23 dB log modulus should be attenuated at the new crossover frequency. Calculating −23 = −20logα, we obtain $\alpha = 14.1$. The pole will then be at $\omega = 0.0032$ (i.e., $1/\alpha T$). Then, the phase lag compensator has the transfer function

$$G_c(s) = 25\frac{s + 0.045}{s + 0.0032}$$

The compensated Bode diagram is drawn as dotted lines in Fig. 5.5.4, where

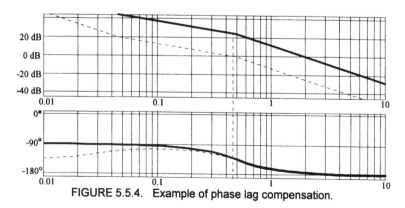

FIGURE 5.5.4. Example of phase lag compensation.

TABLE 5.19.   Phase lead network and phase lag network.

Item	Phase lead network	Phase lag network
Transfer function	$\dfrac{\alpha Ts + 1}{\alpha(Ts + 1)}$	$\dfrac{Ts + 1}{\alpha Ts + 1}$
Location of pole and zero	Zero is nearer to origin.	Pole is nearer to origin.
Method for increasing PM	Offering phase lead	Lowering crossover frequency
Effect on crossover frequency	Moving right	Moving left

the break point corresponding to the pole of the compensator is not included because its location is outside the frequency range of the graph.

The phase lead network and phase lag network are both used to increase phase margin, but are implemented in different ways, as summarized in Table 5.19.

Remark

1. To design phase lead or phase lag compensation networks, the trial and error method with *PCET* can be used.

2. If necessary, a phase lead-lag network may be used, which lowers the gain crossover frequency due to phase lag action, and at the same time offers phase lead at the new gain crossover frequency.

Problems

5.1. (Nyquist plot for first order systems) With *PCET* or by hand, sketch and compare the Nyquist plots for the following three systems with the open loop transfer functions:

a. $G_1(s) = \dfrac{1}{s+1}$,   b. $G_2(s) = \dfrac{3}{s+1}$,   c. $G_3(s) = \dfrac{8}{s+1}$

Determine the stability of each closed loop system.

5.2. (Nyquist plot for second order systems) With *PCET* or by hand, sketch and compare the Nyquist plots for the following two systems with the open loop transfer functions:

a. $G_1(s) = \dfrac{2}{(s+1)(s+2)}$ b. $G_2(s) = \dfrac{5}{(s+1)(s+2)}$

Determine the stability of each closed loop system.

5.3. (Nyquist plot for third order systems) With *PCET* or by hand, sketch and compare the Nyquist plots for the following two systems with the open loop transfer functions:

a. $G_1(s) = \dfrac{3}{(s+1)(s+2)(s+3)}$ b. $G_2(s) = \dfrac{8}{(s+1)(s+2)(s+3)}$

Determine the stability of each closed loop system.

5.4. (Nyquist plot for systems with integral action) With *PCET* or by hand, sketch and compare the Nyquist plots for the following two systems with the open loop transfer functions:

a. $G_1(s) = \dfrac{4}{s(s+1)(s+2)}$ b. $G_2(s) = \dfrac{10}{s(s+1)(s+2)}$

Determine the stability of each closed loop system.

5.5. Consider three unity feedback control systems with open loop transfer functions:

a. $G_1(s) = \dfrac{5}{s(s+1)(s+5)}$ b. $G_2(s) = \dfrac{5}{s^2(s+1)(s+5)}$

c. $G_3(s) = \dfrac{5}{s^3(s+1)(s+5)}$

They have different types. With *PCET*, observe and compare the Nyquist plots of the three systems. Describe the relationship between the type of the system and the low frequency region of the Nyquist plot.

5.6. Consider the following two unity feedback control systems with open loop transfer functions:

a. $G_1(s) = \dfrac{10(0.1s+1)}{(5s+1)(s+30)}$

b. $G_2(s) = \dfrac{10(s+3)}{s(s+2)(s^2+s+2)}$

With *PCET*, plot the system Bode diagrams and determine the stability of the closed loop systems. Give the stability margins if the system is stable.

5.7. The asymptotes of the log modulus curve for a minimum phase system are shown in Fig. 5.P.1. What is the open loop transfer function of the system?

5.8. For a unity feedback control system, if the open loop transfer function is

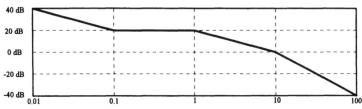

FIGURE 5.P.1. Bode diagram of a system.

$$G(s) = \frac{\omega_0^2}{s(s + 2\zeta\omega_0)}$$

the closed loop system is a standard second order system with a transfer function

$$\frac{Y(s)}{U(s)} = \frac{\omega_0^2}{s^2 + 2\zeta\omega_0 s + \omega_0^2}$$

Prove that the phase margin PM and the damping factor ζ satisfy

$$PM = \text{arctg}\left[2\zeta \middle/ \sqrt{\sqrt{4\zeta^4 + 1} - 2\zeta^2} \right]$$

If damping ratio ζ is small, it reduces to

$$PM = 2\zeta$$

(Hint: Replace s by $j\omega$ in the open loop transfer function, and use the definition of the phase margin in the Nyquist plot.)

5.9. Consider a single loop feedback control system as shown in Fig. 5.P.2, where a P controller is used and the process transfer function is

$$G(s) = \frac{1}{s(s + 1)(s + 2)}$$

Using the trial and error method and subprogram FDA in $PCET$, find all values of K_p that make the system stable. Compare the result with that of Problem 4.7.

5.10. Design a conventional controller for a unity feedback control system to satisfy (a) steady state error = zero; (b) phase margin $\approx 45°$, if the process transfer function is

a. $G_1(s) = \dfrac{10}{(s + 1)(s + 2)}$

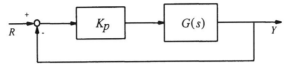

FIGURE 5.P.2.   A single loop feedback control system of Problem 5.9 (5.12).

b. $G_2(s) = \dfrac{12}{(s+1)(s+2)(s+5)}$

5.11. Consider a single loop feedback control system with open loop process transfer function

$$G_p(s) = \frac{12}{(0.4s+1)(0.6s+1)(8s+1)(25s+1)}$$

With *PCET*, design an ideal *PID* controller to satisfy the requirement:

$$GM = 6 \sim 10 \text{ dB} \quad \text{and} \quad PM = 30° \sim 45°$$

5.12. Consider a single loop feedback control system as shown in Fig. 5.P.2. The process transfer function is given as

$$G(s) = \frac{e^{-0.5s}}{(s+1)(2s+1)}$$

Using the subprogram *FDA* of *PCET* and trial and error, design *P*, *PI* and *PID* controllers respectively with the same 10 dB gain margin. Parameters τ_I and τ_D can be selected by the empirical formulas. By *PCET*, draw the Bode diagrams for the three cases and compare their phase margins. Compare the responses to a unit step input of the three cases and find which controller is better for the system.

5.13. For a phase lead network as shown in Fig. 5.5.1a, the maximum phase shift occurs at the frequency ω_m that can be found in (5.5.5). Prove that the contribution of the phase lead network to log modulus in the low frequency period of is $-20\log\alpha$ dB, in the high frequency period is 0 and at ω_m is $-10\log\alpha$ dB.

5.14. Consider a feedback control system with an open loop transfer function

$$G(s) = \frac{0.36}{s(s+1.2)(s+3)}$$

Design a phase lead compensation network to satisfy the condition that a ramp error constant $K_v \geq 6$ and phase margin $M \geq 25°$.

5.15. Design a phase lag compensation network for Problem 5.14.

CHAPTER 6

DISCRETE TIME SYSTEM
ANALYSIS AND DESIGN

6.1. SAMPLING AND SIGNAL CONVERSION

With the dramatic development of digital computers in the past few decades, computer control systems have been used in many applications, such as process control, machine tool control, aircraft control, missile control, space satellite control, robot control, automobile traffic control, and so on. In digital computers, digital signals, rather than continuous signals, are used.

Discrete signals can also be found in practice. For example, the chromatographic composition analyzers for chemical process control collect and produce discontinuous signals. The samples are injected into a chromatographic column and the composition signals are produced only once every few minutes. The input and output signals are discrete.

Discrete signals in discrete time systems

Discrete time systems, or sampled data systems, are dynamic systems in which one or more signals are discontinuous or discrete, and can change only at discrete instants of time. Usually, conversion between continuous signals and discrete signals is necessary in a discrete time system.

A signal (data) obtained from a continuous signal only at discrete instants of time is a discrete signal (sampled data). The continuous signals can be sampled into discrete signals. A sampler is basically a switch that only turns on at discrete instants of time. Fig. 6.1.1a shows a continuous signal $r(t)$. Its discrete signal $r^*(t)$ is shown in Fig. 6.1.1b, where the signal is periodically sampled at every multiple time T, and T is called sampling period. Consequently, $\omega_s = 2\pi/T$ is given the name sampling (angular) frequency. Mathematically, a sampled signal can be written as

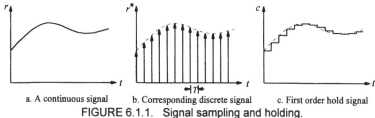

a. A continuous signal   b. Corresponding discrete signal   c. First order hold signal

FIGURE 6.1.1.   Signal sampling and holding.

$$r*(s) = \sum_{k=0}^{\infty} r(t)\delta(t - kT) \qquad (6.1.1)$$

where $\delta(\cdot)$ is a unit impulse function. The integer k is taken from 0 to ∞, because the function considered is zero for $t < 0$.

Zero order hold

The conversion from a discrete signal to a continuous signal is made by a holding device. The simplest holding device is a zero order hold, which holds the discrete signal constant at $r(kT)$ until a new sample occurs at $t = (k + 1)T$. The resulting signal is shown in Fig. 6.1.1c, which is a staircase approximation of the input signal $r(t)$ in Fig. 6.1.1a.

The block diagram of a discrete system with a sampler and a zero order hold is shown in Fig. 6.1.2. The output $c(t)$ of the zero order hold can be represented by

$$c(t) = r(nt) \qquad nT \le t < (n+1)T, \quad n = 0,1,\cdots \qquad (6.1.2)$$

As we already know, the transfer function of an element is the Laplace transform of its impulse response. Using the definition of zero order hold, the impulse response of the zero order hold is shown in Fig. 6.1.3, which can be formulated as

$$c(t) = \begin{cases} 1 & 0 \le t < T \\ 0 & \text{otherwise} \end{cases} \qquad (6.1.3)$$

Therefore, the Laplace transform of (6.1.3), or the transfer function of a zero

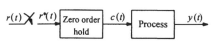

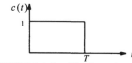

FIGURE 6.1.2.   Block diagram of a   FIGURE 6.1.3.   The impulse
discrete system.   response of a zero order hold.

order hold, will be

$$H_0(s) = \mathcal{L}[c(t)] = \mathcal{L}[u(t) - u(t - T)] = \frac{1 - e^{-Ts}}{s} \qquad (6.1.4)$$

Comparing Fig. 6.1.1a and 6.1.1c, we find that the precision of conversion is limited, and related to the sampling period. The output $c(t)$ of the zero order hold approaches the system input $r(t)$ as the sampling period T approaches zero.

Sampling and holding are summarized in Table 6.1.

The advantage of discrete systems
Since we already have satisfactory continuous control theory, and we just learned that the conversion between discrete signals and continuous signals introduces errors, why are we interested in discrete (digital) systems?

Discrete systems have the following advantages:

1. Discrete systems are expressed by difference equations that are easier to deal with than continuous differential equations.

2. Digital controllers are capable of rapidly performing complex computations with any desired level of accuracy. It paves the way for the application of advanced control theories, such as optimal control, adaptive control and so on.

3. With a digital computer, the control law is implemented by the software, which is very easy to change. This versatility is important should the control law be changed when production requirements, raw materials, economic factors or control objectives are changed.

4. Digital signals are less sensitive to disturbances and more reliable than continuous signals for signal measurement, transmission and processing.

Despite the advantages of digital systems, there are many potential problems inherent in the digitization of analog data. When working with a digital system, the control engineer must take special care to ensure that the system accurately

TABLE 6.1. Sampling and holding.

Item	Sampling	Holding
Signal conversion	Continuous signal to discrete signal	Discrete signal to continuous signal
Device	Sampler	Zero order hold or higher order hold
Mathematical expression	$r^*(t) = \sum\limits_{k=0}^{\infty} r(t)\delta(t - kT)$	$c(t) = r(nT)$ $nT \le t < (n+1)T, \quad n = 0, 1, 2, \cdots$
Laplace transformation	$R(s) = \sum\limits_{k=0}^{\infty} r(kT)e^{-kTs}$	$H_0(s) = \frac{1 - e^{-Ts}}{s}$

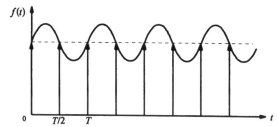

FIGURE 6.1.4. Sampling signal of a sinusoidal wave at a half sampling frequency.

reflects the analog process.

Sampling theorem

The sampling frequency affects the precision of conversion between continuous and discrete signals. What is the minimum sampling frequency required to reconstruct the original signal from the sampled signal?

Shannon's sampling theorem states that the signal $r(t)$ can be reconstructed completely from its sampled signal if and only if the sampling frequency ω_s is twice more than the maximum frequency ω_{max} of the original signal $r(t)$, i.e.,

$$\omega_s > 2\omega_{max} \tag{6.1.5}$$

A more detailed description and the proof of this theorem can be found in the literature [Åström and Wittenmark 1984, Ogata 1990 and 1987].

The sampling theorem can be understood by considering Fig. 6.1.4, which shows that if the sampling frequency is just twice the sinusoidal frequency, the sampled signal of a sinusoidal wave cannot be distinguished from that of a constant.

Remark

The sampling frequency should be chosen reasonably. On the one hand, it should be big enough to ensure that no significant information is lost. On the other hand, high sampling frequency will make it difficult for the computer to handle data flow.

6.2. Z-TRANSFORM

The Laplace transform is very important for continuous system analysis and design. The z-transform plays a similar role in discrete system analysis and design.

Definition

Taking the Laplace transform of a discrete signal of Eq. (6.1.1) yields

$$R*(s) = \mathcal{L}[r*(t)] = \mathcal{L}[\sum_{k=0}^{\infty} r(t)\delta(t - kT)]$$

$$= \mathcal{L}[\sum_{k=0}^{\infty} r(kT)\delta(t - kT)] = \sum_{k=0}^{\infty} r(kT)e^{-kTs} \qquad (6.2.1)$$

By defining

$$z = e^{Ts} \quad \text{or} \quad s = (\ln z)/T \qquad (6.2.2)$$

and denoting $R(z) = R*(s)$, we have

$$Z[r*(t)] = R(z) = R*(s) = \sum_{k=0}^{\infty} r(kT)z^{-k} \qquad (6.2.3)$$

$Z[r*(t)] = R(z)$ is referred to as the z-transform of $r*(t)$. Since only signal values at the sampling instants are considered, we denote

$$Z[r(t)] = Z[r*(t)] \qquad (6.2.4)$$

Example 6.2.1

For a unit step function $u(t)$, its z-transform can be found as

$$Z[u(t)] = \sum_{k=0}^{\infty} z^{-k} = 1 + z^{-1} + z^{-2} + z^{3} + \cdots = \frac{1}{1 - z^{-1}} = \frac{z}{z - 1}$$

Example 6.2.2

Find the z-transform of $r(t) = t \ (t > 0)$.

By definition, we have

$$Z[r(t)] = R(z) = \sum_{k=0}^{\infty} kTz^{-k} = Tz^{-1} + 2Tz^{-2} + 3Tz^{-3} + 4Tz^{-4} + \cdots$$

Considering

$$R(z) - z^{-1}R(z) = Tz^{-1} + Tz^{-2} + Tz^{-3} + Tz^{-4} + \cdots = T\frac{z^{-1}}{1 - z^{-1}}$$

we have

$$R(z) = \frac{z^{-1}T}{(1 - z^{-1})^2} = \frac{Tz}{(z - 1)^2}$$

Example 6.2.3

Prove that

$$Z[\sin\omega t] = \frac{z\sin\omega T}{z^2 - 2z\cos\omega T + 1}$$

From the definition of z-transform, we have

$$Z[\sin\omega t] = \sum_{k=0}^{\infty}(\sin\omega kT)z^{-k}$$

$$= \sum_{k=0}^{\infty}\left(\frac{e^{j\omega kT} - e^{-j\omega kT}}{2j}\right)z^{-k}$$

$$= \frac{1}{2j}\sum_{k=0}^{\infty}e^{j\omega kT}z^{-k} - \frac{1}{2j}\sum_{k=0}^{\infty}e^{-j\omega kT}z^{-k}$$

$$= \frac{1}{2j}\sum_{k=0}^{\infty}(e^{j\omega T}z^{-1})^k - \frac{1}{2j}\sum_{k=0}^{\infty}(e^{-j\omega T}z^{-1})^k$$

$$= \frac{1}{2j}\frac{1}{1-e^{j\omega T}z^{-1}} - \frac{1}{2j}\frac{1}{1-e^{-j\omega T}z^{-1}}$$

$$= \frac{1}{2j}\frac{(e^{j\omega T} - e^{-j\omega T})z^{-1}}{1-(e^{j\omega T} + e^{-j\omega T})z^{-1} + z^{-2}}$$

$$= \frac{z\sin\omega T}{z^2 - 2z\cos\omega T + 1}$$

Some basic function's z-transforms are given in Table 6.2, where the corresponding Laplace transforms are also given for comparison.

Properties of z-transform
For a discrete signal $r^*(t)$ with a fixed sampling period T, we are only concerned about the values at $t = kT$ ($k = 0, 1, 2, ...$). Then, $r(kT)$ is denoted as $r(k)$ for convenience.

The z-transform has some important properties as follows:

1. *Linearity.* It is easy to see from the definition that

$$Z[af(t) \pm bg(t)] = aZ[f(t)] \pm bZ[g(t)] \qquad (6.2.5)$$

where a and b are constants. Equivalently, Eq. (6.2.5) may be rewritten as

$$Z[af(k) \pm bg(k)] = aF(z) \pm bG(z) \qquad (6.2.6)$$

2. *Real translation theorem.* The z-transform of a discrete time function with a time delay equal to the sampling period is given by

TABLE 6.2.   z-transform of basic functions.

Laplace transform $R(s)$	Time function $r(t)$	z-transform $R(z)$
1	$\delta(t)$	1
e^{-kTs}	$\delta(t - kT)$	z^{-k}
$1/s$	$u(t) = \begin{cases} 0 & t < 0 \\ 1 & t \geq 0 \end{cases}$	$\dfrac{z}{z-1}$
$\dfrac{1}{s^2}$	t	$\dfrac{Tz}{(z-1)^2}$
$\dfrac{2}{s^3}$	t^2	$\dfrac{T^2 z(z+1)}{(z-1)^3}$
$\dfrac{1}{s+a}$	e^{-at}	$\dfrac{z}{z-e^{-aT}}$
$\dfrac{1}{(s+a)^2}$	te^{-at}	$\dfrac{Tze^{-aT}}{(z-e^{-aT})^2}$
$\dfrac{\omega}{s^2+\omega^2}$	$\sin\omega t$	$\dfrac{z\sin\omega T}{z^2 - 2z\cos\omega T + 1}$
$\dfrac{s}{s^2+\omega^2}$	$\cos\omega t$	$\dfrac{z(z-\cos\omega T)}{z^2 - 2z\cos\omega T + 1}$

$$Z[f(t-T)] = z^{-1} Z[f(t)] \qquad (6.2.7)$$

$f(t - T)$ shifts $f(t)$ one sampling period behind. From the definition, we have

$$Z[f(t-T)] = \sum_{k=0}^{\infty} f(kT - T)z^{-k} = \sum_{k-1=-1}^{\infty} f[(k-1)T]z^{-(k-1)}z^{-1}$$

Since $f(t) = 0$ for $t < 0$, we have $f(-1) = 0$. Let $p = k - 1$, then

$$Z[f(t-T)] = \sum_{p=0}^{\infty} f(pT)\, z^{-p}z^{-1} = z^{-1} Z[f(t)]$$

Similarly, we have

$$Z[f(t-2T)] = z^{-2} Z[f(t)] \qquad (6.2.8)$$

Generally,

$$Z[f(t-nT)] = z^{-n} Z[f(t)] \qquad (6.2.9)$$

Thus, multiplying the z-transform of a function by z^{-n} has the effect of delaying the time function by time nT.

If $f(t)$ is shifted one sampling period ahead, we have

$$Z[f(t+T)] = \sum_{k=0}^{\infty} f(kT+T)\, z^{-k} = \sum_{k=0}^{\infty} f[(k+1)T]\, z^{-(k+1)}z$$

$$= \sum_{k+1=0}^{\infty} f[(k+1)T]z^{-(k+1)}z - f(0)z$$

$$= zF(z) - zf(0) \qquad (6.2.10)$$

Also, we can obtain that

$$Z[f(t+2T)] = z^2 F(z) - z^2 f(0) - zf(T) \qquad (6.2.11)$$

In general,

$$Z[f(t+nT)] = z^n F(z) - z^n f(0) - z^{n-1} f(T) - \cdots - zf[(n-1)T] \qquad (6.2.12)$$

3. *Multiplication by* e^{-at}. The z-transform of $e^{-at}f(t)$ is

$$Z[e^{-at}f(t)] = F(ze^{aT}) \qquad (6.2.13)$$

From the definition, it follows that

$$Z[e^{-at}f(t)] = \sum_{k=0}^{\infty} f(kT)e^{-akT}z^{-k} = \sum_{k=0}^{\infty} f(kT)(e^{aT}z)^{-k} = F(ze^{aT})$$

Eq. (6.2.13) is proved.

4. *Multiplication by* t. The z-transform of $tf(t)$ is

$$Z[tf(t)] = -zT\frac{d}{dz}F(z) \qquad (6.2.14)$$

Note that

$$-zT\frac{d}{dz}F(z) = -zT\frac{d}{dz}\sum_{k=0}^{\infty} f(kT)z^{-k}$$

$$= -zT\sum_{k=0}^{\infty}(-k)f(kT)z^{-k-1}$$

$$= \sum_{k=0}^{\infty}(kT)f(kT)z^{-k} = [tf(t)]$$

5. *Initial value theorem.* The initial value of a function can be obtained from its z-transform. The initial value theorem says that

$$\lim_{t \to 0} f(t) = \lim_{z \to \infty} F(z) \qquad (6.2.15)$$

As z approaches infinity, the only term on the right side of Eq. (6.2.3) that does not vanish is $f(0)$, which verifies Eq. (6.2.15).

6. *Final value theorem.* The final value theorem says that if $F(z)$ does not have any poles on or outside the unit circle, then

$$\lim_{t \to \infty} f(t) = \lim_{z \to 1}(1 - z^{-1})F(z) \qquad (6.2.16)$$

To prove this theorem, consider the sums S_n and S_{n-1}:

$$S_n = \sum_{k=0}^{n} f(kT)z^{-k}, \qquad S_{n-1} = \sum_{k=0}^{n-1} f(kT)z^{-k}$$

Note that

$$S_n - z^{-1}S_{n-1} = \sum_{k=0}^{n-1}(1 - z^{-1})f(kT)z^{-k} + f(nT)z^{-n}$$

TABLE 6.3.   Properties of z-transform.

Property	Continuous function	Discrete function	z-transform
Linearity	$af(t) + bg(t)$	$af(k) + bg(k)$	$aF(z) + bG(z)$
Real translation	$f(t - T)$	$f(k - 1)$	$z^{-1}F(z)$
	$f(t - 2T)$	$f(k - 2)$	$z^{-2}F(z)$
	$f(t - nT)$	$f(k - n)$	$z^{-n}F(z)$
	$f(t + T)$	$f(k + 1)$	$zF(z) - zf(0)$
	$f(t + 2T)$		$z^2 F(z) - z^2 f(0) - zf(T)$
		$f(k + 2)$	$z^2 F(z) - z^2 f(0) - zf(1)$
	$f(t + nT)$		$z^n F(z) - z^n f(0) - z^{n-1}f(T)$ $- \cdots - zf[(n-1)T]$
		$f(k + n)$	$z^n F(z) - z^n f(0) - z^{n-1}f(1)$ $- \cdots - zf(n-1)$
Multiplication	$e^{-at}f(t)$		$F(ze^{aT})$
		$e^{-ak}f(k)$	$F(ze^{a})$
Initial value theorem	$\lim\limits_{t \to 0} f(t) = \lim\limits_{z \to \infty} F(z)$	$\lim\limits_{k \to 0} f(k) = \lim\limits_{z \to \infty} F(z)$	
Final value theorem	$\lim\limits_{t \to \infty} f(t) = \lim\limits_{z \to 1}(1 - z^{-1})F(z)$	$\lim\limits_{k \to \infty} f(k) = \lim\limits_{z \to 1}(1 - z^{-1})F(z)$	

Taking the limit as z approaches 1 yields

$$\lim_{z \to 1}(S_n - z^{-1}S_{n-1}) = f(nT)$$

Taking the limit as n approaches ∞, we have Eq. (6.2.16).

A summary of z-transform properties is given in Table 6.3.

Inverse z-transform
Usually, z-transforms are more complicated than their corresponding Laplace transforms.   Fortunately, there are some relatively simple techniques for obtaining inverse z-transforms.
We used the partial fraction method to find inverse Laplace transforms in Chapter 2. Here, we can use it to find inverse z-transforms.

Example 6.2.4
The z-transform of a time function is

$$F(z) = \frac{3(1 - e^{-2T})z}{(z-1)(z - e^{-2T})}$$

The partial fraction expansion of $F(z)$ is

$$F(z) = 3[\frac{K_1}{(z-1)} + \frac{K_2}{(z - e^{-2T})}]z$$

$$= 3\frac{K_1z - K_1e^{-2T} + K_2z - K_2}{(z-1)(z - e^{-2T})}z$$

By the comparison of the coefficients, we can obtain $K_1 = 1$ and $K_2 = -1$. Thus,

$$F(z) = 3[\frac{1}{(z-1)} + \frac{-1}{(z - e^{-2T})}]z$$

From Table 6.2, the corresponding time function is

$$f(t) = 3 - 3e^{-2t}$$

The properties of the z-transform can also be used to find the inverse z-transforms.

Example 6.2.5
The z-transform of a time function is

$$F(z) = \frac{3(1 - e^{-2T})e^T}{(ze^T - 1)(ze^T - e^{-2T})}$$

where T is a sampling period. It can be rewritten as

$$F(z) = \frac{3(1 - e^{-2T})ze^T}{(ze^T - 1)(ze^T - e^{-2T})} z^{-1}$$

Along with the properties of multiplication by e^{aT} and real translation, recalling the last example yields

$$f(t) = (3 - 3e^{-2(t-T)})e^{-(t-T)} = 3e^{-(t-T)} - 3e^{-3(t-T)}$$

The long division method is an easier way to find an inverse z-transform than partial fraction expansion. However, the result is usually in series form, which is not as useful as an analytical expression.

Example 6.2.6
    Recalling Example 6.2.4, $F(z)$ may be expressed as

$$F(z) = \frac{3(1 - e^{-2T})z}{(z - 1)(z - e^{-2T})} = \frac{3(1 - e^{-2T})z^{-1}}{(1 - z^{-1})(1 - z^{-1}e^{-2T})}$$

Denote $e^{-2T} = b$ and $1 - e^{-2T} = 1 - b = c$; then

$$F(z) = \frac{3cz^{-1}}{(1 - z^{-1})(1 - z^{-1}b)}$$

By long division, we have

$$
\require{enclose}
\begin{array}{r}
cz^{-1} + c(b+1)z^{-2} + c(b^2+b+1)z^{-3} + \cdots \\[4pt]
1 - (b+1)z^{-1} + bz^{-2} \enclose{longdiv}{\,cz^{-1}\phantom{+c(b+1)z^{-2}+c(b^2+b+1)z^{-3}}} \\
\end{array}
$$

$$cz^{-1} - c(b+1)z^{-2} + bcz^{-3}$$
$$c(b+1)z^{-2} - bcz^{-3}$$
$$c(b+1)z^{-2} - c(b+1)^2 z^{-3} + bc(b+1)z^{-4}$$
$$c(b^2+b+1)z^{-3} - bc(b+1)z^{-4}$$

It follows that

$$F(z)/3 = cz^{-1} + c(b+1)z^{-2} + c(b^2+b+1)z^{-3} + \cdots$$
$$= (1 - e^{-2T})z^{-1} + (1 - e^{-2T})(1 + e^{-2T})z^{-2} + (1 - e^{-2T})(1 + e^{-2T} + e^{-4T})z^{-3} + \cdots$$

$$= (1 - e^{-2T})z^{-1} + (1 - e^{-2 \times 2T})z^{-2} + (1 - e^{-3 \times 2T})z^{-3} + \cdots$$

$$= z^{-1} + z^{-2} + z^{-3} + \cdots - e^{-2T}z^{-1} - e^{-2 \times 2T}z^{-2} - e^{-3 \times 2T}z^{-3} + \cdots$$

$$= \frac{z^{-1}}{1 - z^{-1}} - \frac{e^{-2T}z^{-1}}{1 - e^{-2T}z^{-1}}$$

Hence

$$f(t)/3 = 1 - e^{-2t}$$

or

$$f(t) = 3 - 3e^{-2t}$$

Remarks

In the last three examples, the inverse z-transform of $F(z)$ is not $f(t)$ but $f^*(t)$. It should be noticed that the z-transform of $f(t)$ is unique, but different time functions may have the same z-transform, e.g., $G(z) = F(z)$ when $f(t) \neq g(t)$ but $f^*(t) = g^*(t)$.

6.3. DISCRETE TIME SYSTEM RESPONSES

Before analyzing the responses of discrete systems, we must determine the discrete system models. Discrete systems can be described by difference equations in the time domain or by transfer functions in the z-domain.

Difference equations and pulse transfer functions

For discrete systems, we are only interested in the values of signals at sampling instants. While differential equations are used to describe continuous sys..:m dynamics, difference equations are used for discrete systems.

The difference equation has a general form

$$y(t) + a_1 y(t - 1) + \cdots + a_m y(t - m)$$
$$= b_0 r(t) + b_1 r(t - 1) + \cdots + b_k r(t - k) \tag{6.3.1}$$

Taking z-transform on both sides of (6.3.1) yields

$$Y(z) + a_1 z^{-1} Y(z) + \cdots + a_m z^{-m} Y(z)$$
$$= b_0 R(z) + b_1 z^{-1} R(z) + \cdots + b_k z^{-k} R(z) \tag{6.3.2}$$

or

$$\frac{Y(z)}{R(z)} = \frac{b_0 + b_1 z^{-1} + \cdots + b_k z^{-k}}{1 + a_1 z^{-1} + \cdots + a_m z^{-m}} \tag{6.3.3}$$

This is the transfer function of a discrete system, which represents the relationship between the input and output of the system, and is referred to as a pulse transfer function. The transfer function in the z-domain is analogous to that used in the Laplace domain (s-domain). This input-output model can be used to determine the response of a discrete system to a specified input.

For a continuous system with a transfer function $G(s)$ as shown in Fig. 6.3.1a, a corresponding discrete system is shown in Fig. 6.3.1b by incorporating samplers at input and output. Usually, the sampler at the output is not necessary, as shown in Fig. 6.3.1c, if we consider a fictitious sampler at the output and observe the sequence of values taken by the output only at the sampling instants. The relationship between the discrete input and output of a system may be represented by a transfer function in the z-domain as shown in Fig. 6.3.1d.

Example 6.3.1

A continuous system, as shown in Fig. 6.3.1a, has a transfer function

$$G(s) = \frac{a}{s(s+a)} \tag{6.3.4}$$

After it is sampled as shown in Fig. 6.3.1c, the pulse transfer function can be found as follows:

Rewrite $G(s)$ as

$$G(s) = \frac{1}{s} - \frac{1}{(s+a)}$$

From Table 6.2, the pulse transfer function of the system can be found as

$$\begin{aligned}
G(z) &= \frac{z}{z-1} - \frac{z}{z-e^{-aT}} \\
&= \frac{1}{1-z^{-1}} - \frac{1}{1-e^{-aT}z^{-1}} \\
&= \frac{z^{-1}(1-e^{-aT})}{e^{-aT}z^{-2} - (1+e^{-aT})z^{-1} + 1}
\end{aligned} \tag{6.3.5}$$

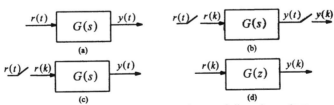

FIGURE 6.3.1.   Continuous system and discrete system.

which means that

$$[e^{-aT}z^{-2} - (1+e^{-aT})z^{-1} + 1]Y(z) = [z^{-1}(1-e^{-aT})]R(z)$$

where $Y(z)$ is the z-transform of the output, and $R(z)$ that of the input. The difference equation of the system is

$$y(k) - (1+e^{-aT})y(k-1) + e^{-aT}y(k-2) = (1-e^{-aT})r(k-1)$$

Sampler configurations
Different sampler configurations for a system will give different transfer functions in the z-domain. Fig. 6.3.2 shows two different sampler configurations. Assume that

$$G_1(s) = \frac{a}{s}, \qquad G_2(s) = \frac{1}{(s+a)}$$

For a system in Fig. 6.3.2a, the transfer function in the s-domain is $G(s) = G_1(s)G_2(s)$, and the transfer function in the z-domain, which we denote $G_1G_2(z)$, has been derived as (6.3.5) in Example 6.3.1. The corresponding transfer functions in the z-domain of the system in Fig. 6.3.2b can be derived from Table 6.2 as

$$G_1(z) = \frac{az}{z-1}, \qquad G_2(z) = \frac{z}{z-e^{-aT}}$$

Then, the overall transfer function will be

$$G(z) = G_1(z)G_2(z) = \frac{az}{z-1} \cdot \frac{z}{z-e^{-aT}}$$

$$= \frac{a}{1-z^{-1}} \cdot \frac{1}{1-e^{-aT}z^{-1}}$$

$$= \frac{a}{e^{-aT}z^{-2} - (1+e^{-aT})z^{-1} + 1}$$

which is different from (6.3.5). Therefore, we must be careful and observe whether there is a sampler between cascaded elements. Note that

$$G_1(z)G_2(z) \neq G_1G_2(z) \qquad\qquad (6.3.6)$$

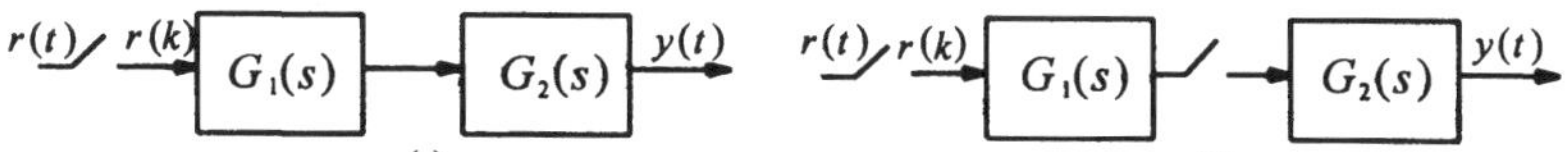

(a) (b)

FIGURE 6.3.2. Sampler configuration.

where $G_1 G_2(z)$ is the z-transform of $G_1(s)G_2(s)$, whereas $G_1(z)G_2(z)$ is the product of the z-transform of $G_1(s)$ and the z-transform of $G_2(s)$.

The general procedure [Raven 1987] for determining the pulse transfer function of a discrete system is shown in Table 6.4.

Example 6.3.2

Consider a feedback control system as shown in Fig. 6.3.3, where the actuating error is sampled. Following Step 1 in Table 6.4, the two outputs are Y and E, and two inputs are E^* and R. Representing the outputs in terms of inputs, we have

$$Y(s) = E^*(s)G(s)$$

$$E(s) = R(s) - G(s)H(s)E^*(s)$$

The corresponding equations in the z-domain are

$$Y(z) = E(z)G(z)$$

$$E(z) = R(z) - GH(z)E(z)$$

Then, the pulse transfer function can be obtained as

$$\frac{Y(z)}{R(z)} = \frac{G(z)}{1 + GH(z)} \tag{6.3.7}$$

Example 6.3.3

Fig. 6.3.4 shows a sampled feedback system. Following the procedure of Table 6.4, the outputs Y and D, in terms of inputs R and D^*, can be found as

TABLE 6.4. The procedure to determine the pulse transfer function.

Steps	Objective	Method
1	Determine the input and output for transfer function.	Consider all sampler outputs as inputs as well as the system input, and consider all sampler inputs as outputs in addition to the system output
2	Determine the transfer function in the s-domain.	Write transfer functions for each output in terms of all the inputs, and deal with the product of continuous transfer functions as a single one.
3	Determine the transfer function in the z-domain.	Use z-transform method to get the transfer function in z-domain.
4	Determine the pulse transfer function.	With the block diagram deduction technique, find the pulse transfer function of the system in z-domain.

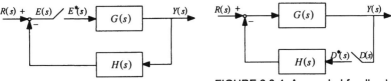

FIGURE 6.3.3 A sampled feedback
control system.

FIGURE 6.3.4 A sampled feedback
control system.

$$Y(s) = G(s)R(s) - H(s)G(s)D(s)$$
$$D(s) = Y(s)$$

Then,

$$Y(z) = GR(z) - HG(z)D(z)$$
$$D(z) = Y(z)$$

The system output in the z-domain is

$$Y(z) = \frac{GR(z)}{1 + HG(z)} \qquad (6.3.8)$$

Remarks

1. In the procedure of Table 6.4, each output equation must be expressed only in terms of the inputs rather than other intermediate variables.

2. As shown in Example 6.3.3, for some sampled systems, the closed loop system cannot be represented by a transfer function, since in the expression of the output in z-domain, the input cannot be separated.

The outputs of some typical sampling systems in the z-domain are shown in Table 6.5.

System responses
The responses of a sampled system to a specified input can be calculated from the difference equation of the system, which can be obtained from the output in the z-domain.

Example 6.3.4
For a system as shown in Fig. 6.3.3, assume that

$$G(s) = \frac{a}{s(s+a)} \qquad \text{and} \qquad H(s) = 1$$

Recalling Example 6.3.1, the closed loop transfer function in the z-domain is

TABLE 6.5. Some systems and their outputs in z-domain.

Block diagram	z-transform of the output
	$$Y(z) = \frac{G(z)R(z)}{1 + GH(z)}$$
	$$Y(z) = \frac{G(z)R(z)}{1 + G(z)H(z)}$$
	$$Y(z) = \frac{G(z)R(z)}{1 + G(z)H(z)}$$
	$$Y(z) = \frac{GR(z)}{1 + GH(z)}$$
	$$Y(z) = \frac{G_2(z)G_1R(z)}{1 + G_1HG_2(z)}$$
	$$Y(z) = \frac{G_2(z)G_1(z)R(z)}{1 + G_1(z)HG_2(z)}$$
	$$Y(z) = \frac{G_2(z)G_1(z)R(z)}{1 + G_1(z)G_2(z)H(z)}$$

$$\frac{Y(z)}{R(z)} = \frac{G(z)}{1 + HG(z)}$$

$$= \frac{z^{-1}(1 - e^{-aT})}{e^{-aT}z^{-2} - (1 + e^{-aT})z^{-1} + 1 + z^{-1}(1 - e^{-aT})}$$

$$= \frac{(1 - e^{-aT})z^{-1}}{e^{-aT}z^{-2} - 2e^{-aT}z^{-1} + 1}$$

Thus,

$$Y(z) - 2e^{-aT}z^{-1}Y(z) + e^{-aT}z^{-2}Y(z) = (1 - e^{-aT})z^{-1}R(z)$$

If $T = 1$ and $a = 1$, its corresponding difference equation is

TABLE 6.6. Data of Example 6.3.4.

k	y(k − 1)	y(k − 2)	r(k − 1)	r(k)
0	0	0	0	0
1	0	0	1	0.6321
2	0.6321	0	1	1.0972
3	1.0972	0.6321	1	1.2067
4	1.2067	1.0972	1	1.1163
5	1.1163	1.2067	1	1.0095
6	1.0095	1.1163	1	0.9642
7	0.9642	1.0095	1	0.9702
8	0.9702	0.9642	1	0.9912
9	0.9912	0.9702	1	1.0045
10	1.0045	0.9912	1	1.0066

$$y(k) = 0.7358 y(k − 1) - 0.3679 y(k − 2) + 0.6321 r(k − 1)$$

The values of $y(k)$ from $k = 0$ to 10 are shown in Table 6.6, and the response is shown in Fig. 6.3.5.

Example 6.3.5

A system is given in Fig. 6.3.6, where

$$G(s) = \frac{2}{s+1}$$

and $H_0(s)$ is a zero order hold with a transfer function

$$H_0(s) = \frac{1 - e^{-sT}}{s}$$

If the sampling period $T = 1$, what is the system response to a unit step input?

The pulse transfer function has been obtained in Example 6.3.2 as

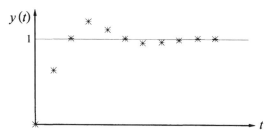

FIGURE 6.3.5. Discrete time response for Example 6.3.4.

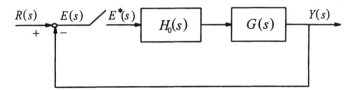

FIGURE 6.3.6. A feedback control system with a zero order hold.

$$\frac{Y(z)}{R(z)} = \frac{G(z)}{1 + GH(z)}$$

where

$$G(z) = Z[H_0(s)G(s)] = Z[\frac{1 - e^{-sT}}{s} \cdot \frac{2}{s+1}]$$

From the translation property of Laplace transforms, e^{-sT} represents a time delay. If the inverse Laplace transform of $G(s)$ is $g(t)$, the inverse Laplace transform of $G(s)e^{-sT}$ is $g(t - T)$. Thus, if the corresponding z-transform of $G(s)$ is $G(z)$, the corresponding z-transform of $G(s)e^{-sT}$ is $z^{-1}G(z)$. Therefore, we have

$$G(z) = 2(1 - z^{-1})Z[\frac{1}{s(s+1)}]$$

$$= 2(1 - z^{-1})Z[\frac{1}{s} - \frac{1}{(s+1)}]$$

$$= 2(1 - z^{-1})[\frac{z}{z-1} - \frac{z}{z-e^{-1}}]$$

$$= 2(1 - z^{-1})[\frac{z^{-1}(1 - e^{-1})}{(1 - z^{-1})(1 - e^{-1}z^{-1})}]$$

$$= 2\frac{z^{-1}(1 - e^{-1})}{(1 - e^{-1}z^{-1})}$$

Because $H(s) = 1$, the system output in the z-domain is

$$Y(z) = \frac{G(z)}{1 + G(z)}R(z)$$

$$= \frac{2z^{-1}(1 - e^{-1})}{1 - e^{-1}z^{-1} + 2z^{-1}(1 - e^{-1})}R(z)$$

$$= \frac{1.2642z^{-1}}{1+0.8963z^{-1}} R(z)$$

The corresponding difference equation is

$$y(k) = -0.8963y(k-1) + 1.2642r(k-1)$$

With respect to a unit step input, the output may be calculated as

$y(0) = 0$	$y(1) = 1.2642$	$y(2) = 0.1311$
$y(3) = 1.1467$	$y(4) = 0.2364$	$y(5) = 1.0523$
$y(6) = 0.3210$	$y(7) = 0.9765$	$y(8) = 0.3890$
$y(9) = 0.9156$	$y(10) = 0.4436$	$y(11) = 0.8666$
$y(12) = 0.4875$	$y(13) = 0.8273$	......

The response of the system to a unit step input is shown in Fig. 6.3.7, where we roughly connect the sampling points instead of drawing a staircase curve.

Remarks
We find that the system above has an oscillatory response to a unit step input, although the system has only a single pole. This situation is not possible for a continuous system.

6.4. SYSTEM ANALYSIS AND DESIGN

As in the case of continuous systems, a block diagram and the system response can serve as the basis for analyzing the dynamic behavior of a discrete system. The stability of a system is essential and must be investigated.

Stability analysis in the z-plane
As we know, a linear continuous system is stable if all poles of the closed loop

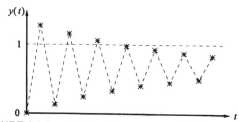

FIGURE 6.3.7. Discrete time response for Example 6.3.5.

transfer function lie in the left half of the s-plane. For a linear discrete system, the closed loop poles may be investigated in the z-plane. From the relationship between the s-plane and the z-plane, we have

$$z = e^{sT} = e^{(\sigma+j\omega)T} = e^{\sigma T}e^{j\omega T} \qquad (6.4.1)$$

where σ is the real part of s, and ω the imaginary part. From (6.4.1), it is clear that $-\infty <\sigma < 0$ (left half s-plane) corresponds to $|z| < 1$ (inside the unit circle), and the imaginary axis of the s-plane corresponds to the unit circle in the z-plane as shown in Fig. 6.4.1.

Therefore, we can conclude that a discrete system is stable if and only if all the poles of the discrete closed loop transfer function lie within the unit circle.

The corresponding relationship between the s-plane and the z-plane is given in Table 6.7.

Remarks

1. The corresponding relationship between the s-plane and z-plane is not 'one to one mapping'. Many points in the s-plane correspond to a given point in the z-plane, since e^{Ts} is a periodic function with respect to ω.

2. The effect of real poles in the z-plane on the system response is more complicated than that in the s-plane. This can also be found in Table 6.7.

Modified Routh criterion

The Routh criterion has been used to find how many poles of a continuous system are located on the right half plane. With modification, it can also be used to analyze the stability of a discrete system.

Consider a bilinear transformation

$$w = \frac{z+1}{z-1} \qquad (6.4.2)$$

Equivalently, it can be rewritten as

$$z = \frac{w+1}{w-1} \qquad (6.4.3)$$

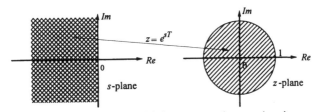

FIGURE 6.4.1. Relationship between s-plane and z-plane.

TABLE 6.7. Relationship between s-plane and z-plane.

s-plane	z-plane	Comment
Right half plane	Outside the unit circle	Unstable region
Left half plane	Inside the unit circle	Stable region
Origin	$z = 1$	Integral action
Negative real axis	$0 < z < 1$	Nonoscillatory convergent response
Positive real axis	$1 < z < \infty$	Monotonic divergent response
Points $\sigma + j(\omega+2\pi k/T)$ $k = 0, \pm1, \cdots$	One point $e^{(\sigma + j\omega)T}$	Many points in s-plane are mapped to one point in z-plane
$\sigma + j(\omega+2\pi k/T)$ $\sigma < 0, \ k = 0, \pm1, \cdots$	$-1 < z < 0$	Oscillatory convergent response
$\sigma + j(\omega+2\pi k/T)$ $\sigma > 0, \ k = 0, \pm1, \cdots$	$-\infty < z < -1$	Oscillatory divergent response

This transformation maps the unit circle in the z-plane to the vertical imaginary axis of the w-plane. The poles outside the unit circle in the z-plane are mapped into the right half of the w-plane. Therefore, when the characteristic equation is expressed in the w-plane by the transformation of (6.4.2) or (6.4.3), the Routh criterion may be used in the same way as for continuous systems.

Example 6.4.1

Recalling Example 6.3.4, the characteristic equation is

$$1 - 0.7358z^{-1} + 0.3679z^{-2} = 0$$

or

$$z^2 - 0.7358z + 0.3679 = 0$$

Substituting (6.4.3) into this yields

$$0.6321w^2 + 1.2642w + 2.1037 = 0$$

The Routh array can be written as

$$
\begin{array}{ll}
0.6321 & 2.1037 \\
1.2642 & 0 \\
2.1037 &
\end{array}
$$

Since all the signs in the first column are positive, the system is stable.

The system stability also can be checked by the location of system poles. The roots of the characteristic polynomial can be solved as $z = 0.3679 \pm j0.4822$. Because of $|z| = 0.607 < 1$, the system is stable.

Controllers in discrete systems

Both continuous and discrete controllers can be used in discrete control systems. In Fig. 6.4.2a, a continuous controller is used, while in Fig. 6.4.2b a discrete controller is used, both sides of which are sampled. In the case of a continuous controller used in a discrete system, the discrete transfer function of the controller is meaningless. The continuous transfer function of the controller should be combined with the process transfer function. Only the discrete transfer function of their combination is meaningful.

In most discrete systems, discrete rather than continuous controllers are used. In a computer controlled system, the discrete (digital) controller is an algorithm implemented in the computer.

Discrete (digital) controller design

Usually, there are two ways to design discrete (digital) controllers:

1. Design a continuous controller for a continuous system first, then implement the controller digitally. This way is especially suitable for situations where the continuous controller gives satisfactory performance for a continuous system, but a discrete (digital) controller is suggested for gaining more advantages [Kuo 1991].

2. Design a discrete (digital) controller directly with discrete time system theory for a discrete time system. Before designing the controller, the pulse transfer function of the process (with a zero order hold) in the z-domain must be obtained.

Digital implementation for a *PID* controller

Recalling Eq. (3.4.13), a continuous *PID* controller can be described as

$$c(t) = K_c(e(t) + \frac{1}{\tau_I}\int_0^t e(t)dt + \tau_D \frac{de(t)}{dt}) \qquad (6.4.4)$$

where $c(t)$ is the output deviation variable of the controller, and $e(t)$ is the actuating error. The transfer function of a *PID* controller is

$$\frac{C(s)}{E(s)} = K_c(1 + \frac{1}{\tau_I s} + \tau_D s) \qquad (6.4.5)$$

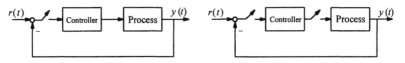

a. Continuous controller   b. Discrete controller

FIGURE 6.4.2. Continuous and discrete controllers in the discrete systems.

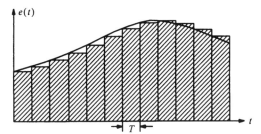

FIGURE 6.4.3. Numerical evaluation of an integral.

Through a numerical approximation as shown in Fig. 6.4.3, the integral in (6.4.4) can be evaluated by

$$\int_0^t e(t)dt \approx T \sum_{k=1}^{n} e(kT) \tag{6.4.6}$$

where $nT = t$, and T is the sampling period. As shown in Fig. 6.4.4, the derivative at $t = nT$ can be approximated by

$$\frac{de}{dt} \approx \frac{e(nT) - e[(n-1)T]}{T} \tag{6.4.7}$$

Therefore, the numerical approximation of a *PID* controller is

$$c(nT) = K_c e(nT) + \frac{K_c T}{\tau_I} \sum_{k=1}^{n} e(kT) + \frac{K_c \tau_D}{T} \{e(nT) - e[(n-1)T]\} \tag{6.4.8}$$

This is also referred to as the position form of the *PID* control algorithm since the actual controller output is calculated.

An alternative form for *PID* control algorithms is the velocity form, in which the change of the controller output from the preceding period is calculated rather than the actual controller output itself. Taking the preceding period from (6.4.8),

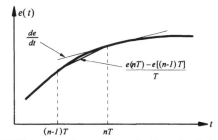

FIGURE 6.4.4. Numerical evaluation of a derivative.

we have

$$c[(n-1)T] = K_c e[(n-1)T] + \frac{K_c T}{\tau_I} \sum_{k=1}^{n-1} e(kT) +$$

$$+ \frac{K_c \tau_D}{T} \{ e[(n-1)T] - e[(n-2)T] \} \qquad (6.4.9)$$

Subtracting (6.4.9) from (6.4.8) gives

$$c(nT) - c[(n-1)T] = (K_c + \frac{K_c T}{\tau_I} + \frac{K_c \tau_D}{T}) e(nT)$$

$$- (K_c + \frac{2K_c \tau_D}{T}) e[(n-1)T] + \frac{K_c \tau_D}{T} e[(n-2)T] \qquad (6.4.10)$$

Similarly, the velocity form of a *PI* controller is

$$c(nT) - c[(n-1)T] = (K_c + \frac{K_c T}{\tau_I}) e(nT) - K_c e[(n-1)T] \qquad (6.4.11)$$

Selection of sampling period

The selection of the sampling period is an important but complicated issue. A number of empirical rules for this selection have been reported. Shannon's sampling theorem, mentioned in section 6.1, can be used to determine the sampling period. However, it is more of an art than a science. The response of a system may react unfavorably to an unsuitably selected sampling period.

When an integral action or a derivative action is included in the system, the selection of a sampling period becomes more complicated. Adjusting the parameters of a *PID* controller may require adjusting the sampling period. Selected empirical formulas for sampling period selection can be found in Table 6.8.

Deadbeat controller design

The general transfer function of a discrete controller can be written as

TABLE 6.8. Selection of sampling period.

Case	Sampling period T	Description
Process model $\dfrac{Ke^{-\tau s}}{T_1 s + 1}$	$0.2 < T/\tau < 1$	If τ is too small, $T < 0.1T_I$
PI controller	$T > \tau_I / 100$	
PID controller	$0.1\tau_D < T < 0.5\tau_D$	Consider $T > \tau_I / 100$

$$G_c(z) = \frac{(z - z_1)(z - z_2)\cdots(z - z_m)}{(z - p_1)(z - p_2)\cdots(z - p_n)} \tag{6.4.12}$$

where z_i are the zeros of the controller, and p_i the poles. For a system in Fig. 6.4.2b, if the discrete transfer functions of the process and controller are $G(z)$ and $G_c(z)$, respectively, the pulse transfer function of the system is

$$\frac{Y(z)}{R(z)} = \frac{G(z)G_c(z)}{1 + G(z)G_c(z)} \tag{6.4.13}$$

Assuming the output and input of the system are known, from (6.4.13) the controller can be designed as

$$G_c(z) = \frac{Y(z)}{G(z)[R(z) - Y(z)]} \tag{6.4.14}$$

The performance of a system reflects the relationship between the input and output of the system. Therefore, we can define the output response to a specified input to design the controller. Usually, a unit step input is assigned as the specified input, i.e.,

$$R(z) = \frac{z}{z - 1} \tag{6.4.15}$$

In order to obtain the robust response from the system, we may require the output to reach the steady state in only one sampling period. This control strategy is called deadbeat control. Then, the output response to a unit step input should be $y(0) = 0, y(1) = 1, y(2) = 1, \cdots, y(n) = 1$. From the definition of the z-transform (6.2.3), the z-transform of the output is

$$
\begin{aligned}
Y(z) &= \sum_{k=0}^{\infty} y(k)z^{-k} \\
&= y(0) + y(1)z^{-1} + y(2)z^{-2} + \cdots + y(n)z^{-n} + \cdots \\
&= z^{-1} + z^{-2} + \cdots + z^{-n} + \cdots \\
&= \frac{z^{-1}}{1 - z^{-1}} = \frac{z}{z - 1}
\end{aligned} \tag{6.4.16}
$$

Substituting (6.4.15) and (6.4.16) into (6.4.14) yields

$$G_c(z) = \frac{1}{G(z)(z - 1)} \tag{6.4.17}$$

Example 6.4.2

For the system in Example 6.3.5 (Fig. 6.3.6), we can design a controller so that the output response of the system to a unit step input is a one sampling period delay of the output.

From Example 6.3.5, the transfer function $G(z)$ is

$$G(z) = 2\frac{z^{-1}(1-e^{-1})}{1-e^{-1}z^{-1}}$$

The required controller from (6.4.17) is

$$G_c(z) = \frac{1}{G(z)(z-1)} = \frac{1}{2\dfrac{z^{-1}(1-e^{-1})}{1-e^{-1}z^{-1}}(z-1)}$$

$$= \frac{1}{2\dfrac{1-e^{-1}}{z-e^{-1}}(z-1)} = \frac{1}{2(1-e^{-1})} \cdot \frac{z-e^{-1}}{z-1}$$

Remarks

1. The discrete transfer function $G(z)$ represents all the elements between the controller output sampler and the output of the system, which includes the zero order hold, final control element and process.

2. As shown in Fig. 6.4.5, with the deadbeat design method, large overshoots or highly oscillatory behavior may occur between sampling instants. This is a drawback of deadbeat systems [Stephanopoulos 1984].

3. Clearly, the order of denominator of $G_c(z)$ must be greater than or equal to that of the numerator for realization. It follows from (6.4.17) that the order difference between the denominator and numerator of $G(z)$ cannot be greater than or equal to two. For this kind of system, a time delay of the output to the

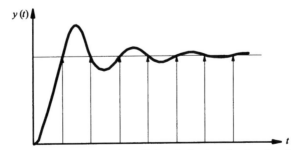

FIGURE 6.4.5.   Overshoot and oscillatory behavior of a deadbeat system.

input of two or more sampling periods may be required.

If a two sampling period time delay of the output to a unit step input is required, i.e., the system output reaches the steady state in two sampling periods, we can assign the output to be $y(0) = 0$, $y(1) = a$, $y(2) = 1$, $\cdots$, $y(n) = 1$. From the definition of the z-transform (6.2.3), the z-transform of the output is

$$Y(z) = \sum_{k=0}^{\infty} y(k)z$$

$$= y(0) + y(1)z^{-1} + y(2)z^{-2} + \cdots + y(n)z^{-n} + \cdots$$

$$= az^{-1} + z^{-2} + \cdots + z^{-n} + \cdots$$

$$= az^{-2} + \frac{z^{-2}}{1-z^{-1}}$$

$$= \frac{az^{-1} - az^{-2} + z^{-2}}{1-z^{-1}}$$

$$= \frac{az - a + 1}{z(z-1)} \tag{6.4.18}$$

Then the controller is

$$G_c(z) = \frac{\dfrac{az - a + 1}{z(z-1)}}{G(z)(\dfrac{z}{z-1} - \dfrac{az - a + 1}{z(z-1)})} = \frac{az - a + 1}{G(z)(z^2 - az + a - 1)} \tag{6.4.19}$$

If $a = 1$, this expression becomes (6.4.17). If $a = 0$, it is the case of a two sampling period delay, and the controller will be

$$G_c(z) = \frac{1}{G(z)(z^2 - 1)} \tag{6.4.20}$$

In such a case, the denominator of the discrete transfer function $G(z)$ may be two orders higher than the numerator.

More case studies are summarized in Table 6.9.

Discrete time system analysis with *PCET*

Incorporated in the software package *PCET* is the subprogram *DSA*, which will enable the reader to examine the dynamics of a unit feedback discrete time system. A typical process model is used, which has the transfer function

TABLE 6.9.   Deadbeat controller design.

Input	Output requirement	Controller	Comments
A unit step input	One sampling period delay	$\dfrac{1}{G(z)(z-1)}$	The denominator of $G(z)$ can be one order higher than the numerator.
	Two sampling period delay	$\dfrac{1}{G(z)(z^2-1)}$	The denominator of $G(z)$ can be two orders higher than the numerator.
	Reach the steady state in two sampling periods	$\dfrac{az-a+1}{G(z)(z^2-az+a-1)}$	a is the output value at T, where T is sampling period.
	Three sampling period delay	$\dfrac{1}{G(z)(z^3-1)}$	The denominator of $G(z)$ can be three orders higher than the numerator.
	n sampling period delay	$\dfrac{1}{G(z)(z^n-1)}$	The denominator of $G(z)$ can be n orders higher than numerator.

$$G(s) = \frac{K_0}{(T_1s+1)(T^2s^2+2T\zeta s+1)}e^{-\tau s}$$

The system configuration is shown in Fig. 6.4.6, where a digital *PID* controller and a zero order hold are used.

The sampling period can be assigned by the user. One must be aware that if the sampling period is too small, or the dimension scaled is too large, simulation error may happen, since too many calculation steps produce a large accumulated error.

Similar to *TDA* for time domain analysis, the subprogram *DSA* used in the z-domain can be applied for open loop process analysis, controller analysis, closed loop system analysis and controller tuning. The discrete time response of the process or closed loop system to a unit step input or disturbance can be drawn on the screen or with a printer. The keystroke sequences required to obtain the responses are illustrated in Table 6.10.

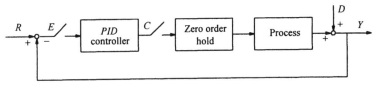

FIGURE 6.4.6.   The discrete time system used in *PCET*.

TABLE 6.10. Plotting the discrete time unit step response using *PCET*.

Step	User input	Function
1	Type *PCET*	Start program
2	Select *DSA*	Discrete time system analysis
3	Press 1, 2, 3 or 4	1. Analysis of open loop process 2. Analysis of controller action 3. Analysis of closed loop response 4. Change setpoint/disturbance
4	Input system model, controller parameters and sampling period	Specify system dynamics.
5	Revise input if necessary	Verify input data
6	Select autoscaling, or input figure dimension	Autoscaling computes axis dimension from response data
7		Desired response is displayed on the screen
8	Press a function key to continue	F1-Exit, back to main Menu F2-Restart, restart *DSA* F3- Load, load a saved data file F4-Save, save the data in a file F5-Scale, change the dimension F6-Modify, change the model F7-Tuning, change controller

Problems

6.1. A sampling period T is assumed. By the definition of the z-transform, find the z-transforms of the following time functions:

a. $f(t) = e^{-at}$ $(a > 0)$ b. $f(t) = \cos\omega t$

c. $f(t) = 2te^{-at}$

6.2. Find the pulse transfer function of the system from its definition, if the corresponding continuous transfer function is

a) $G(s) = \dfrac{1}{s}$ b) $G(s) = \dfrac{a}{s+a}$

c) $G(s) = \dfrac{a}{s(s+a)}$ d) $G(s) = \dfrac{a(1-e^{-sT})}{s(s+a)}$

6.3. If the Laplace transform of a function is given by

$$F(s) = \frac{2(1-e^{-s})}{s^2(s+1)}$$

what is the z-transform of the function for $T = 1$? (A z-transform table can be used.)

6.4. If the z-transform of function $f(t) = t$ is known as

$$Z(t) = \frac{Tz}{(z-1)^2}$$

what is the z-transform of function $f(t) = t^2$?

6.5. Find the pulse transfer functions or output transfer functions for the systems shown in Fig. 6.P.1.

6.6. Consider a system with an open loop discrete transfer function

$$G(z) = \frac{z^2 - 1.25z - 0.6}{z^3 - 2.5z^2 + z + 1}$$

How many closed loop poles are located outside the unit circle?

6.7. Consider a system as shown in Fig. 6.P.2. Find all the values of gain K so that the system is stable.

6.8. For the system shown in Fig. 6.4.2b, find the discrete transfer function of

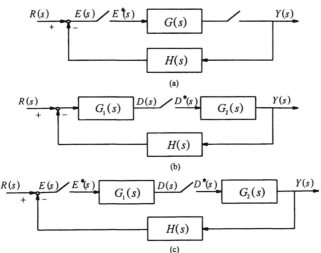

FIGURE 6.P.1. Some sampler configurations.

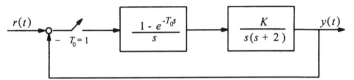

FIGURE 6.P.2. Discrete system of Problem 6.7.

the controller so that the response of the system to a unit step function is $y(0)=0$, $y(1)=0$, $y(2)=0$, $y(3)=1$, $y(4)=1$, $\cdots$, $y(n)=1$. In other words, the output is a three sampling period delay of the input.

6.9. For the system shown in Fig. 6.4.2b, find the discrete transfer function of the controller so that the response of the system to a unit step function is $y(0)=0$, $y(1)=a$, $y(2)=b$, $y(3)=1$, $y(4)=1$, $\cdots$, $y(n)=1$. In other words, the output reaches the steady state in three sampling periods.

6.10. (Effect of sampling period) For a system as shown in 6.P.3 with the process transfer function

$$G(s) = \frac{1}{(s+1)(4s^2 + 1.2s + 1)}$$

use *PCET* to do the following:

a) With subprogram *DSA*, draw the responses of the system to a unit step input with sampling periods 0.1, 1 and 10 on the same screen.

b) With the subprogram *FDA*, draw the Bode diagram of the system. From the log modulus, find the frequency ω_d at which the log modulus is 3 dB lower than that as $\omega \to 0$. (The band $0 \le \omega \le \omega_d$ is called the frequency bandwidth.)

c) Assume that any signal that has maximum frequency $\le \omega_d$ can pass through the system without considerable distortion. With Shannon's sampling theorem, which sampling period in the three is unreasonable? Give the reason.

6.11. (Closed loop discrete time response) With subprogram *DSA* in *PCET*, draw the response of the system of Example 6.3.5 to a unit step input. Is Fig. 6.3.7 correct?

6.12. (Controller design for a first order system) Consider a discrete system as shown in Fig. 6.4.6. The first order process has the transfer function

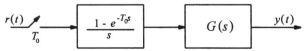

FIGURE 6.P.3. An open loop discrete system.

$$G(s) = \frac{2}{5.2s + 1}$$

For a 0.5 second sampling period, do the following:

a) Find the open loop response to a unit step input.

b) Examine the closed loop response obtained using a proportional (P) controller with $K_c = 1.5$. Does offset appear?

c) Design a proportional controller by trial and error so that the closed loop response satisfies the following performance requirements:

Offset ≤ 0.1

Settling time ≤ 4 sec

d) Design a proportional-integral (PI) controller by trial and error so that the closed loop system meets the performance specifications:

Overshoot ≤ 0.3

Settling time ≤ 5 sec

e) Compare the results obtained in parts a), b), c) and d) to the results obtained in Problem 3.17 of Chapter 3, and give your conclusion.

6.13. (Controller design for a third order system) Consider a discrete system as shown in Fig. 6.4.6. The third order process has the transfer function

$$G(s) = \frac{4}{(3s + 1)(4s^2 + 4s + 1)}$$

Use subprogram DSA to design a PID controller by trial and error so that the closed loop system satisfies the time domain performance criteria:

Overshoot ≤ 0.25

Settling time ≤ 50 seconds

Select a suitable sampling period T that satisfies the formulas in Table 6.8. If T is too large or too small, what will happen in the simulation? Explain.

CHAPTER 7

PROCESS CONTROL SYSTEM DESIGN

7.1. SELECTION OF CONTROLLED VARIABLES AND MANIPULATED VARIABLES

After studying the basic theory of control system analysis and controller design, we come to discuss how to design a system, i.e., how to select controlled and manipulated variables, how to select sensors and final control elements, how to select controllers and so on.

To construct an effective control system, it is very important to select the proper controlled variables and manipulated variables. Different selections of controlled variables will bring about different results. How many controlled variables or manipulated variables should be selected? What principles should we follow to select these variables? We discuss these issues in the following lines.

Number of controlled variables and manipulated variables
In Section 2.1 we studied the degree of freedom concept, which can be expressed in the following formula:

$$N_f = N_v - N_e \qquad (7.1.1)$$

where N_f is the degree of freedom, N_v is the number of process variables including outputs and unspecified inputs, and N_e is the number of independent equations relating N_v variables.

To have an exactly specified process, the degree of freedom should be zero. For an unspecified process, the degree of freedom is greater than zero, and it can be decreased by selecting manipulated variables, or by identifying process variables that are determined by the process environment. It follows that

$$N_f = N_m + N_s \qquad (7.1.2)$$

201

where N_m is the number of manipulated variables and N_s the number of process variables specified by the environment.

Eq. (7.1.2) indicates that the number of manipulated variables cannot exceed the degree of freedom. Normally, the number of required independent manipulated variables is equal to that of independent controlled variables.

Example 7.1.1

From a mass balance equation and an energy balance equation, the following differential equations of a stirred tank heater were obtained in Example 2.1.4:

$$A\frac{dh}{dt} = F_i - F \tag{7.1.3}$$

$$Ah\frac{dT}{dt} = F_i(T_i - T) + \frac{Q}{\rho c_p} \tag{7.1.4}$$

where A is cross sectional area; h is the height of liquid level; F_i and F are volumetric flow rates in the inlet and outlet, respectively; T_i and T are the temperatures of the liquid in the inlet and tank, respectively; Q is the amount of heat supplied by the stream per unit of time; ρ is the density of the liquid and c_p is the heat capacity.

The cross sectional area A, density ρ and heat capacity c_p are constants, and 6 independent variables are h, F_i, F, T_i, T and Q. Since there are two independent equations, the degree of freedom is $6 - 2 = 4$. T_i and F_i would be specified by upstream units, i.e., $N = 2$. Thus, there should be $4 - 2 = 2$ manipulated variables. For instance, we choose two manipulated variables F and Q to control two controlled variables h and T, respectively.

Selection of controlled variables

The selection of controlled variables depends on the considerations of economics, safety, constraints, availability and reliability of sensors, etc. Engineering judgment is the main tool for determining the controlled variables. A good understanding of the process is essential for the selection. The selection of controlled variables from the available output variables could follow the suggested guidelines:

1. Select output variables that are the direct measures of product quality, or the main factors that affect the quality. This selection will make the product quality easier to control.

2. Select output variables that may exceed equipment and operating constraints. This selection may keep the controlled variables with the operating constraints.

3. Usually do not select output variables that are self-regulating. This selection coincides with Guideline 2. Usually, a self-regulating variable does not exceed the operating constraints.

4. Select output variables so that the transfer function between the selected variable and one of manipulated variables has favorable dynamic and static characteristics.

5. Select output variables that can be measured easily, rapidly and reliably. The time constant of measuring instrumentation should be small enough to satisfy system requirements.

Remarks

If the ultimately controlled variables cannot be simply measured, other variables may be selected as controlled variables. The ultimately controlled variables that cannot be easily measured are called indirectly controlled variables [D'azzo 1988]. These indirectly controlled variables may be computed from selected controlled variables. The common examples of ultimately controlled variables are heat removal rates, mass flow rates, ratio of flow rates, and so on.

Selection of manipulated variables

Once the controlled variables have been chosen, the control configuration may be decided by the selection of manipulated variables. For some simple systems, the variables that should be selected as manipulated variables may be obvious after the controlled variables have been determined. However, the selection may not be so easy for complex systems. The following guidelines for the selection of manipulated variables are suggested:

1. Select input variables that have large effects on the specified controlled variables. This means that the proportional gain between the selected manipulated variable and the specified controlled variable should be large.

2. Select input variables that have large ranges. Such a selection will make the controlled variables easier to control.

3. Select input variables that rapidly affect the controlled variables. This indicates that the dynamic responses should be as quick as possible.

4. Select input variables that affect the specified controlled variable directly rather than indirectly, and have a minimum effect on other controlled variables. This suggests that the interaction of different loops of the system should be as small as possible.

Remark

Sometimes the guidelines for the selection of controlled variables or manipulated variables conflict with each other. An appropriate compromise may be made by considering operation safety. More detailed discussion can be found in the literature [Seborg et al. 1989, Newell and Lee 1988].

7.2. SENSORS, TRANSMITTERS AND TRANSDUCERS

After the controlled variables and manipulated variables have been selected, all measurement, manipulation and transmission instruments may be chosen. These instruments can convert, amplify and transmit physical or chemical signals.

Transducers
Devices that convert physical or chemical variables from one form to another with a desired strength are referred to as transducers. A transducer for measurement typically consists of a sensing element (sensor) and a driving element (transmitter) as shown in Fig. 7.2.1. The sensing element converts a physical or chemical signal (information) into a more appropriate physical form. For instance, a thermocouple converts physical information (temperature) into a physical signal — electromotive force (emf) — that is easily amplified for signal processing. The transmitter usually converts the sensor output to a signal level appropriate for the controller input. At this level, the signal may drive the transmission lines connecting the transducer and the controller. The transmission lines are often very long since the controllers are usually located in a distant control room. Since the electromotive force obtained from a thermocouple is very weak, it is often necessary to amplify and convert the signal to one that is compatible with the controller input, e.g., a digital signal for a digital controller.

Measured variables
In process control systems, the measured physical and chemical variables may be among the following list:
Temperature. One of the most common temperature sensors is a thermocouple, as shown in Fig. 7.2.2, which consists of two wires of different material that produce a millivolt signal varying with the hot junction

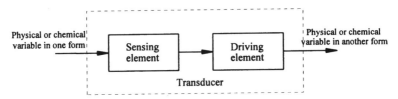

FIGURE 7.2.1. The composition of a transducer.

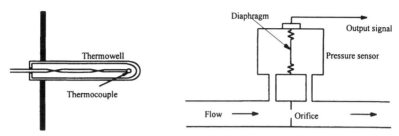

FIGURE 7.2.2. Schematic diagram of FIGURE 7.2.3. Flow rate measurement
typical thermocouple.   with a pressure sensor.

temperature.   Filled bulb sensors and resistance thermometers are also
commonly used.   The filled bulb temperature sensor is filled with a gas that
produces different pressures at different temperatures.   The resistance
thermometer works on the principle that the resistance of an element changes
with its temperature.

Usually, a sensor needs to be protected with a thermowell wall, although it
deteriorates the dynamic characteristics of the sensor. The time constant of a
thermocouple and a heavy thermowell may be more than 30 seconds. This may
degrade control performance to a considerable degree.

The transfer function of a temperature sensor may be modeled as a first order
system or an overdamped second order system.

Pressure and differential pressure. There are two kinds of pressure sensors:
pneumatic and electronic.   The pressure or differential pressure may be a
measure of a liquid level, or a flow rate.   For a typical device, a pressure
difference causes a small displacement of a sensing diaphragm. A force balance
leads to a second order model given by

$$\tau^2 \frac{d^2 z}{dt^2} + 2\zeta\tau \frac{dz}{dt} + z = K\Delta p \qquad (7.2.1)$$

where z is a displacement of the sensing diaphragm, Δp is an actuating pressure
difference, and τ, ζ and K are parameters depending on the structure of the
sensor.

Liquid or gas flow rate. Liquid or gas flow may be measured with a turbine
flow meter, which uses the number of turbine revolutions per unit time to
compute the flow rate.

Most flow rate sensors measure flow rates indirectly, for example, through
pressure drop. Fig. 7.2.3 shows a typical orifice plate. The volumetric flow rate
is proportional to the square root of the pressure drop across the orifice. Flow
rate sensors have very fast dynamic characteristics, thus a simple algebraic
model may be used as

$$f = \alpha \sqrt{\Delta p} \qquad (7.2.2)$$

where f is the volumetric flow rate, Δp is the pressure difference between the stable flows before and after the orifice, and α is a constant dependent on the structure.

Liquid level. The three most common ways to detect liquid levels are: 1) following the position of a float; 2) weighing a fixed buoy on a heavy cylinder; 3) measuring static pressure difference between two points, one in the vapor above the liquid, another under the liquid surface.

pH. The acidity or basicity of a process fluid may be measured by a specially designed electrode for measuring hydrogen ion concentration. The solid accumulation on the surface of the measuring electrode may cause an increase of time delay [Shinskey 1988]. Continuous cleaning with a device, e.g., an ultrasonic device, may be required to eliminate this variability.

Chemical composition. Many measurements of chemical composition are difficult and expensive to obtain. Measurements with gas chromatography and various types of spectroscopic analyzers fall in this category. The dominant dynamic feature of composition analyzers is the large time delay, which will result in ineffective control.

Some typical measuring devices are listed in Table 7.1, and more detailed devices may be found in other references [De Silva 1989, Stephanopoulos 1984].

Time delays
Time delay (transportation lag, or dead time) is commonly encountered in

Table 7.1. Typical measuring devices.

Measured variable	Measuring devices	Comments
Temperature	Thermocouples	First or second order
	Filled bulb sensors	system with large time
	Resistance thermometers	constant
Pressure	Diaphragm	Second order system
Liquid or gas flow rate	Turbine flow meters	Fast dynamics $\quad f \propto \omega$
	Pressure drop sensors	$f = \alpha \sqrt{\Delta p}$
Liquid level	Float actuated devices	Transmitting to pressure
	Displacer devices	or
	Liquid head devices	float force to measure
pH	pH meters (measuring electrodes)	Mechanical or ultrasonic device to clean electrode to avoid large time delay
Chemical composition	Gas chromatography	Large time delay
	Spectroscopic analyzers	and expensive

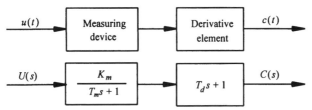

FIGURE 7.2.4.   Reduction of measurement lag with derivative action.

chemical process control systems. The use of some measuring devices, such as gas chromatography to measure concentration in gas stream samples, will introduce time delays.   The installation of measuring devices in different locations in a plant will also give rise to time delays. Measuring devices should be installed at a location that minimizes the time delay.

The presence of time delays in a process limits the performance of a conventional control system, and may cause system instability. In the view of frequency response, the time delay contributes phase shift to the open loop transfer function, which will decrease the phase margin. In order to maintain the same phase margin as the system without time delay, the controller gain must be reduced. Consequently, the response of the closed loop system will be sluggish, compared with the system having the same phase margin without time delay.

The technique of time delay compensation [Seborg et al. 1989] can be used to improve the performance of a system with time delay.

Measurement lags
Measurement lag refers to the phase lag (phase shift) and gain decrease caused by the measuring device which has a transfer function with a significant time constant. As we have studied, a thermocouple is a first order or second order element, and may produce phase lag in a signal. The transfer function of a transducer may be viewed as a constant only under special conditions.

In general, measurement and transmission elements should be selected so that the time constants of the elements are less than one tenth of the largest process time constant. The less the measurement and transmission time constant is, the smaller the measurement error will be. The installation location may affect the measurement or transmission time constant; thus choosing the optimum locations to install measuring devices is important.

Derivative action may reduce the measurement lag. If a derivative element is connected in series with a measuring device as shown in Fig. 7.2.4, and the transfer function of the measuring device is $K_m /(T_m s + 1)$, the transfer function obtained by introducing derivative action $T_d s + 1$ will be

Table 7.2. Time delay and measurement lag.

	Time delay	Measurement lag
Transfer function	$e^{-\tau s}$	$\dfrac{K}{Ts+1}$
Source	Measurement instrument with time delay or phase lag and incorrect installation locations	
Method to reduce their effect on system behavior	1. Use of instrument with small time delay 2. Installing instrument at the best location 3. reduction of controller gain 4. Time delay compensation	1. Use of fast dynamic measuring device 2. Installation of instrument at the best location 3. Use of derivative action

$$\frac{C(s)}{U(s)} = \frac{K_m(T_d s + 1)}{T_m s + 1} \tag{7.2.3}$$

If $T_d = T_m$ is chosen, then the pole and zero of (7.2.3) are canceled, and the output of the derivative element will be proportional to the measured variable. The dynamic measurement error vanishes.

Recalling *PID* controller design, we moved derivative action to the feedback path from the feedforward path to overcome derivative kick as shown in Fig. 3.4.4. Obviously, the movement of derivative action is also beneficial in reducing dynamic measurement error.

Time delays and measurement lags are summarized in Table 7.2.

7.3. FINAL CONTROL ELEMENTS

Final control elements, which enable processes to be manipulated, are important in control systems,. A final control element may be a speed drive of a pump, a screw conveyor, or a blower. In a chemical process control system, the most common final control element is a control valve.

Classification of control valves

Control valves may be driven electrically or pneumatically. Electrically driven control valves are driven by DC motors for continuous control systems, or stepping motors for digital control systems. Some are driven by electromagnetic mechanisms. In chemical process control systems, pneumatically driven control valves, called pneumatic control valves, are widely used, in which a diaphragm that senses the pneumatic signals moves a stem up or down to control the flow passing through an orifice.

Control valves may be classified by their actions in failure states. An air-to-close valve is shown in Fig. 7.3.1a, where the plug on the stem will close the orifice to stop the fluid flow when the air pressure input signal above the diaphragm increases. If the air pressure input signal is lost, the plug will open the orifice. These kinds of valves are also called fail-open valves. Correspondingly, the pneumatic control valves with opposite actions are called air-to-open or fail-closed valves. The plug on the stem will stop the fluid flow when the air pressure input signal is lost. An air-to-open valve is shown in Fig. 7.3.1b.

Flow rates of valves
For a nonflashing liquid, the flow rate through a valve may be obtained from the following equation:

$$F = Cf(l)\sqrt{\Delta p/\rho} \tag{7.3.1}$$

where F is the flow rate of the fluid, C is the valve size coefficient, $f(l)$ is the valve flow characteristic curve, l is the stem position, Δp is the pressure drop over the valve, and ρ is the specific gravity of the flowing liquid (relative to water).

The control engineer typically desires a control valve with a predictable (often linear) response of flow rate to stem position. The flow rate response in this case is governed by $f(l)$.

Characteristics of control valves
The flow characteristic curve $f(l)$ may be nondimensionalized as

$$F / F_{max} = f(l / L) \tag{7.3.2}$$

where F_{max} is the maximum flow rate, F/F_{max} is the relative flow rate (i.e., the ratio of the flow rate to the maximum flow rate), L is the maximum stem position, and l/L is the ratio of stem position to maximum stem position.

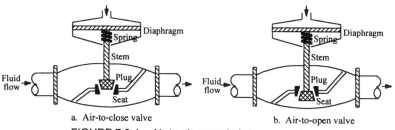

a. Air-to-close valve   b. Air-to-open valve

FIGURE 7.3.1. Air-to-close and air-to-open valves.

The valve flow characteristic curve $f(l/L)$ depends on the geometrical shape of the plug and seat surfaces of the valve. There are several valve flow characteristic curves commonly encountered in process control systems. These correspond to linear, equal percentage, and quick opening valves.

The linear trim valve has a flow characteristic curve of

$$\frac{d(F/F_{max})}{d(l/L)} = C \qquad (7.3.3)$$

Integrating (7.3.3) yields

$$F/F_{max} = C(l/L) + C_0 \qquad (7.3.4)$$

where C and C_0 are constants. Assuming that

$$F = \begin{cases} F_{min} & \text{when} \quad l = 0 \\ F_{max} & \text{when} \quad l = L \end{cases} \qquad (7.3.5)$$

then $C_0 = F_{min}/F_{max}$ and $C = 1 - F_{min}/F_{max}$ are obtained. It follows that

$$F/F_{max} = \frac{1}{R}[1 + (R-1)\frac{l}{L}] \qquad (7.3.6)$$

where $R = F_{max}/F_{min}$ is an adjustable range, and is usually about 20 to 50.

The flow characteristic curve of an equal percentage trim valve is

$$\frac{d(F/F_{max})}{d(l/L)} = C(F/F_{max}) \qquad (7.3.7)$$

which indicates that the slope of the F/F_{max} versus l/L curve (or F versus l curve) is a constant fraction of F/F_{max} (or F), leading to an equal percentage change in flow for a specified change of l anywhere in the range. With the same method and under the same conditions as that for a linear trim valve, we may obtain

$$F/F_{max} = R^{(l/L-1)} \qquad (7.3.8)$$

Similarly, for a quick trim valve, also called a square root valve, the flow characteristic curve is

$$\frac{d(F/F_{max})}{d(l/L)} = C(F/F_{max})^{-1} \qquad (7.3.9)$$

and the flow equation under the condition of (7.3.5) is

$$F/F_{max} = \frac{1}{R}[1 + (R^2 - 1)\frac{l}{L}]^{1/2} \qquad (7.3.10)$$

Table 7.3. Flow characteristic curves of valves.

Name	Characteristic formula		Feature
	Derivative form	Algebraic form	
Linear trim valve	$\dfrac{d(F/F_{max})}{d(l/L)} = C$	$\dfrac{F}{F_{max}} = \dfrac{1}{R}[1+(R-1)\dfrac{l}{L}]$	Gain is constant.
Equal percentage trim valve	$\dfrac{d(F/F_{max})}{d(l/L)} = C(\dfrac{F}{F_{max}})$	$F/F_{max} = R^{(l/L-1)}$	Gain $\uparrow$ while $l\uparrow$
Quick trim valve	$\dfrac{d(F/F_{max})}{d(l/L)} = C(\dfrac{F}{F_{max}})^{-1}$	$\dfrac{F}{F_{max}} = \dfrac{1}{R}[1+(R^2-1)\dfrac{l}{L}]^{1/2}$	Gain $\downarrow$ while $l\uparrow$

The flow characteristic curves of these three commonly used valves are summarized in Table 7.3, and are shown in Fig. 7.3.2 for $R = 30$.

Selection of valve
Different control trim valves are made to meet the requirement of keeping the steady state and dynamic characteristics of the control loop stationary for different control systems. The characteristics of a valve may compensate those of a process (an object) or a measurement. Complete compensation is not easy. Therefore, the only consideration for characteristic compensation is that the product of the gains of process (object), measuring device and valve remains constant as the flow rate changes between F_{max} and F_{min}, i.e., only steady state compensation is considered.

Very often, if the pressure drop over the control valve is fairly constant, then a linear trim valve is required. When the pressure drop over the control valve is not constant, equal-percentage valves are often used.

Another consideration for the selection of valves is the maximum flow

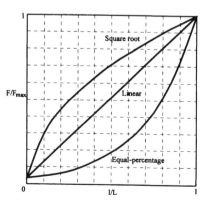

FIGURE 7.3.2. Flow characteristic curve of valves.

capacity, which should meet the production requirements. The maximum flow rate through a valve depends on the size of the valve.

In the view of control theory, the size of a valve should be chosen so that the most pressure drop in the loop is across the valve, thus allowing the process to be easily controlled by the valve. On the other hand, in the view of the most economical operating conditions, the valve should take the smallest possible pressure drop, because a small pressure drop means that a lower-head pump can be used, resulting in lower energy costs, i.e., less power consumed by the driving motor.

Sizing of control valves is a controversial subject in process control. A compromise is to size a valve to take 1/4 to 1/3 of the total pressure drop in the loop at design flow rate [Seborg et al. 1989]. Another suggestion for sizing a control valve is that the designing pressure drop over the control valve be 50 percent of the total pressure drop in the loop [Luyben 1990].

The selection of valves between "air-to-close" and "air-to-open" depends on safety considerations, more precisely, on which way the valve should operate for a failure. For example, an air-to-open valve should be used for steam pressure in a reactor heating coil. This selection guarantees that overheat will not be brought about by a transmitter failure. An overheat situation may be more serious than operation at a low temperature.

The valve selection principles are summarized in Table 7.4.

7.4. CONTROLLER SELECTION AND TUNING

In Chapter 3, the characteristics and principal functions of the conventional controllers are discussed in detail. The most popularly used conventional controllers are on-off, *P*, *PI* and *PID* controllers.

Table 7.4. Valve selection.

Selection item	Selection consideration	Selected content
Action	Safety of the system	Air-to-close (fail-open) Air-to-open (fail-closed)
Maximum flow capacity	Production capability	Matching the maximum flow rate of valve to maximum production capability
Flow characteristic	Stationary gain of the loop	Linear trim valve Equal-percentage trim valve Square root trim valve
Size	1. Manipulating ability 2. Economical operating conditions	Taking 1/4 to 1/3 pressure in the loop

Review of effects of conventional controllers on system

The effects of *P*, *I* and *D* control actions on a system response may be summarized as follows:

The increase of a proportional (*P*) control action may accelerate the system response, but an offset will persist in the system (steady state error ≠ 0).

Integral (*I*) control action can eliminate any offset in the system (steady state error = 0), but may increase overshoot and bring about a long oscillatory system response.

Derivative (*D*) control action may increase the stabilizing effect on the system response, and make the response less oscillatory.

Theoretically, *PID* controllers should be the best among conventional controllers. In practice, however, care should be taken when incorporating a derivative control action, especially in a noisy environment. The derivative action may amplify noise, and turn a *PID* controller into a Proportional Integral Disaster. Since a *PID* controller has three parameters to adjust in order to compromise various effects, the controller tuning is difficult.

Choosing a suitable controller

There is a compromise between the system response and tuning problem. The following guidelines are suggested:

1. The *P* controller, the simplest one, should always be considered first. If the offset of the process is tolerable, or the integral action is already included in the process, i.e., there is a term of 1/*s* in the process transfer function, then a *P* controller is likely the most suitable device.

2. The *PI* controller may be considered if a *P* controller is not suitable. The *PI* controller can eliminate offset, but the integral action makes the system more sensitive, and may lead to instability. If the offset must be eliminated, the *PI* controller may be a good choice.

3. The *PID* controller should be considered for a high quality control requirement, if a *PI* controller is not satisfactory. The introduction of derivative control action brings a stabilizing effect to the system, while the presence of integral control action makes the system more oscillatory.

Selection of conventional controllers for various processes

Different kinds of processes have different characteristics, and need different kinds of controllers.

Flow loops. A flow control system is shown in Fig. 7.4.1. The turbine flow meter is used to measure the flow rate. Both the flow meter and valve have fast dynamic characteristics, and a *PI* controller with small integral action and low

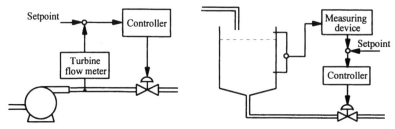

FIGURE 7.4.1.   A flow loop.   FIGURE 7.4.2.   A liquid level loop.

gain is satisfactory for eliminating offset, reducing the effect of the noisy flow signals and maintaining a better response.

Liquid level loops. A liquid level in an industrial process, as shown in Fig. 7.4.2, indicates the liquid storage in a tank. Thus, it is satisfactory to keep the level within a certain range around a desired setpoint. The real control objective is not to keep the liquid level exactly at a setpoint, but to prevent abrupt changes of flow rate at the outlet. A *P* controller is often used to give a smooth change in flow rate. If a *PI* controller is used rather than a *P* controller, then the controller may control the liquid level exactly at the setpoint. This means that the flow rate at the outlet may be controlled such that it is always equal to that at the inlet. This situation is useless for maintaining a smooth flow rate at the outlet. If the control objective is to make the flow rate at the outlet equal to that at the inlet, then connecting the inlet and outlet must be better than, or at least the same as, the perfect liquid level control system. Therefore, a *PI* controller is not a good choice for maintaining a smooth flow rate at the outlet. Of course, if the control objective is to keep the liquid level at a setpoint, the *PI* controller may be used.

Pressure loops. Pressure loops are very different from other loops; therefore different kinds of controllers may be chosen for different pressure loops. However, most of them have a very fast response, almost like a flow rate control. In this case, *PI* controllers are normally used with a small amount of integral control action, i.e., τ_I is large.

Fig. 7.4.3 is the case of fast response, where the manipulated vapor flow rate affects the vapor pressure directly and quickly. A *PI* controller is satisfactory.

A different case is shown in Fig. 7.4.4, where the vapor pressure is affected by the flow of cooling water for condensation, and the heat transfer process has a slow response. A *PID* controller should be selected rather than a *PI* controller, which may slow the system response.

Temperature loops. Because of measurement lags and process heat transfer lags, the response of a temperature loop is always slow. Therefore, a *PID* controller is a common choice.

Conventional controller selection is summarized in Table 7.5.

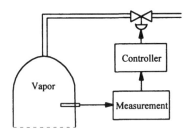

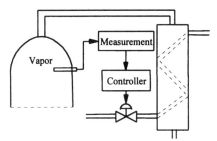

FIGURE 7.4.3. Fast response
pressure loop.

FIGURE 7.4.4. Slow response pressure
loop.

Controller tuning
The controller tuning problem involves deciding what values should be taken for the parameters of the selected conventional controllers. There are two ways to perform controller tuning: theoretical and semi-empirical. As we discussed in previous chapters, the parameters of a *PID* controller may be determined by time domain or frequency domain design methods if the transfer function of the process is given. Some semi-empirical methods are popularly used in practice, since the process transfer function may not be exactly known.

Trial and error tuning method
One of the most popular semi-empirical controller tuning methods is the trial and error method, which may be described as follows [Seborg et al. 1989, Luyben 1990]:

TABLE 7.5. Controller selection.

Controller	Suitable process	Process characteristics			Control quality
		Time delay	Disturbance	Open loop response	
On-off	Temperature Level height	Small	Very small	Sluggish	Low
P	Level height pressure	Small	Small	Sluggish	Medium
PI	Level height Pressure Temperature Flow rate	Small or medium	May be large	Sluggish with offset	Without offset
PD	Pressure Temperature	Small or medium	Small	Fast with overshoot	Less with overshoot
PID	Level height Pressure Temperature	Small or large	Small or large	Sluggish or fast	High

1. Take all integral and derivative actions out of the controller, i.e., set the integral time τ_I at its maximum value and the derivative time τ_D at its minimum value.

2. Set the proportional gain K_c at a low value (then the system response is sluggish), and put the controller on automatic.

3. Increase the proportional gain in small increments to find a proportional gain K_{cu}, called an ultimate gain, at which the system response is a sustained oscillation with constant amplitude.

4. Set the proportional gain K_c at $0.5K_{cu}$. Of course, the sustained oscillation disappears.

5. Decrease the integral time in small increments to find a critical integral time τ_I at which the sustained oscillation returns.

6. Set the integral time τ_I equal to 3 times the critical integral time. The oscillation should disappear.

7. A critical derivative time may be found by increasing derivative time in small increments until a sustained oscillation occurs again.

8. Set the derivative time τ_D to one-third of the critical value. The parameters K_c, τ_I and τ_D are now tuned.

Remarks

1. The reason for a small increment or decrement each time is to keep the process unsaturated. When the proportional gain is greater than the ultimate gain, the system response will be theoretically unbounded. In practice, sustained oscillation appears in the system because of saturation. Thus the ultimate gain is the minimum proportional gain at which the system response is a sustained oscillation.

2. Since disturbances always exist, it is not necessary to make a setpoint change to evaluate whether or not the response is a sustained oscillation; any disturbance will result in a sustained oscillating response if the proportional gain is greater than the ultimate one.

3. The steps of the trial and error tuning method for a *P* controller or *PI* controller are the same as that for a *PID* controller except for deletion of the steps that deal with the absent control actions.

4. This method may be detrimental to the safety of the tuned system.

Process reaction curve method
Another popular semi-empirical controller tuning method is the process reaction curve method.

Most physical processes encountered in a chemical plant are first order processes, or higher order overdamped ones. The oscillatory underdamped response may be found after the addition of a feedback controller.

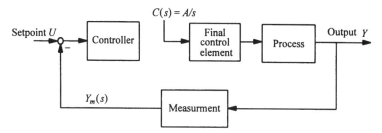

FIGURE 7.4.5. A system for getting process reaction curve.

If we cut off the connection between controller and final control element and introduce a step input to the final control element, as shown in Fig. 7.4.5, the response at measurement output is usually overdamped. This response is referred to as the process reaction curve, and may be approximated by a first order system with a dead time, as shown in Fig. 7.4.6. As we have already studied in Chapter 3, the approximated transfer function may be written as

$$G(s) = \frac{Ke^{-\tau s}}{Ts + 1} \qquad (7.4.1)$$

where K and T are the gain and time constant of the first order system, respectively, and τ is the dead time. The parameters K and T may be calculated as follows:

$$K = B/A \qquad (7.4.2)$$

$$T = B/S \qquad (7.4.3)$$

where A is the amplitude of the step input, B the amplitude of the measurement output, and S the slope of the process reaction curve, as shown in Fig. 7.4.6. The parameter τ may be measured directly from the process reaction curve.

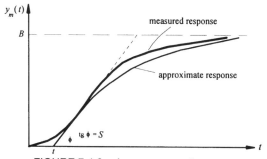

FIGURE 7.4.6. A process reaction curve.

Table 7.6. Cohen and Coon controller design algorithms.

Controller	Parameter	Formula
P	K_c	$\dfrac{1}{K}\dfrac{T}{\tau}[1+\dfrac{\tau}{3T}]$
PI	K_c	$\dfrac{1}{K}\dfrac{T}{\tau}[0.9+\dfrac{\tau}{12T}]$
	τ_I	$\dfrac{\tau[30+3\tau/T]}{9+20\tau/T}$
PID	K_c	$\dfrac{1}{K}\dfrac{T}{\tau}[1.25+\dfrac{\tau}{4T}]$
	τ_I	$\dfrac{\tau[32+6\tau/T]}{13+8\tau/T}$
	τ_D	$\dfrac{4\tau}{11+2\tau/T}$

Cohen and Coon [Cohen and Coon 1953] reported the design parameters of the conventional controller with the parameters K, T and τ based on the closed loop response with a decay ratio of 1/4. Their design algorithms are shown in Table 7.6.

Example 7.4.1

The measured output response of a temperature control open loop system to a unit step input acting on the final control element, i.e., the process reaction curve, is shown in Fig. 7.4.6. From this figure, we find that $B = 2$, $\phi = 45°$ and $\tau = 0.3$. With the Cohen and Coon method, a *PID* controller may be designed as follows:

From (7.4.2) and (7.4.3), it can be seen that

$$K = B/A = 2 \qquad T = B/\tan45° = 2$$

The approximated open loop transfer function of the system (7.4.1) is

$$G(s) = \frac{Ke^{-\tau s}}{Ts+1} = \frac{2e^{-0.3s}}{2s+1}$$

The parameters of the *PID* controller may then be determined as follows:

$$K_c = \frac{1}{K}\frac{T}{\tau}[1.25+\frac{\tau}{4T}] = \frac{1}{2}\frac{2}{0.3}[1.25+\frac{0.3}{8}] = 4.29$$

$$\tau_I = \frac{\tau[32+6\tau/T]}{13+8\tau/T} = \frac{0.3[32+6\cdot0.3/2]}{13+8\cdot0.3/2} = 0.695$$

and

$$\tau_D = \frac{4\tau}{11+2\tau/T} = \frac{4 \cdot 0.3}{11+2 \cdot 0.3/2} = 0.106$$

Tuning a controller in the frequency domain with *PCET*

As we studied in Chapter 5, the trial and error method can be used in system simulation using *PCET* to tune a *PID* controller. The procedure for determining the controller parameters has been shown in Table 5.16. With *PCET*, we can obtain the Bode diagram of the open loop system. By adjusting the controller's parameters to obtain a satisfactory Bode diagram with suitable phase margin and gain margin, we can set the values of the controller parameters.

The 8 steps of the trial and error tuning method mentioned earlier in this section can be used to find a suitable controller. With this method, the proportional gain, reset rate (integral time) and derivative time are adjusted one after another. The procedure is as follows:

1. After doing the steps in Table 5.5, we obtain the Bode diagram of the system with the controller parameters $K_c = 1$, $1/\tau_I = 0$ and $\tau_D = 0$.

2. Repeat Steps 3 and 4 in Table 5.16 to find the ultimate gain K_{cu} to make phase margin and gain margin equal to zero (or close to zero).

3. Take the proportional gain K_c as $0.5K_{cu}$.

4. Repeat Steps 3 and 4 in Table 5.16 to find the critical reset rate $1/\tau_I$ at which the phase margin and gain margin are zero.

5. Take the reset rate $1/\tau_I$ equal to one third of the critical value.

6. A critical derivative time may be found by repeating Steps 3 and 4 in Table 5.16.

7. Set the derivative time τ_D at one-third of the critical value.

8. If necessary, small adjustment of parameters can be made to get better system performance.

7.5. SELECTION OF A CONTROLLER'S ACTION DIRECTION

When designing a controller, aside from choosing suitable control laws, we need to choose a controller's action direction, which reflects the sign of the controller and is related to the action direction of the process and that of the final control element. The traditional method of determining the controller's action direction relies on experiments and experience, which are time consuming and error-prone. Thus a new design criterion has been developed [Rao, Jiang and Tsai 1988].

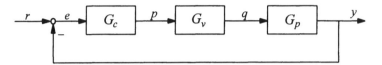

FIGURE 7.5.1. Block diagram for a single loop control system.

Definition of Action Direction Function

Consider the simple (single loop) feedback control system in Fig. 7.5.1. In this block diagram, the three blocks represent the controlled process, G_p, the final control element G_v and the controller G_c, respectively.

In general, the action direction function for a block is defined in terms of the input-output relation of this block. When the output of the block increases as its input becomes larger, the action direction function for this block is assigned as "positive," and "negative" for the opposite case.

For a controller, the action direction function, R_c, is defined as "positive" if the output signal p of the controller increases as the error signal e becomes larger, and vice versa. For an air-to-open valve, when the input signal fails, the output orifice of the valve will be closed. In other words, when input signal increases, the output orifice will increase. This means that the action direction function R_v for an air-to-open valve is "positive." Similarly, the action direction function for an air-to-close valve is "negative." The action direction function for the process block, R_p, is defined as "positive" if the process output y increases when the manipulated variable becomes larger, and vice versa. In the case of positive R_p, the action direction of the process is referred to as "positive."

The action direction function for a loop, R_l, is defined as "positive" if the process output y increases as the setpoint r increases. It can be shown that the following heuristic relation exists:

$$R_L = R_c * R_v * R_p \qquad (7.5.1)$$

where the sign product $R_1 * R_2$ of two action direction functions with identical signs is defined as "positive"; otherwise, it is "negative."

Design Criterion

Since we always need the output to increase (decrease) while the input increases (decreases), the action direction function of the controller can be decided by the following criterion.

For a control system as shown in Fig. 7.5.1, the action direction functions of the controllers should be so chosen that the action direction function for the loop is always positive. That is,

$$R_L = R_c * R_v * R_p = \text{"positive"} \qquad (7.5.2)$$

Table 7.7. Selection of the action direction function of controller.

Action direction function of process	Action direction function of valve	Action direction function of controller
Positive	Positive (air-to-open)	Positive
Positive	Negative (air-to-close)	Negative
Negative	Positive (air-to-open)	Negative
Negative	Negative (air-to-close)	Positive

According to the criterion, Table 7.7 can be obtained, which shows correct action directions of the controller. In industry, the terminology of "direct" and "inverse" is used for "positive" and "negative."

Example 7.5.1
    If a process has a "positive" action function and an air-to-close valve is used, the action direction function of the controller should be "negative", since

$$\text{"positive"} = R_L = R_c * R_v * R_p = R_c * \text{"negative"} * \text{"positive"}$$

requires R_c to be "negative."

Principle of Equivalence
For a multiloop control system as described by Fig. 7.5.2, the correct action direction of all controllers can be selected in the same way as a single loop control system by the principle of equivalence.
    As we can see, the inner-most loop is a single loop system, similar to that of Fig. 7.5.1. In this loop, the action direction of the controller can be decided by the above criterion. The inner-most loop can be treated as a valve block with a positive action direction function in the second inner-most loop. Therefore, the action direction function of the controller in the second inner most loop can be determined by the above criterion.   In the same way, the action direction functions of all controllers in the system can be determined. Based on the above methodology, we can develop the principle of equivalence as follows:
    Choosing the direction of the controller's action in a loop of a multiloop control system is equivalent to that for a single loop system if we treat the inner loop as a "valve block" with a "positive" action direction function.

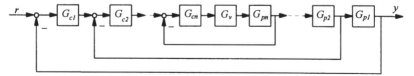

FIGURE 7.5.2. Block diagram for a multiloop control system.

Problems

7.1. When designing a two-tank system as shown in Fig. 2.P.3 (in Chapter 2), if only the heights of liquid level in two tanks should be maintained, then determine

a. the system model (differential equations),

b. the number of degrees of freedom for the system, and

c. the controlled variables and manipulated variables.

7.2. What kind of valves (air-to-open valve or air-to-close valve) should be used for the following manipulated variables? Give the reasons.

a. Steam pressure in a reactor heating coil.

b. Flow of cooling water to a distillation condenser.

c. Flow of effluent from a waste water treatment holding tank into a river.

7.3. A process reaction curve to an input $1.5u(t)$ acting on the final control system is shown in Fig. 7.4.6, where $u(t)$ is a unit step signal, and it can be read that $B = 3$, $\phi = 55°$ and $\tau = 0.2$. If a *PID* controller is selected, determine the parameters of the controller with the Cohen and Coon method.

7.4. (*P* Controller design and tuning for a first order system)   A first order system with a time delay has an open loop transfer function

$$G(s) = \frac{e^{-s}}{5.2s+1}$$

a. Design a proportional (*P*) controller with the Cohen and Coon method.

b. With *PCET*, plot the closed loop response of the system to a unit step input using the controller you designed in (a.).   Estimate the overshoot, offset and settling time from the response.

c. Design a proportional (*P*) controller by means of a trial and error method starting with the initial value of K_c you obtained above to make the closed loop system response satisfy the following performance requirements:

$$\text{overshoot} < 0.3$$
$$\text{settling time} < 8 \text{ sec}$$
$$\text{offset} < 0.25$$

Give the parameters of the designed controller and plot the closed loop response curve of the system with this controller to a unit step input.

7.5. (*PI* Controller design and tuning for a first order system) For the same process used in Problem 7.4, do the following:

a. Design a proportional-integral (*PI*) controller with the Cohen and Coon method.

b. With *PCET*, plot the closed loop response curve of the system to a unit step input using the controller you designed in (a.). Estimate the overshoot, offset and settling time from the system response curve.

c. Design a proportional-integral (*PI*) controller by means of a trial and error method with the initial values of the controller parameters you obtained above to make the closed loop system response satisfy the following performance requirements:

overshoot < 0.3

settling time < 10 sec

Give the parameters of the designed controller and plot the closed loop response curve of the system with this controller to a unit step input.

d. Compare the results of Problems 7.4 and 7.5. What is the effect of integral action on the response?

7.6. (*PID* Controller design and tuning for a first order system) For the same process used in Problem 7.4, with *PCET* do the following:

a. Design a proportional-integral-derivative (*PID*) controller with the Cohen and Coon method.

b. Observe and plot the closed loop response curve of the system to a unit step input using the controller you designed in (a.). Estimate the overshoot, decay ratio, offset and settling time from the system response curve.

c. Design a proportional-integral-derivative (*PID*) controller by means of a trial and error method with the initial values of the controller parameters you obtained above to make the closed loop system response satisfy the following performance requirements:

overshoot < 0.15

settling time < 5 sec

Give the parameters of this controller and plot the closed loop response curve of the system with the designed controller to a unit step input.

d. Compare the results of Problems 7.4, 7.5 and 7.6. What conclusion can you make?

7.7. (*PID* Controller design and tuning for a higher order system) For a unity feedback control system with the process transfer function

$$G(s) = \frac{10}{(s+1)(s+2)(s+10)}$$

with *PCET* do the following:

a. Plot the process response to a unit step, and from this approximate the process model as a first order system with time delay.

b. Plot and compare the process responses of the original model and the approximated model on the same screen. If the approximation are not good enough, find a better first order model with the trial and error method.

c. Design a proportional-integral-derivative (*PID*) controller for the system with the approximated model and the Cohen and Coon method.

d. Observe and plot the closed loop response curve of the approximated system to a unit step input using the controller you designed above. Estimate the overshoot, offset and settling time from the response curve.

e. Design a proportional-integral-derivative (*PID*) controller for the system with the approximated model by means of a trial and error method with the initial values of the controller parameters you have already calculated to make the closed loop system response satisfy the following performance requirements:

$$\text{overshoot} < 0.15$$
$$\text{settling time} < 4 \text{ sec}$$

Give the parameters of the designed controller and plot the closed loop response curve of the system with this controller to a unit step input.

f. Compare the responses of step c and e, and give the conclusion.

7.8 With the Frequency Domain Analysis (*FDA*) package of *PCET* and the trial and error method, find a *PID* controller for the system with the process transfer function

$$G(s) = \frac{1000}{(s+1)(s+2)(s+10)(s+100)}$$

Find the approximated model first. The performance requirements are

$$\text{Phase margin} = 30° \sim 45°$$
$$\text{Gain margin} = 6 \sim 10 \text{ dB}$$

(If the autoscaling introduces overflow, use the manual scaling with small ranges.)

7.9. For a three loop control system as shown in Fig. 7.5.2, if the action direction functions of process 1 and 3 are "positive", and that of process 2 is "negative," and the valve used in loop 3 is an air-to-close valve, determine the action direction functions of all three controllers.

CHAPTER 8

ADVANCED CONTROL SYSTEMS

8.1. CASCADE CONTROL

Advanced control strategies are often employed in process control systems. In order to eliminate the effect of disturbances and improve control system's dynamic performance, a common control configuration is taken with two controllers. When the output of one controller is used to manipulate the setpoint of another controller, then we say the two controllers are cascaded, and the system is referred to as a cascade control system.

An example of a cascade control system

Consider an oxidation oven as shown in Fig. 8.1.1a. The chemical reaction of ammonia and oxygen occurs in the oven. The reaction can be represented as

$$4NH_3 + 5O_2 \rightarrow 4NO + 6H_2O + Q \tag{8.1.1}$$

where Q represents the heat released during the reaction. The temperature in the oven is supposed to be controlled at $840 \pm 5°C$.

A simple feedback control configuration is shown in Fig. 8.1.1a, where the temperature in the oven is measured and the flow rate of ammonia is manipulated. Because of existing disturbances and the large time constant of the process, the temperature in the oven is difficult to control at $840 \pm 5°C$ with this simple feedback control configuration. The block diagram of this simple control configuration is shown in Fig. 8.1.1b; for later comparison the process is divided into two parts, one for the flow process and the other for the temperature process.

The main disturbance in the system acts on the flow process, and affects the ammonia flow rate directly. Because of the large time constant of the temperature process, the temperature in the oxidation oven will not be affected for a considerable time. Only after the temperature is affected can the controller finally reduce the effect of the disturbance on the system. The slow control

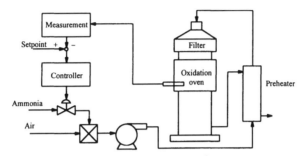

a. An oxidation oven temperature control system

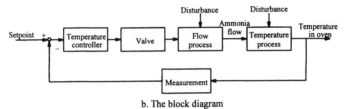

b. The block diagram

FIGURE 8.1.1.   A simple temperature control system.

action is the main reason for poor control performance. If the ammonia flow rate can be controlled, i.e., the effect of the disturbance on the ammonia flow rate can be reduced, the system performance will be improved. A new controller for reducing the effect of this disturbance before it affects the oven temperature is incorporated to construct an inner feedback control loop. This cascade control configuration is shown in Fig. 8.1.2, where the temperature controller is used to manipulate the setpoint of the ammonia flow rate controller.

Note that there are two control loops with two different measurements and a

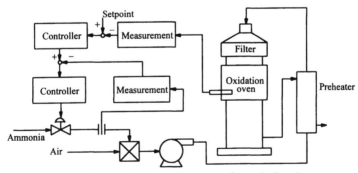

FIGURE 8.1.2.   A temperature cascade control system.

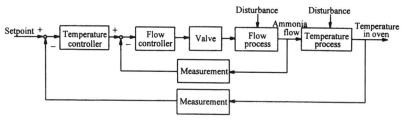

FIGURE 8.1.3. The block diagram of a temperature cascade control system.

common manipulated variable. The block diagram of the cascade system is shown in Fig. 8.1.3.

TABLE 8.1. Terminology of cascade control systems.

Terminology	Definition	Corresponding part in example
Primary (controlled) variable	The variable we want to control.	Temperature in the oven
Secondary (controlled) variable	The variable we are not interested in controlling, but which affects the primary controlled variable.	Ammonia flow rate
Primary process	The process that has the primary variable as its output.	Temperature process
Secondary process	The process that has the secondary controlled variable as its output.	Ammonia flow process
Primary controller (Master controller)	The controller that has the difference between setpoint and the measurement of primary variable as an input, and whose output is the reference input of another controller.	Temperature controller
Secondary controller (slave controller)	The controller that has the output of primary controller as the reference input, and has the difference between the reference input and measurement of secondary variable as its input.	Ammonia flow controller
Primary loop	The loop composed of a primary controller, a secondary loop, a primary process and the measuring device of a primary variable.	Temperature control loop
Secondary loop	The loop containing of a secondary controller, a final control element, a secondary process and the measuring device of a secondary variable.	Ammonia flow control loop

Terminology in a cascade control system
It is convenient to introduce terminology before proceeding further. The terms primary controlled variable (primary variable), secondary controlled variable (secondary variable), primary process, secondary process, primary controller (master controller), secondary controller (slave controller), primary loop and secondary loop are defined in Table 8.1.

Block diagram of a cascade system
The block diagram of a general cascade system with the terminology above is shown in Fig. 8.1.4, which may be redrawn with corresponding transfer

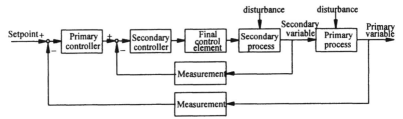

FIGURE 8.1.4.   A block diagram of a general cascade control system.

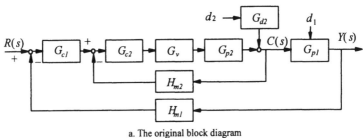

a. The original block diagram

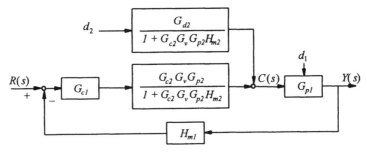

b. Block diagram transformation

FIGURE 8.1.5.   Block diagram transformation for a cascade system.

functions as shown in Fig. 8.1.5a, where the subscript 1 stands for the primary loop, 2 for the secondary loop, p for process, v for valve, c for controller, m for measurement and d for disturbance. For example, $G_{d2}(s)$ is the disturbance transfer function in the secondary loop.

By using the block diagram transformation techniques discussed in Chapter 2, Fig. 8.1.5b can be obtained. The transfer function of the secondary loop is

$$G_2(s) = \frac{G_{c2}(s)G_v(s)G_{p2}(s)}{1 + G_{c2}(s)G_v(s)G_{p2}(s)H_{m2}(s)} \tag{8.1.2}$$

and the equivalent disturbance transfer function for the secondary loop is

$$G_3(s) = \frac{G_{d2}(s)}{1 + G_{c2}(s)G_v(s)G_{p2}(s)H_{m2}(s)} \tag{8.1.3}$$

Comparing Fig. 8.1.4 with Fig. 8.1.1b, we find that the diagram in Fig. 8.1.5 is valid for a simple feedback system if $H_{m2}(s) = 0$ and $G_{c2}(s) = 1$. Thus, the transfer functions of $G_2(s)$ and $G_3(s)$ are reduced to $G_v(s)G_{p2}(s)$ and $G_{d2}(s)$ for the corresponding simple feedback system, respectively.

Advantage of a cascade control system
The principal advantages of cascade control are as follows:

1. Disturbances occurring in the secondary loop are reduced by the secondary controller before they can influence the primary variable. For the example above, the unexpected change in ammonia flow rate is measured and fed back to the secondary controller, and thus reduced by the secondary controller before it affects the oven temperature.

As mentioned above, the equivalent disturbance transfer function for the secondary loop of a cascade control system is (8.1.3), which is $1/(1 + G_{c2}G_vG_{p2}H_{m2})$ times the corresponding transfer function of a simple feedback control system. In other words, the offset of the secondary variable of a cascade control system caused by the disturbance will be $1/(1 + K_{c2}K_vK_{p2}K_{m2})$ times that of the corresponding simple feedback control system, where K_{c2}, K_v, K_{p2} and K_{m2} are the static gains of the transfer functions G_{c2}, G_v, G_{p2} and H_{m2}, respectively.

2. The dynamic characteristics of the secondary process are improved by incorporating a secondary loop. If the secondary process is approximated by a first order system as

$$G_{p2}(s) = \frac{K_{p2}}{T_{p2}s + 1} \tag{8.1.4}$$

and other elements in the secondary loop are taken as

$$G_{c2}(s) = K_{c2} \qquad G_v(S) = K_v \qquad H_{m2}(s) = K_{m2} \tag{8.1.5}$$

then, the transfer function of the secondary loop is

$$G_2(s) = \frac{G_{c2}(s)G_v(s)G_{p2}(s)}{1 + G_{c2}(s)G_v(s)G_{p2}(s)H_{m2}(s)} = \frac{K_{c2}K_v \dfrac{K_{p2}}{T_{p2}s + 1}}{1 + K_{c2}K_v \dfrac{K_{p2}}{T_{p2}s + 1} K_{m2}}$$

$$= \frac{K_{c2}K_vK_{p2}}{T_{p2}s + K_{c2}K_vK_{p2}K_{m2} + 1} = \frac{\hat{K}}{\hat{T}s + 1} \tag{8.1.6}$$

where

$$\hat{K} = \frac{K_{c2}K_vK_{p2}}{1 + K_{c2}K_vK_{p2}K_{m2}} \tag{8.1.7}$$

and

$$\hat{T} = \frac{T_{p2}}{1 + K_{c2}K_vK_{p2}K_{m2}} < T_{p2} \tag{8.1.8}$$

The time constant of the secondary loop is smaller than that of the corresponding simple feedback control system. Thus, the phase lag is reduced, and the response speed of the primary loop is improved.

3. The secondary controller reduces the effect of gain variations in the secondary loop on the system response. It may also be seen from (8.1.7) that if $K_{c2}K_vK_{p2}$ is large enough, the effect of any gain variation in K_{c2}, K_v or K_{p2} may be neglected. For example, if the gain of a secondary process changes to $K_{p2} + \Delta K_{p2}$, the gain variation of the closed secondary loop is

$$\Delta\hat{K} = \frac{K_{c2}K_v(K_{p2} + \Delta K_{p2})}{1 + K_{c2}K_v(K_{p2} + \Delta K_{p2})K_{m2}} - \frac{K_{c2}K_vK_{p2}}{1 + K_{c2}K_vK_{p2}K_{m2}}$$

$$= \frac{K_{c2}K_v\Delta K_{p2}}{[1 + K_{c2}K_v(K_{p2} + \Delta K_{p2})K_{m2}][1 + K_{c2}K_vK_{p2}K_{m2}]}$$

$$\cong \frac{K_{c2}K_v\Delta K_{p2}}{[1 + K_{c2}K_vK_{p2}K_{m2}]^2} \tag{8.1.9}$$

which may be neglected, because $K_{c2}K_vK_{p2}$ is assumed to be large enough.

The performance or parameter comparison between a cascade control system and a simple feedback control system may be found in Table 8.2.

TABLE 8.2. Performance comparison in secondary loops.

	Offset of secondary variable	Time constant of secondary loop	Closed loop gain variation
Simple feedback control system	$K_{d2}d$	T_{p2}	ΔK_{p2}
Cascade control system	$\dfrac{K_{d2}d}{1 + K_{c2}K_v K_{p2}K_{m2}}$	$\dfrac{T_{p2}}{1 + K_{c2}K_v K_{p2}K_{m2}}$	$\dfrac{\Delta K_{p2}}{[1 + K_{c2}K_v K_{p2}K_{m2}]^2}$
Effect of cascade control	Decrease	Decrease	Decrease

Example 8.1.1

For a cascade control system as shown in Fig. 8.1.5a, the transfer functions in the secondary loop are assumed to be

$$G_{p2}(s) = \frac{6}{12s+1} \qquad G_{d2}(s) = \frac{4}{2s+1}$$

$$G_v = 1 \qquad G_{c2} = 10 \qquad H_{m2} = 1$$

The closed loop transfer function of the secondary loop may be found as

$$G_2(s) = \frac{K}{Ts+1}$$

where

$$K = \frac{K_{c2}K_v K_{p2}}{1 + K_{c2}K_v K_{p2}K_{m2}} = \frac{10 \times 1 \times 6}{1 + 10 \times 1 \times 6 \times 1} = \frac{60}{61} \cong 1$$

and

$$T = \frac{T_{p2}}{1 + K_{c2}K_v K_{p2}K_{m2}} = \frac{12}{1 + 10 \times 1 \times 6 \times 1} = \frac{12}{61} \cong 0.2$$

Comparing $G_2(s)$ with $G_{p2}(s)$, the dynamic characteristics of the secondary loop are improved. Furthermore, if the gain of the secondary process varies from 6 to 9, i.e., 50% increase, the closed loop static gain is

$$K = \frac{K_{c2}K_v K_{p2}}{1 + K_{c2}K_v K_{p2}K_{m2}} = \frac{10 \times 1 \times 9}{1 + 10 \times 1 \times 9 \times 1} = \frac{90}{91} \cong 1$$

It shows that the static gain of the closed secondary loop remains almost unchanged.

If the disturbance is a unit step function, then from (8.1.3), the offset of the secondary variable caused by the disturbance will be

$$c_{ss} = \lim_{s \to 0} sG_3(s)\frac{1}{s} = \frac{G_{d2}(0)}{1 + G_{c2}(0)G_v(0)G_{p2}(0)H_{m2}(0)}$$

$$= \frac{4}{1 + 10 \cdot 1 \cdot 6 \cdot 1} = \frac{4}{61} = 0.066$$

By contrast, the offset of the same variable for a simple feedback control system without the secondary loop is

$$c_{ss} = \lim_{s \to 0} sG_d(s)\frac{1}{s} = G_d(0) = 4$$

Principles of the selection of primary and secondary variables
The selection of primary variables is just the same as that of controlled variables discussed before.

The selection of the secondary variable is related to the advantages of cascade control. Namely the secondary loop should include the main disturbances in the system so as to reduce their effect. Also, the secondary loop should include those elements that have poor characteristics or have gain variations.

Selection of primary and secondary controller
It can be proved that adding an integral action to either of the two controllers can eliminate the offset (steady state error) caused by disturbances in the secondary loop. Usually, integral action is only added in the primary controller, since the proportional gain in the secondary loop is very large, and the offset of the secondary loop is small, as we observed in Example 8.1.1. An offset in the secondary loop is not important, since controlling the output of the secondary process is not our objective. Adding integral action in the secondary loop may bring about reset windup, which may degrade the secondary loop performance.

A secondary controller is usually a proportional controller, whereas a primary

TABLE 8.3. Cascade controllers and their function.

	Primary loop	Secondary loop
Controller	*PID*	*P* with large gain
Functions	1. Eliminate the system offset. 2. Speed up the system response 3. Reduce the effect of disturbance in both loops.	1. Reduce the effect of disturbance in the secondary loop. 2. Improve the dynamic response of the secondary loop. 3. Reduce the effect of gain variation in the secondary loop.

controller often adopts a *PID* structure. The integral control action of the primary controller will eliminate the offset caused by disturbances not only in the secondary loop, but also in the primary loop. The derivative control action is used to compensate the process phase lag and to speed up the system response.

The functions of primary and secondary controllers are summarized in Table 8.3.

Analysis of a cascade system with *PCET*

Subprogram *IAC* (Industrial Application Case) in *PCET* deals with industrial processes. A paper machine headbox process is typically incorporated. With this package, single loop control and cascade control systems can be analyzed.

The procedure for analyzing the cascade control systems is described in Table 8.4. The subprogram *TDA* is incorporated with *IAC*. The secondary loop response is obtained first. The secondary (closed) loop is automatically dealt with as an element block in the primary loop, when the primary loop response is drawn. When drawing the response, we can follow the same procedure used for closed loop analysis in *TDA* described before The input can be selected as a setpoint or disturbance by pressing "4" in Step 4 of Table 3.8. The default values are R(setpoint) = 1 and D(disturbance) = 0. When observing the responses to the disturbance in the secondary loop, the setpoint should be set to zero and disturbance can be set to 1 for secondary loop analysis, and both the setpoint and disturbance in the primary loop should be set to zero for primary loop analysis.

TABLE 8.4. Plotting response of a cascade control system with *PCET*.

Step	User input		Function
1	Type *PCET*		Start program
2	Select *IAC*		Start industrial application case Headbox control system introduction
3	Select F3	"F1-Exit"	Return to main menu
		"F2-Single"	Go into *TDA*
		"F3-Cascade"	Get headbox cascade control diagram
		"F4-Next"	Get the nest page of the introduction
4	Select F2	"F1-Exit"	Return to previous display
		"F2-Response"	Go to *TDA* to get secondary loop response
		"F3-Help"	Get help
		"F4-Block"	Display the Block diagram
5	As procedure in *TDA*		Get the secondary loop response
6	Press F1 to exit		Go to *TDA* to get primary loop response
7	As procedure in *TDA*		Get the primary loop response

8.2. RATIO CONTROL

In many chemical processes, the control objective is to maintain the ratio of two variables at a specified value, rather than the numerical values of the separate variables. Ratio control is used to achieve this objective.

The most popular applications of the ratio control include: holding a constant reflux ratio in a distillation column; keeping a fuel-air ratio to a furnace at an optimum value for efficient combustion; maintaining the ratio between a feeding flow rate and a stream rate in the reboiler of a distillation column and so on. Usually in the chemical industry, the variables whose ratio is controlled are flow rates.

Configuration of ratio control

A simple open loop ratio control system is shown in Fig. 8.2.1, which is the simplest ratio control configuration. One of the two flow rates, called the wild flow rate or wild variable, is measured and used to control the other flow rate, called the adjusted flow rate. Since no effort is taken to reduce the disturbances in the system, the only systems suitable for ratio control are those with low quality control requirements and few disturbances.

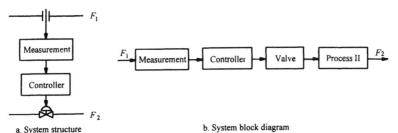

FIGURE 8.2.1.   A simple open loop ratio control system.

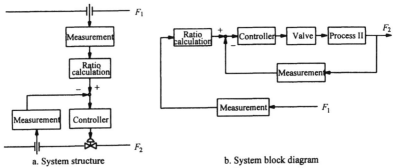

FIGURE 8.2.2.   A single loop ratio control system.

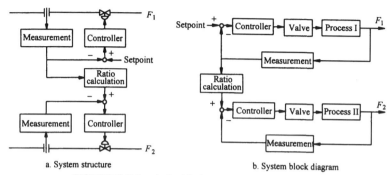

FIGURE 8.2.3. A double loop ratio control system.

A single loop ratio control system is shown in Fig. 8.2.2, where the adjusted flow rate control loop is incorporated. The wild flow rate is measured and fed to the ratio calculation device. The output of the ratio calculation device is proportional to the measured value by a specified ratio, and is the setpoint of the adjusted flow rate control loop.

If we want to maintain the ratio of two flow rates at a constant, while holding the values of the flow rates constant, a double loop ratio control configuration can be used as shown in Fig. 8.2.3. The wild flow rate and the adjusted flow rate are controlled by their own feedback control loops. Thus, the disturbances in the two loops are reduced by two controllers. The ratio of two flow rates is determined by the ratio calculation device that connects the two loops.

The single loop ratio control configuration is a typical setup, and one widely used in process industry.

Three different ratio control configurations are summarized in Table 8.5.

Ratio calculation devices

There are two basic ratio calculation devices: divider and multiplier (ratio station). They effect different ratio control schemes.

A single loop ratio control configuration with a divider element as a ratio

TABLE 8.5. Configuration of ratio control system.

Item	Adjustable variable	Configuration	Performance
Simple open loop ratio control	Adjusted flow	Open loop	Simple, but affected greatly by disturbances
Single loop ratio control	Adjusted flow	One closed loop	Disturbance to adjusted flow reduced
Double loop ratio control	Wild flow and adjusted flow	Two closed loops	Maintain not only the ratio but also the flow rates

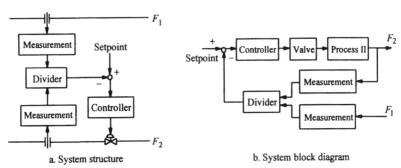

a. System structure   b. System block diagram

FIGURE 8.2.4.   A ratio control scheme with divider.

calculation device is shown in Fig. 8.2.4. The measured values of wild flow rate and adjusted flow rate are the inputs of the divider. The ratio of the two divider inputs is calculated and sent out as the output of the divider. The difference between the setpoint and the divider output is fed to the ratio controller, which usually is a *PI* controller for eliminating the steady state error. It is obvious that the setpoint of the system is the desired ratio.

With a divider as a ratio calculation device, the ratio of two flow rates may be read directly. However, this scheme has a serious disadvantage:

It is obvious from Fig. 8.2.4 that the divider is in an adjusted flow rate control loop. If two measuring devices have the same gain, then the ratio R is written as

$$R = F_2/F_1 \tag{8.2.1}$$

where F_1 and F_2 are the wild flow rate and adjusted flow rate, respectively. The variation rate of the gain of the divider to the adjusted flow rate F_2 is

$$K_d = (\frac{\partial R}{\partial F_2}) = 1/F_1 \tag{8.2.2}$$

which varies with the wild flow rate F_1. If the wild flow rate is small, a small change in its value may result in a big change in the divider gain.

Using the inverse of (8.2.1) as the desired ratio also poses a problem. With R now rewritten as

$$R = F_1/F_2 \tag{8.2.3}$$

the variation rate of the gain of the divider then will be

$$K_d = (\frac{\partial R}{\partial F_2}) = -F_1/F_2^2 = -R/F_2 \tag{8.2.4}$$

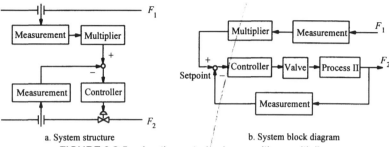

a. System structure   b. System block diagram
FIGURE 8.2.5.   A ratio control scheme with a multiplier.

which is a nonlinear function of F_2. A small change in the adjusted flow rate will result in not only a big change in the divider gain, but also a serious nonlinear characteristic.

Another possible scheme is one with a multiplier element as shown in Fig. 8.2.5, where the multiplier is used as a ratio calculation device. The measured value of wild flow rate is fed to the multiplier. The output of the multiplier is the setpoint of the adjusted flow rate feedback control loop. The ratio calculation device in this scheme is called a ratio station, where the measured wild flow rate is multiplied by a specified gain. With this scheme, the open loop gain remains constant. A multiplier is a widely used ratio calculation device.

Two kinds of ratio calculation devices are compared in Table 8.6.

Gains of ratio calculation devices

The desired ratio of two flow rates may be different from the gain of a ratio calculation device. For a single loop ratio control system with a ratio station as shown in Fig. 8.2.5, if the transfer functions of the controller, valve, process and measurement in the adjusted flow rate control loop are $G_{c2}(s)$, $G_{v2}(s)$, $G_{p2}(s)$ and $H_{m2}(s)$, respectively, the transfer function of the measurement in the wild flow rate control loop is $H_{m1}(s)$, and that of the ratio station is K, then the transfer function between the two flow rates may be found as

$$\frac{F_2(s)}{F_1(s)} = \frac{H_{m1}(s)KG_{p2}(s)G_{v2}(s)G_{c2}(s)}{1 + H_{m2}(s)G_{p2}(s)G_{v2}(s)G_{c2}(s)}$$

(8.2.5)

TABLE 8.6.   Basic schemes for ratio calculation.

Item	Input	Output	Location	Comment
Divider	Adjusted and wild flow rates	Input of controller	In the closed loop	1. Easy to read ratio 2. Nonlinear or varied gain
Ratio station	Wild flow rate	Setpoint of closed loop	Out of closed loop	Keeps the gain unchanged

Since the gain of the transfer function $G(s)$ is equal to $G(0)$, from (8.2.5), the desired ratio of the two flow rates, R, should be

$$R = \frac{F_2(0)}{F_1(0)} = \frac{H_{m1}(0)KG_{p2}(0)G_{v2}(0)G_{c2}(0)}{1 + H_{m2}(0)G_{p2}(0)G_{v2}(0)G_{c2}(0)} \qquad (8.2.6)$$

If $G_{c2}(0)G_{v2}(0)G_{p2}(0)H_{m2}(0) \gg 1$, (8.2.6) may be approximated as

$$K = \frac{H_{m2}(0)F_2(0)}{H_{m1}(0)F_1(0)} \qquad (8.2.7)$$

which indicates that the gain of the ratio station is the ratio of the measured values of the two flow rates rather than the two flow rates themselves.

In fact, this conclusion may be obtained directly from the block diagram shown in Fig. 8.2.5b, in view of the fact that the ratio station input is the measured value of the wild flow rate and its output is compared with the measured value of the adjusted flow rate.

Obviously, the desired ratio and the gain of the ratio station have the following relationship

$$K = \frac{H_{m2}(0)}{H_{m1}(0)} R \qquad (8.2.8)$$

A similar situation appears in a ratio control system with a divider as the ratio calculation device. In Fig. 8.2.4b, the inputs of the divider are the measured values of the two flow rates, thus, the setpoint of the adjusted flow rate loop should be the desired ratio of the measured values of the two flow rates rather than the desired ratio of the two flow rates. If the setpoint in Fig. 8.2.4b is denoted as K, then (8.2.8) is still valid.

Dynamic compensation
Generally speaking, the ratio of two flow rates remains constant only at steady state. Some chemical process ratio control systems constantly require a precise ratio of two flow rates for operational safety and high product quality. Therefore, a dynamic compensation may be introduced into the system to overcome the transient imbalance of the two flow rates.

A double loop ratio control system with a dynamic compensation element is diagrammed in Fig. 8.2.6a, where the dynamic compensation element is incorporated in front of the ratio station. Fig. 8.2.6b is the corresponding block diagram.

The transfer functions of controller, valve, process and measurement in the wild flow rate control loop are $G_{c1}(s)$, $G_{v1}(s)$, $G_{p1}(s)$ and $H_{m1}(s)$, and those in the adjusted flow rate control loop are $G_{c2}(s)$, $G_{v2}(s)$, $G_{p2}(s)$ and $H_{m2}(s)$, respectively.

The transfer function of the ratio station is K, and that of the dynamic compensation element is $G_z(s)$.

The transfer function between the two flow rates is

$$\frac{F_2(s)}{F_1(s)} = \frac{H_{m1}(s)G_z(s)KG_{p2}(s)G_{v2}(s)G_{c2}(s)}{1 + H_{m2}(s)G_{p2}(s)G_{v2}(s)G_{c2}(s)} \qquad (8.2.9)$$

To overcome transient imbalance it is required that

$$F_2(s) = RF_1(s) \qquad (8.2.10)$$

where R is the desired ratio. Therefore, we have

$$R = \frac{H_{m1}(s)G_z(s)KG_{p2}(s)G_{v2}(s)G_{c2}(s)}{1 + H_{m2}(s)G_{p2}(s)G_{v2}(s)G_{c2}(s)} \qquad (8.2.11)$$

This yields

$$G_z(s) = \frac{1 + H_{m2}(s)G_{p2}(s)G_{v2}(s)G_{c2}(s)}{H_{m1}(s)G_{p2}(s)G_{v2}(s)G_{c2}(s)} \cdot \frac{R}{K} \qquad (8.2.12)$$

or

$$G_z(s) = \frac{1 + H_{m2}(s)G_{p2}(s)G_{v2}(s)G_{c2}(s)}{H_{m1}(s)G_{p2}(s)G_{v2}(s)G_{c2}(s)} \cdot \frac{H_{m1}(0)}{H_{m2}(0)} \qquad (8.2.13)$$

With such a compensation, the system can keep the ratio of flow rates constant dynamically.

From (8.2.13), it is clear that the transfer function of the dynamic compensation element has nothing to do with the ratio of the two flow rates, which is determined by the ratio calculation device. Therefore, if we want to

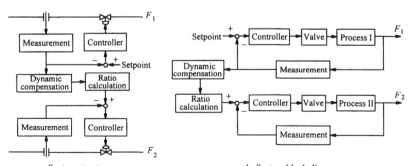

a. System structure   b. System block diagram

FIGURE 8.2.6.   A dynamic compensation ratio control system.

change the ratio of the flow rates, we can change the coefficient of the ratio
station without affecting the dynamic compensation element.

Example 8.2.1

A single loop ratio control system as shown in Fig. 8.2.5 has the transfer
functions

$$G_{p2}(s) = \frac{1}{0.5s+1} \qquad G_{c2}(s) = 1 + \frac{1}{0.2s}$$

$$G_{v2}(s) = 1 \qquad H_{m1}(s) = 0.7 \qquad H_{m2}(s) = 0.8$$

where $G_{p2}(s)$, $G_{c2}(s)$, $G_{v2}(s)$, $H_{m1}(s)$ and $H_{m2}(s)$ are transfer functions of the
adjusted flow process, *PI* controller, valve, transducer of the wild flow rate and
transducer of the adjusted flow rate, respectively.

From (8.2.13), the transfer function of the dynamic compensation element
may be calculated as

$$G_z(s) = \frac{1 + H_{m2}(s)G_{p2}(s)G_{v2}(s)G_{c2}(s)}{H_{m1}(s)G_{p2}(s)G_{v2}(s)G_{c2}(s)} \cdot \frac{H_{m1}(0)}{H_{m2}(0)}$$

$$= \frac{1 + 0.8 \cdot \dfrac{1}{0.5s+1} \cdot 1 \cdot \dfrac{0.2s+1}{0.2s}}{0.7 \cdot \dfrac{1}{0.5s+1} \cdot 1 \cdot \dfrac{0.2s+1}{0.2s}} \cdot \frac{0.7}{0.8}$$

$$= \frac{0.01s^2 + 0.36s + 0.8}{0.7(0.2s+1)} \cdot \frac{0.7}{0.8}$$

$$= \frac{0.0125s^2 + 0.45s + 1}{0.2s+1}$$

$$= \frac{(0.03s+1)(0.42s+1)}{0.2s+1} \qquad (8.2.14)$$

which is an unrealizable transfer function, since the order of the numerator is
higher than that of the denominator. In order to make it realizable, Eq. (8.2.14)
may be approximated in several ways. One is to add the two time constants
together, which yields

$$G_z(s) = \frac{(0.45s+1)}{0.2s+1} \qquad (8.2.15)$$

8.3. FEEDFORWARD CONTROL

The importance of feedback control has been demonstrated in the preceding chapters. With feedback control, the deviation of the controlled variable from the desired value in the system will be detected and a suitable control action will be produced to reduce the deviation, no matter which disturbance causes the deviation. Because the feedback controller produces a control action based on the actuating errors, it is not necessary to measure process disturbances. Even if the process model is not precisely known, or if the parameters of the process have changed, the feedback controller may still work well. However, is a feedback control configuration ideal for all cases?

Disadvantages of feedback control
Feedback control has some inherent disadvantages including the following:

1. No actions are taken until the deviation of the controlled variable occurs, i.e., it does not provide any predictive control action to reduce the effects of measurable disturbances. The controller needs a measurable error to generate a restoring action.

2. If a disturbance occurs at a high frequency relative to the time constant of the system, the system output may never reach steady state. For example, if a first order closed loop system has a time constant T, and the disturbance appears at intervals less than $3T$, then the system output will always be disturbed and no steady state will ever be reached. Thus, feedback control may not be satisfactory for processes with large time constants and/or long time delays.

3. Feedback control may create instability in the closed loop response, because a feedback controller solves control problems by trial and error, which is characteristic of the oscillatory response of a feedback loop.

What is feedforward control
If the disturbance acting on a system can be measured, it should be corrected in some way before it upsets the process. This is the objective of a feedforward control approach.

A feedforward control system is an open loop control system whose control action starts immediately after a change in the disturbance has been detected. Feedforward control is theoretically capable of perfect control, i.e., in theory it can eliminate all the effects of measurable disturbances on the system output.

It should be emphasized that the disturbances entering the system must be measured on line. If not, feedforward control is impossible. The performance of a feedforward control system depends on the accuracy and speed of the disturbance measurement and control action computation.

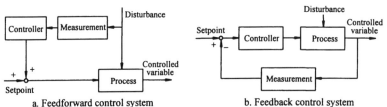

a. Feedforward control system   b. Feedback control system

FIGURE 8.3.1.   Comparison between feedforward and feedback control systems.

The block diagram of a feedforward control system is shown in Fig. 8.3.1a. For comparison, a feedback control system is shown in Fig. 8.3.1b.

Disadvantages of feedforward control

Feedforward control can overcome some disadvantages of feedback control, but it also loses the advantages of feedback control. Feedforward control also introduces the following shortcomings:

1. As mentioned, the disturbances entering the system must be measurable on line. In many cases, however, the disturbances are unmeasurable.

2. After disturbances are measured, the effectiveness of feedforward control to eliminate the disturbance effect mainly depends on the accuracy of control action computation, which requires a precise process model. Feedforward control is very sensitive to process parameter variation.

3. An ideal feedforward controller designed for perfect control may not be physically realizable. In this case, only an approximate controller may be practically designed.

Feedforward feedback control systems

After discussing the advantages and disadvantages of feedforward and feedback control, we can combine feedforward control with a feedback scheme to design a system with superior performance. We may use feedforward control to reduce the effect of measured disturbances immediately before they disturb the system. At the same time, the deviations of controlled variables caused by unmeasured disturbances, the inaccuracy of disturbance measurement and the error of process modeling or variation of process parameters, will be reduced by the feedback control action.

In fact, we have studied some feedforward control systems and the feedforward feedback control systems in the previous section. The simple open loop ratio control system as shown in Fig. 8.2.1 is a feedforward control system. The single loop ratio control system as shown in Fig. 8.2.2 is a feedforward feedback control system. The single loop ratio control system has a better performance than the simple open loop ratio control system. In other words, the feedforward feedback control system gives better performance than the pure feedforward control system. The double loop ratio control system is also a combined feedforward feedback control system.

Design of a feedforward controller for an open loop system
The design method for a feedforward controller depends on its objective. Usually, the objective of a feedforward controller is to reject disturbances, i.e., to eliminate the effect of disturbances on a controlled variable. Therefore, the principle of designing a feedforward controller is to balance the direct effect of disturbances on a controlled variable with the effect of the feedforward controller. In other words, the feedforward controller should have a similar but opposite effect sign on the controlled variable as that produced by the disturbances.

For an open loop system shown in Fig. 8.3.2a, where G_v, G_p and G_d are the transfer functions of the valve, main process and disturbance process, respectively, the feedforward control configuration may be constructed as shown in Fig. 8.3.2b, where G_{md} is the transfer function of the disturbance measurement; G_{cf} is a feedforward controller, which is used to produce a control signal to balance the direct effect of the disturbance on the system output; and G_s is a compensation controller, which is incorporated for setpoint tracking.

The system output may be expressed as

$$Y(s) = G_p G_v G_s R + [G_d - G_p G_v G_{cf} G_{md}]D \qquad (8.3.1)$$

To completely eliminate the effect of the disturbance, it is required that

$$G_d - G_p G_v G_{cf} G_{md} = 0$$

or

$$G_{cf}(s) = \frac{G_d}{G_p G_v G_{md}} \qquad (8.3.2)$$

Sometimes, for the sake of easy implementation, a steady state feedforward controller is suggested. In such a case, the designed controllers are simplified to

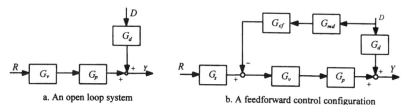

a. An open loop system   b. A feedforward control configuration
FIGURE 8.3.2. Feedforward control system.

$$G_{cf}(s) = \frac{K_d}{K_p K_v K_{md}} \qquad (8.3.3)$$

where K_{md}, K_v, K_p and K_d are the static gains of the disturbance measurement, valve, main process and disturbance process, respectively. Thus the controller may be constructed as a proportional controller.

Example 8.3.1

For an open loop system and the feedforward control configuration shown in Fig 8.3.2, if $G_{md}(s)= 1$, $G_v(s)= 1$ and

$$G_p(s) = \frac{8}{1.6s+1} \qquad G_d(s) = \frac{4}{2s+1}$$

then the controllers for disturbance rejection may be designed as

$$G_{cf}(s) = \frac{G_d(s)}{G_p(s)} = \frac{0.5(1.6s+1)}{2s+1} \qquad (8.3.4)$$

Controller (8.3.4) is a lag-lead element, which has been discussed in Chapter 5.

If only steady state feedforward control is desired, the controller may be simplified to

$$G_c(s) = 0.5$$

This is just a proportional gain.

Design of a feedforward controller for a closed loop system

For a closed loop system, if the disturbance can be measured, we can incorporate a feedforward controller for disturbance rejection. The block diagram of such a feedforward feedback system configuration is given in Fig. 8.3.3. In this closed loop configuration, the feedforward controller $G_{cf}(s)$ and feedback controller $G_c(s)$ are in different loops.

The system output caused by the disturbance may be represented as

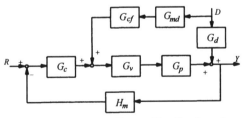

FIGURE 8.3.3. A feedforward feedback system.

$$\frac{Y(s)}{D(s)} = \frac{G_d + G_p G_v G_{cf} G_{md}}{1 + G_c G_v G_p H_m}$$

(8.3.5)

To eliminate the effect of the disturbance on the output, it is required that

$$G_d + G_p G_v G_{cf} G_{md} = 0$$

(8.3.6)

or

$$G_{cf} = -\frac{G_d}{G_p G_v G_{md}}$$

(8.3.7)

Remarks

1. From (8.3.5), the characteristic equation is

$$\cdot \quad 1 + G_c G_v G_p H_m = 0$$

where the transfer function of the feedforward controller does not appear. Therefore, the feedforward controller has no effect on the stability of the control system.

2. The design of the feedforward controller depends on not only the control objective, but also the system configuration and process model.

Example 8.3.2

For a system configuration as shown in Fig. 8.3.3, where

$$G_p(s) = \frac{2}{0.5s + 1} \qquad G_d(s) = \frac{1}{0.7s + 1}$$

$$G_{md} = 0.8 \qquad G_v = 0.9$$

the feedforward controller is designed from (8.3.7) as

$$G_{cf} = -\frac{G_d}{G_p G_v G_{md}} \approx -\frac{0.7(0.5s + 1)}{0.7s + 1}$$

Example 8.3.3

If a time delay exists in the disturbance process of the system in Example 8.3.2, that is,

$$G_p(s) = \frac{2}{0.5s + 1} \qquad G_d(s) = \frac{1}{0.7s + 1} e^{-0.3s}$$

$$G_{md} = 0.8 \qquad\qquad G_v = 0.9$$

then the feedforward controller will be designed as

$$G_{cf} = -\frac{G_d}{G_p G_v G_{md}} \approx -\frac{0.7(0.5s+1)}{0.7s+1} \cdot e^{-0.3s}$$

With the approximation formula that can be found in Appendix B

$$e^{-\tau s} = \frac{1 - 0.5\tau s}{1 + 0.5\tau s} \qquad\qquad (8.3.8)$$

the feedforward controller may be designed as

$$G_{cf} = \frac{0.7(0.5s+1)(0.15s-1)}{(0.7s+1)(0.15s+1)}$$

If the order of the disturbance process is less than that of the main process, or the measurement and valve cannot be approximated by a constant, then the designed feedforward controller may be physically unrealizable, and an approximation should be used.

Example 8.3.4

If a feedforward control system as shown in Fig. 8.3.3 has the transfer functions

$$G_p = \frac{2}{(0.5s+1)(0.4s+1)} \qquad\qquad G_d = \frac{1}{0.7s+1}$$

$$G_{md} = 0.8 \qquad\qquad G_v = 0.9$$

then the designed controller will be

$$G_{cf} = \frac{0.7(0.5s+1)(0.4s+1)}{0.7s+1}$$

This is physically unrealizable. By means of approximation, we just simply add the two time constants in the numerator together, yielding

$$G_{cf} = -\frac{0.7(0.9s+1)}{0.7s+1}$$

Example 8.3.5

If a time delay exists in the main process of the system in Example 8.3.2, i.e.,

$$G_p(s) = \frac{2}{0.5s+1} \cdot e^{-0.3s} \qquad G_d(s) = \frac{1}{0.7s+1}$$

$$G_{md} = 0.8 \qquad G_v = 0.9$$

then the feedforward controller will be designed as

$$G_{cf} = -\frac{G_d}{G_p G_v G_{md}} \approx -\frac{0.7(0.5s+1)}{0.7s+1} \cdot e^{0.3s}$$

which is also physically unrealizable, because the feedforward controller is required to predict the disturbance. For this system, we may still use the controller designed in Example 8.3.2.

Comparison between feedforward and feedback controls

Having studied feedforward control, we can compare it with feedback control in detail.

The design of a feedback controller is based on feedback control theory, but the design method of a feedforward controller for disturbance rejection is based on the balance principle: The direct effect of the disturbance on the controlled variable is balanced by the offsetting effect of the feedforward controller.

The detected variable for feedback control is a controlled variable while the detected variable for feedforward control is a disturbance variable. Thus, a feedback controller acts based on the difference between the setpoint and the measured controlled variable, whereas a feedforward controller acts based on the measured disturbance. Therefore, if the disturbance is unmeasurable, feedforward control for disturbance rejection is impossible.

The conventional feedback controller may be chosen from on-off, P, PI, PD

TABLE 8.7. Comparison between feedforward control and feedback control.

Item	Feedback control	Feedforward control
Design principle	Feedback control theory	Balance principle
Measured variable	Controlled variable	Disturbance variable
Controller input	Error between controlled variable and setpoint	Measured disturbance
Controller implementation	Possible and economical	Sometimes approximation is needed for realization
Control system configuration	Closed loop	Open loop
Typical controller	On-off, P, PI, PD, PID	Lead-lag controller
Action	After process is affected by disturbance	Before process is affected by disturbance

and *PID* controllers, which are physically realizable. However, not all of these controllers can be used as a feedforward controller that relies on the process model. Sometimes, feedforward controllers are physically unrealizable. In such cases, an approximation is recommended.

Table 8.7 is a summary of the comparison between feedforward and feedback control.

8.4. OTHER ADVANCED CONTROL SYSTEMS

In the following section we will briefly introduce some other advanced control strategies, such as internal model control, Smith predictor and selective control.

Internal model control (*IMC*)

Since the controller of a control system is designed based on the system model as well as required system performance, model uncertainty may bring about some problems, especially for a system with a feedforward controller. Model uncertainty is explicitly taken into account in an internal model control strategy.

Fig. 8.4.1a is a classical feedback control system, where transfer function $G(s)$ is considered to be the actual controlled plant including process, final control element and measurement. An internal model control strategy is constructed by incorporating a model for the actual plant. The model $G_I(s)$ is connected in parallel to the plant as shown in Fig. 8.4.1b. The difference between the real plant output and the model output is fed back to a controller that has a transfer function $G_{cI}(s)$.

The model uncertainty is included in the feedback information, thus the effect of the uncertainty on the system output will be reduced. The difference between the outputs of the actual plant and internal model is caused not only by the model uncertainty, but also by disturbances acting on the plant. Therefore, internal model control can also reduce the disturbance effect on the system

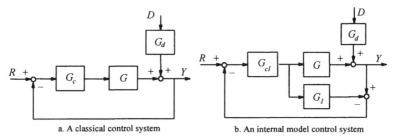

a. A classical control system   b. An internal model control system

FIGURE 8.4.1. Feedback control strategies.

output.

The transfer function of the system output can be expressed as

$$Y(s) = \frac{G_{cl}G}{1 + G_{cl}(G - G_I)}R + \frac{1 - G_{cl}G_I}{1 + G_{cl}(G - G_I)}G_d D \qquad (8.4.1)$$

If the model $G_I(s)$ is perfect, i.e., $G_I(s) = G(s)$, Eq. (8.4.1) reduces to

$$Y(s) = G_{cl}GR + (1 - G_{cl}G)G_d D \qquad (8.4.2)$$

If the model $G_I(s)$ is a real rational function with all its poles and zeros in the left half s-plane, then the controller may be designed as

$$G_{cl}(s) = 1 / G_I \qquad (8.4.3)$$

With this controller, the disturbance effect on the system output is completely eliminated and (8.4.1) is reduced to

$$Y(s) = \frac{G_{cl}G}{1 + G_{cl}(G - G_I)}R = R \qquad (8.4.4)$$

This means that perfect setpoint tracking occurs. Unfortunately, it case is uncommon, because (8.4.3) is usually unrealizable. In order to make the controller realizable, a low-pass filter with a steady state gain of one is incorporated, which has the form

$$F(s) = \frac{1}{(T_c s + 1)^r} \qquad (8.4.5)$$

where T_c is the desired closed loop time constant, and r is a positive integer. r is selected so that the controller $G_{cl}(s)$ is realizable, i.e., the denominator of the controller has an order higher than or equal to that of the numerator. Therefore, the controller will be

$$G_{cl}(s) = \frac{1}{G(s)(T_c s + 1)^r} \qquad (8.4.6)$$

Time delay compensation — Smith predictor

Example 8.3.5 shows that time delay in a process may bring about some problems. Time delay is often encountered in industrial processes. The time delay may be taken into account explicitly using a time delay compensation technique.

The Smith predictor is one of the best known techniques for improving the performance of systems with time delay [Smith 1957]. The control configuration with a Smith predictor is shown in Fig. 8.4.2, where the transfer

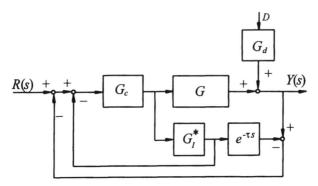

FIGURE 8.4.2. Smith predictor configuration.

function $G(s)$ represents the plant with time delay including the process, final control element and measurement, and the model of the plant is represented by $G_I(s) = G_I^*(s)e^{-\tau s}$. We divide the model into two parts, isolating the time delay in order to predict the effect of control action on the plant output without time delay. The signal is taken from the transfer function $G_I^*(s)$ in the block diagram, and is fed to the controller. The signal is also delayed by the amount of time delay τ for comparison with the actual plant output.

With the block diagram reduction technique, the control configuration in Fig. 8.4.2 may be transformed into Fig. 8.4.3, which may be viewed as a cascade system. The transfer function of the system can be easily calculated as

$$\frac{Y(s)}{R(s)} = \frac{GG_c}{1 + G_I^*(1 - e^{-\tau s})G_c + GG_c} \tag{8.4.7}$$

If the model is perfect, i.e., $G(s) = G_I^*(s)e^{-\tau s}$, then (8.4.7) is reduced to

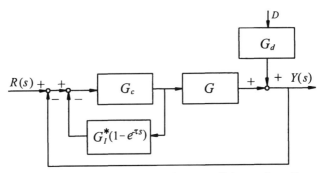

FIGURE 8.4.3. An alternative Smith predictor configuration.

$$\frac{Y(s)}{R(s)} = \frac{G_c G_I^* e^{-\tau s}}{1 + G_I^* G_c} \tag{8.4.8}$$

By contrast, the corresponding conventional feedback control system, as shown in Fig. 8.4.1a, has the transfer function

$$\frac{Y(s)}{R(s)} = \frac{G_c G_I^* e^{-\tau s}}{1 + G_I^* G_c e^{-\tau s}} \tag{8.4.9}$$

By comparing (8.4.8) with (8.4.9), we can see that the Smith predictor eliminates the time delay term from the characteristic equation of the system.

Selective control systems
Usually, the number of controlled variables is equal to the number of manipulated variables in a control system. If there are more controlled variables than manipulated variables in a system, a selector should be used so that a manipulated variable can be shared by two or more controlled variables. Selective control is typically used in four application areas [Shinskey 1988]:
      1. Equipment protection,
      2. Auctioneering,
      3. Redundant instrumentation,
      4. Variable structuring.
    The block diagram of a compressor system with a selector for equipment protection is shown in Fig. 8.4.4, where the motor speed is a unique manipulated variable, while the discharge pressure and the flow rate are two controlled variables. The discharge of a compressor is controlled by a flow rate control system, i.e., controller II works in a normal situation. Since the discharge pressure must be under an upper limit, a high selector switch (*HS*) is introduced for this safety consideration. *HS* switches the flow rate control action to the pressure control action whenever the discharge pressure exceeds the upper limit,

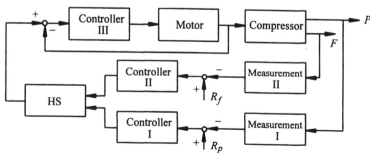

FIGURE 8.4.4. A selective temperature control system.

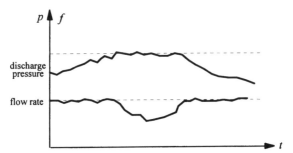

FIGURE 8.4.5. Curves of the controlled variables.

i.e., controller I comes into action and controller II stops working. Controller III is an inner loop controller for the motor's speed. It should be noticed that cascade control systems are found in Fig. 8.4.4. In a normal situation, a flow rate control system (loop II) is a primary loop while the motor speed control loop is a secondary loop. If the discharge pressure exceeds the upper limit, the pressure control system (loop I) is a primary loop while the motor speed control loop is still a secondary loop.

If the discharge pressure, under disturbances, goes up, the flow rate control loop is still connected until the discharge pressure reaches the upper limit. During this period, the flow rate is controlled at a specified value determined by the setpoint. After the control action is changed, the discharge pressure is controlled at the upper limit, with the increasing tendency of the discharge pressure causing the motor speed to decrease. Decreasing motor speed will reduce the increasing tendency of the discharge pressure as well as the flow rate. If the discharge pressure goes down, the flow rate control action takes hold, while controller I stops working. The process may be represented by Fig. 8.4.5.

The selection of the highest signal from among several inputs is also called

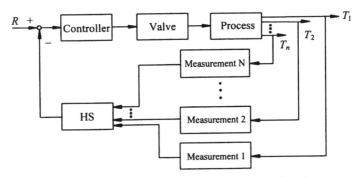

FIGURE 8.4.6. An auctioneering selective control system.

auctioneering. For example, the temperature of a fixed-bed chemical reactor is measured at N different locations, and each value is required to be under an upper limit. A selective control configuration may be constructed as shown in Fig. 8.4.6, where the N temperature measurements are compared with each other, and the highest one is selected and sent to the controller that will adjust the flow rate of the coolant to decrease the temperature.

To protect against instrument failure, N instruments may be installed to take the same measurement. The selector may choose the median measurement and reject the others to make the reading more reliable. This redundant instrumentation is another application of selective control.

Variable structure, which is used to transfer a controller from one valve to another, is only occasionally used and will not be examined here.

Problems

8.1. For a cascade control system as shown in Fig. 8.1.5a, the transfer functions in the secondary loop are assumed to be

$$G_{p2}(s) = \frac{14}{8s+1} \qquad G_{d2}(s) = \frac{7}{9s+1}$$

$$G_v = 1 \qquad G_{c2} = 50 \qquad H_{m2} = 0.8$$

What is the transfer function of the closed secondary loop? Is the characteristic of the secondary loop improved compared with the secondary process? Discuss with detail.

8.2. Recalling Problem 8.1, if the transfer functions in the primary loop are

$$G_{c1} = 20(1+\frac{1}{30s}) \qquad G_{p1} = \frac{5}{2s+1} \qquad H_{m1} = 0.7$$

what are the steady state values of the primary variable and secondary variable if the disturbance in the secondary loop is a unit step signal and the setpoint is zero?

8.3. Recalling Problem 8.2, if we choose

$$G_{c2} = 20(1+\frac{1}{30s}) \qquad G_{c1} = 50$$

what are the steady state values of the primary variable and secondary variable if the disturbance in the secondary loop is a unit step signal and the setpoint is zero? Compare the result with that in Problem 8.2 and give your conclusion.

8.4. A double loop ratio control system is shown in Fig. 8.P.1. Assume that

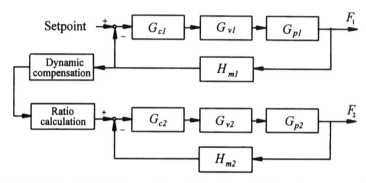

FIGURE 8.P.1. A double loop ratio control system with dynamic compensation.

$$G_{p2}(s) = \frac{3}{8s+1} \qquad G_{p1}(s) = \frac{2}{4s+1}$$

$$G_{v1} = 1 \qquad G_{c1} = 3 \qquad H_{m1} = 0.8$$

$$G_{v2} = 0.9 \qquad G_{c2} = 15(1+\frac{1}{2s}) \qquad H_{m2} = 0.7$$

If the specified ratio is

$$R = F_2 / F_1 = 2$$

design the ratio station and dynamic compensator. (If the compensator is physically unrealizable, make its approximation for realization.) If the steady state $F_1 = 20$ is required, find the setpoint.

8.5. If a feedforward control system as shown in Fig. 8.3.3 has the transfer functions

$$G_p = \frac{0.7}{(0.8s+1)(1.4s+1)} \qquad G_d = \frac{1}{0.9s+1}$$

$$G_{md} = 0.8 \qquad G_v = 0.9$$

$$G_c = 30(1+\frac{1}{150s}) \qquad H_m = 1$$

then,

a. design a feedforward controller to eliminate the effect of disturbance on the output. If the controller is physically unrealizable, make its approximation for realization.

b. calculate the steady state error of the system while $r(t)$ is a unit step input and $d = 0$.

c. calculate the steady state error of the system while $r(t) = 0$ and $d(t)$ is a unit step disturbance.

d. calculate the steady state error of the system while $r(t)$ and $d(t)$ are unit step functions.

8.6. (Cascade control) For a cascade control system, the secondary process is given by

$$G(s) = \frac{1}{1.5s + 1}$$

and the primary process is assumed to be

$$G(s) = \frac{1}{s + 1} e^{-s}$$

With the software package IAC in $PCET$, do the following:

a. If both secondary and primary controllers are P controllers, observe the system response to a unit step input (setpoint). Does offset exist?

b. If both secondary and primary controllers are PI controllers, observe the system response to a unit step input (setpoint). Does offset exist?

c. If the secondary controller is a P controller and the primary controller is a PI controller, observe the system response to a unit step input (setpoint). Does offset exist?

d. If the secondary controller is a PI controller and the primary controller is a P controller, observe the system response to a unit step input (setpoint). Does offset exist?

e. Compare the results you obtain, and give your conclusion. Does the conclusion conflict with that in Section 8.1?

f. If both the secondary and primary controllers are P controllers, observe the system response to a disturbance in the secondary loop. Does offset exist?

g. If both the secondary and primary controllers are PI controllers, observe the system response to a disturbance in the secondary loop. Does offset exist?

h. If the secondary controller is a P controller and the primary controller is a PI controller, observe the system response to a disturbance in the secondary loop. Does offset exist?

i. If the secondary controller is a PI controller and the primary controller is a P controller, observe the system response to a disturbance in the secondary loop. Does offset exist?

j. Compare the results you obtain, and give your conclusion.

CHAPTER 9

LINEAR STATE SPACE ANALYSIS AND DESIGN

9.1. MATRIX ALGEBRA

Linear state space theory is mathematically based on matrix theory. In this section, the mathematical background of matrix theory is reviewed.

For classification, from now on we will use capital bold letters to represent matrices and lowercase bold letters for vectors.

Matrix and matrix algebra

A matrix is a rectangular array of numbers or functions. An $n \times m$ matrix $\mathbf{A}$ (n rows and m columns) can be denoted as

$$\mathbf{A} = \begin{pmatrix} a_{11} & a_{12} & \cdots & a_{1m} \\ a_{21} & a_{22} & \cdots & a_{2m} \\ \vdots & \vdots & & \vdots \\ a_{n1} & a_{n2} & \cdots & a_{nm} \end{pmatrix} \tag{9.1.1}$$

where a_{ij} is the element in the ith row and jth column. Briefly, we denote $\mathbf{A} = (a_{ij})$. If $n = m$, $\mathbf{A}$ is called a square matrix.

With the notation above, some basic matrix calculations are listed in Table 9.1.

Looking at the last item in Table 9.1, if $\mathbf{A}$ and $\mathbf{B}$ are $n \times m$ and $m \times l$ matrices, respectively, the product $\mathbf{C} = \mathbf{AB}$ is an $n \times l$ matrix. $\mathbf{BA}$ is meaningless if $l \neq n$. Even if $\mathbf{A}$ and $\mathbf{B}$ are conformable for $\mathbf{AB}$ and $\mathbf{BA}$ (i.e., $n = l$), usually $\mathbf{AB} \neq \mathbf{BA}$.

The associative, commutative, and distributive laws are still valid, and are tabulated in Table 9.2.

The null matrix $\mathbf{0}$ is one whose elements are all equal to zero. It follows that $\mathbf{A} + \mathbf{0} = \mathbf{A}$ and $\mathbf{0A} = \mathbf{0}$. Clearly, $\mathbf{0A} = \mathbf{0}$ is different from $\mathbf{0A} = \mathbf{0}$. For example if

TABLE 9.1. Basic matrix calculation.

Item	Formula	Comments
Addition of two matrices	$\mathbf{A} + \mathbf{B} = (a_{ij} + b_{ij})$	**A** and **B** have the same dimension.
Subtraction of two matrices	$\mathbf{A} - \mathbf{B} = (a_{ij} - b_{ij})$	**A** and **B** have the same dimension.
Product of a matrix and a scalar	$k\mathbf{A} = (ka_{ij})$	k is a scalar.
Product of two matrices	$\mathbf{AB} = \mathbf{C} = (c_{ij}) = (\sum_{k=1}^{m} a_{ik}b_{kj})$	**A** is an $n{\times}m$ matrix. **B** is an $m{\times}l$ matrix. **AB** is an $n{\times}l$ matrix.

A is a 2×3 matrix and **0** is a 3×2 null matrix, then **0A** is a 3×3 null matrix, but 0**A**, the product of a scalar 0 and a matrix **A**, is a 2×3 null matrix.

A diagonal matrix is a square matrix with all elements equal to zero except those on the diagonal (i.e., the elements with $i = j$). The identity, or unit matrix **I** is a diagonal matrix with all diagonal elements equal to one. It should be noted that **IA** = **AI** = **A**, but two identities may have different dimensions. For example, if **A** is a 2×3 matrix, **I** in **IA** is a 2×2 matrix while **I** in **AI** is a 3×3 matrix. We denote the $n{\times}n$ unit matrix as $\mathbf{I}_n$ for distinction.

The transpose of a matrix $\mathbf{A} = (a_{ij})$ is $\mathbf{A}^T = (a_{ji})$. It is easy to prove that

$$(\mathbf{AB})^T = \mathbf{B}^T\mathbf{A}^T, \qquad (\mathbf{ABC})^T = \mathbf{C}^T\mathbf{B}^T\mathbf{A}^T \qquad (9.1.2)$$

If $\mathbf{A} = \mathbf{A}^T$, i.e., $a_{ij} = a_{ji}$ for all i and j, then **A** is a symmetric matrix.

A vector is a matrix that has only one column or one row. Usually, a small bold letter represents a column vector, e.g., vector **x**, whereas a row vector is denoted by $\mathbf{x}^T$, that is the transpose of a column vector **x**.

Determinant and the rank of a matrix

A determinant is a scalar value for a square matrix, denoted as det **A**, or |**A**|. The value of a determinant is determined by obtaining the minors and cofactors of all elements in the determinant. For an $n{\times}n$ determinant, the minor of an element a_{ij} is a determinant of order $n - 1$ obtained by removing the ith row and jth column

TABLE 9.2. Matrix calculation laws.

Law	Formula
Commutative law of addition	$\mathbf{A} + \mathbf{B} = \mathbf{B} + \mathbf{A}$
Associative law of addition	$\mathbf{A} + (\mathbf{B} + \mathbf{C}) = (\mathbf{A} + \mathbf{B}) + \mathbf{C}$
Associative law of multiplication	$(\mathbf{AB})\mathbf{C} = \mathbf{A}(\mathbf{BC})$
Distributive law of a scalar to matrices	$\alpha(\mathbf{A} + \mathbf{B}) = \alpha\mathbf{A} + \alpha\mathbf{B} = (\mathbf{A} + \mathbf{B})\alpha$
Distributive law of matrices	$\mathbf{A}(\mathbf{B} + \mathbf{C}) = \mathbf{AB} + \mathbf{AC}$ $(\mathbf{B} + \mathbf{C})\mathbf{A} = \mathbf{BA} + \mathbf{CA}$

of the original determinant. The cofactor of an element a_{ij} in a determinant is its minor with a sign $(-1)^{i+j}$. Therefore,

$$\alpha_{ij} = (-1)^{i+j} M_{ij}$$

where α_{ij} is the cofactor of a_{ij}, and M_{ij} is the minor of a_{ij}.

The determinants of order 1 to 3 are listed as follows:

$$n = 1 \qquad |\mathbf{A}| = a_{11}$$

$$n = 2 \qquad |\mathbf{A}| = a_{11}a_{22} - a_{12}a_{21}$$

$$n = 3 \qquad |\mathbf{A}| = a_{11}a_{22}a_{33} + a_{12}a_{23}a_{31} + a_{13}a_{21}a_{32}$$
$$- a_{13}a_{22}a_{31} - a_{12}a_{21}a_{33} - a_{11}a_{32}a_{23}$$

The rank of $\mathbf{A}$, designated as rank($\mathbf{A}$), is defined as the size of the largest nonzero determinant that can be formed from $\mathbf{A}$. Obviously, the maximum possible rank of an $n \times m$ matrix is $\min(n,m)$. For an $n \times n$ square matrix, the maximum possible rank is n. A matrix with maximum rank is said to be of full rank. A square matrix with full rank is said to be nonsingular. Otherwise the square matrix is a singular matrix. Clearly, for a square matrix $\mathbf{A}$, if $|\mathbf{A}| = 0$, $\mathbf{A}$ is singular.

Example 9.1.1

Find the rank of the following matrix:

$$\mathbf{A} = \begin{pmatrix} 1 & 3 & 0 & 0 \\ 2 & 6 & 0 & 0 \\ 2 & 4 & 1 & 2 \end{pmatrix}$$

Because of

$$\begin{vmatrix} 1 & 3 & 0 \\ 2 & 6 & 0 \\ 2 & 4 & 1 \end{vmatrix} = 0, \quad \begin{vmatrix} 1 & 3 & 0 \\ 2 & 6 & 0 \\ 2 & 4 & 2 \end{vmatrix} = 0, \quad \begin{vmatrix} 1 & 0 & 0 \\ 2 & 0 & 0 \\ 2 & 1 & 2 \end{vmatrix} = 0, \quad \begin{vmatrix} 3 & 0 & 0 \\ 6 & 0 & 0 \\ 4 & 1 & 2 \end{vmatrix} = 0$$

but

$$\begin{vmatrix} 2 & 6 \\ 2 & 4 \end{vmatrix} = -4 \neq 0$$

we have

$$\text{Rank}(\mathbf{A}) = 2$$

Matrix inversion

If square matrices $\mathbf{A}$ and $\mathbf{B}$ satisfy

$$AB = BA = I \tag{9.1.3}$$

then **B** is the inverse of **A**, or vice versa. The inverse of **A**, denoted A^{-1}, can be calculated from

$$A^{-1} = \frac{\text{adj}\,A}{\det A} \tag{9.1.4}$$

where adj **A** is the adjoint matrix of **A**, which is defined as

$$\text{adj}\,A = \begin{pmatrix} \alpha_{11} & \alpha_{12} & \cdots & \alpha_{1m} \\ \alpha_{21} & \alpha_{22} & \cdots & \alpha_{2m} \\ \vdots & \vdots & & \vdots \\ \alpha_{n1} & \alpha_{n2} & \cdots & \alpha_{nm} \end{pmatrix} \tag{9.1.5}$$

where α_{ij} is the cofactor of the element a_{ij} of the square matrix **A**. It is easy to find that for a 2×2 matrix

$$A = \begin{pmatrix} a_{11} & a_{12} \\ a_{21} & a_{22} \end{pmatrix}$$

the inverse of **A** is

$$A^{-1} = \frac{\text{adj}\,A}{\det A} = \frac{1}{(a_{11}a_{22} - a_{12}a_{21})} \begin{pmatrix} a_{22} & -a_{12} \\ -a_{21} & a_{11} \end{pmatrix} \tag{9.1.6}$$

The inverse of a matrix can be obtained by matrix row transforms. If

$$[A|I] \xrightarrow{\;row\ transforms\;} [I|B] \tag{9.1.7}$$

then $B = A^{-1}$. We illustrate this method in the following example:

Example 9.1.2

Find the inverse of the square matrix **A**

$$A = \begin{pmatrix} 1 & 3 \\ 2 & -1 \end{pmatrix}$$

Denote row(i) as the ith row of the matrix. The procedure used to obtain the inverse is given as follows:

$$[A|I] = \begin{pmatrix} 1 & 3 & | & 1 & 0 \\ 2 & -1 & | & 0 & 1 \end{pmatrix} \xrightarrow{\;row(2) + row(1) \times (-2) \to row(2)\;}$$

$$= \begin{pmatrix} 1 & 3 & | & 1 & 0 \\ 0 & -7 & | & -2 & 1 \end{pmatrix} \xrightarrow{\;row(2) \div (-7) \to row(2)\;}$$

$$= \begin{pmatrix} 1 & 3 & | & 1 & 0 \\ 0 & 1 & | & 2/7 & 2/7 \end{pmatrix} \xrightarrow{\;\;row(1) + row(2)\times(-3) \rightarrow row(1)\;\;}$$

$$= \begin{pmatrix} 1 & 0 & | & 1/7 & 3/7 \\ 0 & 1 & | & 2/7 & -1/7 \end{pmatrix} = [\mathbf{I} | \mathbf{A}^{-1}]$$

The inverse of **A** has been found as

$$\mathbf{A}^{-1} = \begin{pmatrix} 1/7 & 3/7 \\ 2/7 & -1/7 \end{pmatrix}$$

To find the inverse of a product of matrices, we can use the following formula similar to the formula of the transpose of a product of matrices:

$$(\mathbf{ABC})^{-1} = \mathbf{C}^{-1}\mathbf{B}^{-1}\mathbf{A}^{-1} \qquad (9.1.8)$$

A comparison between transposes and inverses can be found in Table 9.3.

Eigenvalues and eigenvectors
For an $n \times n$ square matrix **A**, if a scalar λ and a nonzero $n \times 1$ vector **x** satisfy

$$\lambda \mathbf{x} = \mathbf{A}\mathbf{x} \qquad (9.1.9)$$

then λ is referred to as an eigenvalue of matrix **A**, and the vector **x** as an eigenvector associated with the eigenvalue λ. Alternatively, (9.1.9) can be written as

$$(\lambda \mathbf{I} - \mathbf{A})\mathbf{x} = \mathbf{0} \qquad (9.1.10)$$

where **I** is an identity matrix. Thus a nonzero solution **x** exists if and only if

$$\det (\lambda \mathbf{I} - \mathbf{A}) = 0 \qquad (9.1.11)$$

This is referred to as a characteristic equation from which the eigenvalues of the matrix **A** can be found. The eigenvector associated with λ can be calculated from (9.1.10).

Obviously, if **x** is an eigenvector, $k\mathbf{x}$ is also an eigenvector associated with the same eigenvalue, where $k \neq 0$ is a constant.

TABLE 9.3.   Transpose and inverse.

Item	Transpose	Inverse
Definition	If $\mathbf{A} = (\alpha_{ij})$, then $\mathbf{A} = (\alpha_{ji})$	If $\mathbf{AB} = \mathbf{BA} = \mathbf{I}$, then $\mathbf{A}^{-1} = \mathbf{B}$, or $\mathbf{A} = \mathbf{B}^{-1}$
Addition	$(\mathbf{A} + \mathbf{B})^T = \mathbf{A}^T + \mathbf{B}^T$	$(\mathbf{A} + \mathbf{B})^{-1} \neq \mathbf{A}^{-1} + \mathbf{B}^{-1}$
Multiplication	$(\mathbf{AB})^T = \mathbf{B}^T\mathbf{A}^T$ $(\mathbf{ABC})^T = \mathbf{C}^T\mathbf{B}^T\mathbf{A}^T$	$(\mathbf{AB})^{-1} = \mathbf{B}^{-1}\mathbf{A}^{-1}$ $(\mathbf{ABC})^{-1} = \mathbf{C}^{-1}\mathbf{B}^{-1}\mathbf{A}^{-1}$

Example 9.1.3

Find the eigenvalues and eigenvectors for the matrix

$$A = \begin{pmatrix} -6 & -11 & -6 \\ 1 & 0 & 0 \\ 0 & 1 & 0 \end{pmatrix}$$

From (9.1.11), the characteristic equation is

$$\begin{vmatrix} \lambda + 6 & 11 & 6 \\ -1 & \lambda & 0 \\ 0 & -1 & \lambda \end{vmatrix} = 0$$

or

$$\lambda^3 + 6\lambda^2 + 11\lambda + 6 = 0$$

The eigenvalues of **A** are

$$\lambda_1 = -1, \qquad \lambda_2 = -2, \qquad \lambda_3 = -3$$

From (9.1.10), the eigenvector associated with λ_1 can be obtained by solving

$$\begin{pmatrix} -1 + 6 & 11 & 6 \\ -1 & -1 & 0 \\ 0 & -1 & -1 \end{pmatrix} \begin{pmatrix} x_1 \\ x_2 \\ x_3 \end{pmatrix} = 0$$

or

$$5x_1 + 11x_2 + 6x_3 = 0$$

$$-x_1 - x_2 = 0$$

$$-x_2 - x_3 = 0$$

If $x_1 = 1$ is chosen, $(1, -1, 1)^T$ is the eigenvector of **A** associated with the eigenvalue $\lambda_1 = -1$. Similarly, $(4, -2, 1)^T$ is the eigenvector for $\lambda_2 = -2$, and $(9, -3, 1)^T$ for $\lambda_3 = -3$.

Diagonalization of matrices and Jordan canonical form

If an $n \times n$ matrix **A** has n distinct eigenvalues, n distinct eigenvectors can be found. A modal matrix of matrix **A** consists of the eigenvectors of **A**. If the eigenvectors of an $n \times n$ matrix **A** are $\mathbf{w}_1, \mathbf{w}_2, ..., \mathbf{w}_n$, then, the modal matrix **W** is

$$\mathbf{W} = [\mathbf{w}_1, \mathbf{w}_2, ..., \mathbf{w}_n] \qquad (9.1.12)$$

A matrix **A** can be diagonalized by the modal matrix and its inverse:

$$\mathbf{J} = \mathbf{W}^{-1} \mathbf{A} \mathbf{W} \qquad (9.1.13)$$

where the diagonal matrix **J** has the eigenvalues of **A** on the diagonal.

If $\mathbf{B}^{-1}\mathbf{AB} = \mathbf{C}$, the relationship between $\mathbf{A}$ and $\mathbf{C}$ is called a similarity transformation, and $\mathbf{A}$ and $\mathbf{C}$ are similar matrices. Two similar matrices have the same eigenvalues.

If a square matrix $\mathbf{A}$ has multiple eigenvalues, a Jordan canonical form matrix instead of a diagonal matrix can be obtained, which has the form

$$\mathbf{J} = \begin{pmatrix} \mathbf{J}_{11} & 0 & \cdots & 0 \\ 0 & \mathbf{J}_{22} & \cdots & 0 \\ \vdots & \vdots & & \vdots \\ 0 & 0 & \cdots & \mathbf{J}_{rr} \end{pmatrix} \tag{9.1.14}$$

where

$$\mathbf{J}_{ii} = \begin{pmatrix} \lambda_i & 1 & 0 & \cdots & 0 \\ 0 & \lambda_i & 1 & \cdots & 0 \\ \vdots & \vdots & \vdots & & \vdots \\ 0 & 0 & 0 & \cdots & 1 \\ 0 & 0 & 0 & \cdots & \lambda_i \end{pmatrix}_{n_i \times n_i} \tag{9.1.15}$$

and λ_i is the ith eigenvalue of the matrix $\mathbf{A}$. The modal matrix $\mathbf{W}$ consists of the eigenvectors and the general eigenvectors that can be obtained from the formulas

$$(\lambda_i \mathbf{I} - \mathbf{A})\mathbf{x}_{i1} = \mathbf{0} \tag{9.1.16}$$

and

$$\mathbf{x}_{ij} = (\lambda_i \mathbf{I} - \mathbf{A})\mathbf{x}_{i(j-1)} \tag{9.1.17}$$

where $\mathbf{x}_{i1}$ is the eigenvector associated with the eigenvalue λ_i and $2 \le j \le n_i$ [Chen 1984].

The procedure for obtaining the Jordan canonical form or a diagonal matrix is shown in Table 9.4.

Example 9.1.4

Recalling Example 9.1.3, the modal matrix $\mathbf{W}$ and its inverse can be found as

TABLE 9.4. Procedure to obtain Jordan canonical form.

Step	Objective	Formulas
1	To find eigenvalues	(9.1.11)
2	To find eigenvectors	(9.1.10)
3	To find general eigenvectors for multiple eigenvalues	(9.1.16), (9.1.17)
4	To form model matrix	(9.1.12)
5	To calculate the inverse of the model matrix	(9.1.7)
6	To obtain the Jordan canonical form	(9.1.13)

$$W = \begin{pmatrix} 1 & 4 & 9 \\ -1 & -2 & -3 \\ 1 & 1 & 1 \end{pmatrix}, \qquad W^{-1} = \begin{pmatrix} -0.5 & 2.5 & 3 \\ -1 & -4 & -3 \\ -0.5 & 1.5 & 1 \end{pmatrix}$$

Then, the matrix A can be diagonalized by

$$J = W^{-1}AW$$

$$= \begin{pmatrix} -0.5 & 2.5 & 3 \\ -1 & -4 & -3 \\ -0.5 & 1.5 & 1 \end{pmatrix} \begin{pmatrix} -6 & -11 & -6 \\ 1 & 0 & 0 \\ 0 & 1 & 0 \end{pmatrix} \begin{pmatrix} 1 & 4 & 9 \\ -1 & -2 & -3 \\ 1 & 1 & 1 \end{pmatrix}$$

$$= \begin{pmatrix} -1 & 0 & 0 \\ 0 & -2 & 0 \\ 0 & 0 & -3 \end{pmatrix} = \text{diag}\{-1, -2, -3\}$$

The elements on the diagonal are the eigenvalues of the matrix A.

Quadratic form

A Quadratic form of a vector x is defined as

$$V(x) = x^T P x$$

$$= [x_1, x_2, \cdots, x_n] \begin{pmatrix} p_{11} & p_{12} & \cdots & p_{1n} \\ p_{21} & p_{22} & \cdots & p_{2n} \\ \vdots & \vdots & & \vdots \\ p_{n1} & p_{n2} & \cdots & p_{nn} \end{pmatrix} \begin{pmatrix} x_1 \\ x_2 \\ \vdots \\ x_n \end{pmatrix}$$

$$= \sum_{i,j}^{n} p_{ij} x_i x_j \qquad (9.1.18)$$

where P is a real and symmetric matrix. As we will see, the quadratic form is very useful in the analysis of system stability and the optimal control problems.

The quadratic form $V(x)$, or the symmetric matrix P, is said to be positive definite if $V(x) > 0$ for all $x \neq 0$ and $V(x) = 0$ for $x = 0$. $V(x)$ or matrix P is said to be negative definite if $-V(x)$ is positive definite. The quadratic form $V(x)$ or the symmetric matrix P is said to be positive semidefinite if $V(x) \geq 0$ for all x. $V(x)$ or matrix P is said to be negative semidefinite if $-V(x)$ is positive semidefinite.

The Sylvester criterion can be used to check positive definiteness of a quadratic form or its corresponding symmetric matrix. It states that the necessary and sufficient conditions for a quadratic form being positive definite are that all the successive principal minors of the corresponding symmetric matrix are positive, i.e.,

TABLE 9.5.   Quadratic form and its definiteness.

Definiteness	Positive	Negative	Semipositive	Seminegative
Definition	$V(x)>0$ for $x\neq0$ $V(x)=0$ for $x=0$	$V(x)<0$ for $x\neq0$ $V(x)=0$ for $x=0$	$V(x)\geq0$ for $x\neq0$ $V(x)=0$ for $x=0$	$V(x)\leq0$ for $x\neq0$ $V(x)=0$ for $x=0$
Eigenvalues λ_i of P	$\lambda_i>0$ $i=1,2,\cdots,n$	$\lambda_i<0$ $i=1,2,\cdots,n$	$\lambda_i\geq0$ $i=1,2,\cdots,n$	$\lambda_i\leq0$ $i=1,2,\cdots,n$
Sylvester Criteria	All principal minors of P are greater than 0.	All principal minors of $-P$ are greater than 0.	All principal minors of P are greater than or equal to 0.	All principal minors of $-P$ are greater than or equal to 0.

$$p_{11}>0, \qquad \begin{vmatrix} p_{11} & p_{12} \\ p_{21} & p_{22} \end{vmatrix}>0, \qquad \cdots$$

$$\begin{vmatrix} p_{11} & p_{12} & \cdots & p_{1n} \\ p_{21} & p_{22} & \cdots & p_{2n} \\ \vdots & \vdots & & \vdots \\ p_{n1} & p_{n2} & \cdots & p_{nn} \end{vmatrix}>0 \qquad (9.1.19)$$

The concepts of definiteness (positive definite, negative definite, positive semidefinite and negative semidefinite) can be extended to a general function. The Sylvester criteria for definiteness, and the relationship between eigenvalues and definiteness are shown in Table 9.5.

9.2.   State, State Space and State Space Representation

We have studied conventional control theory that, generally speaking, is applicable only to linear time-invariant systems having a single input and a single output.   In conventional control theory, only the inputs, outputs and actuating errors are considered important.   In this chapter, the state space control theory is studied.   This is a useful tool for studying systems with multiple inputs and multiple outputs and/or time-variant systems.

State space
Besides inputs, outputs and actuating errors, there are other variables in a control system which can also reflect the system behavior or states, and which may be described in a system model.

The state of a dynamic system is the smallest set of system variables that, together with the inputs, can completely determine the system behavior.   With this definition, the state of the system at time t is determined by the system state at time t_0 and the input for $t \geq t_0$.   The variables that determine the system state are referred to as state variables. All the state variables of a dynamic system, which can completely describe the system behavior, can construct a vector that is called a state vector.

If a system can be completely described by n state variables, the n-dimensional space whose n coordinate axes represent the n state variables is called a state space. The state of the system at any time can be represented by a point (a state vector) in the state space. The system state will trace a trajectory in the state space while time t varies.

State space representation

A linear time-invariant system may be described by a linear constant differential equation as

$$y^{(n)} + a_{n-1}y^{(n-1)} + \cdots + a_1\dot{y} + a_0 y = u \tag{9.2.1}$$

We can define n state variables as

$$\begin{aligned} x_1 &= y \\ x_2 &= \dot{y} \\ &\cdots\cdots \\ x_n &= y^{(n-1)} \end{aligned} \tag{9.2.2}$$

Taking the derivative of both sides of (9.2.2) and substituting (9.2.1) into it yields

$$\begin{aligned} \dot{x}_1 &= x_2 \\ \dot{x}_2 &= x_3 \\ &\cdots\cdots \\ \dot{x}_{n-1} &= x_n \\ \dot{x}_n &= -a_0 x_1 - a_1 x_2 - \cdots - a_{n-1} x_n + u \end{aligned} \tag{9.2.3}$$

Rewriting this set of equations in vector form, we have

$$\dot{x} = Ax + bu \tag{9.2.4}$$

$$y = c^T x \tag{9.2.5}$$

where

$$x = \begin{pmatrix} x_1 \\ x_2 \\ \vdots \\ x_{n-1} \\ x_n \end{pmatrix}, \quad A = \begin{pmatrix} 0 & 1 & 0 & \cdots & 0 \\ 0 & 0 & 1 & \cdots & 0 \\ \vdots & \vdots & \vdots & & \vdots \\ 0 & 0 & 0 & \cdots & 1 \\ -a_0 & -a_1 & -a_2 & \cdots & -a_{n-1} \end{pmatrix}, \quad b = \begin{pmatrix} 0 \\ 0 \\ \vdots \\ 0 \\ 1 \end{pmatrix} \tag{9.2.6}$$

and

$$\mathbf{c}^T = [1, 0, \cdots, 0] \tag{9.2.7}$$

The bold lower case letters $\mathbf{b}$ and $\mathbf{c}^T$ are used instead of capital case letters, since only a single input u and single output y appear in the system. In other words, the system is a single-input and single-output (*SISO*) system. Eq. (9.2.4) is referred to as the state equation, while (9.2.5) is called the output equation. Eqs. (9.2.4) and (9.2.5) together compose a state space representation.

For a multi-input and multi-output (*MIMO*) system, the state space representation of the system takes the form

$$\dot{\mathbf{x}} = \mathbf{A}\mathbf{x} + \mathbf{B}\mathbf{u} \tag{9.2.8}$$

$$\mathbf{y} = \mathbf{C}\mathbf{x} + \mathbf{D}\mathbf{u} \tag{9.2.9}$$

where $\mathbf{x} \in \Re^n$, $\mathbf{y} \in \Re^l$, $\mathbf{u} \in \Re^m$ and $\mathbf{A}$, $\mathbf{B}$, $\mathbf{C}$, $\mathbf{D}$ are compatible constant matrices. The system can be expressed graphically as shown in Fig. 9.2.1. In this book, we mainly discuss single-input and single-output systems. Also, usually $\mathbf{D} = \mathbf{0}$.

A particular set $\{\mathbf{A}, \mathbf{B}, \mathbf{C}, \mathbf{D}\}$ is called a system representation. For a *SISO* system, it can be written as $\{\mathbf{A}, \mathbf{b}, \mathbf{c}^T, \mathbf{0}\}$.

The system representation is not unique. A system represented by $\{\mathbf{A}, \mathbf{b}, \mathbf{c}^T, \mathbf{0}\}$, i.e., Eqs. (9.2.4) and (9.2.5), can assume a transformation $\mathbf{x} = \mathbf{T}\mathbf{z}$ or $\mathbf{z} = \mathbf{T}^{-1}\mathbf{x}$, where $\mathbf{T}$ is a nonsingular square matrix. Then,

$$\dot{\mathbf{x}} = \mathbf{T}\dot{\mathbf{z}} = \mathbf{A}\mathbf{x} + \mathbf{b}u = \mathbf{A}\mathbf{T}\mathbf{z} + \mathbf{b}u$$

or

$$\dot{\mathbf{z}} = \mathbf{T}^{-1}\mathbf{A}\mathbf{T}\mathbf{z} + \mathbf{T}^{-1}\mathbf{b}u \tag{9.2.10}$$

and

$$y = \mathbf{c}^T\mathbf{x} = \mathbf{c}^T\mathbf{T}\mathbf{z} \tag{9.2.11}$$

If we use (9.2.10) and (9.2.11) to replace (9.2.4) and (9.2.5), i.e., to select state vector $\mathbf{z}$ instead of $\mathbf{x}$ to represent the system, we find that the input and output of the system do not change. This result implies that we have freedom to choose

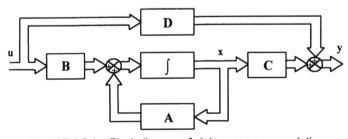

FIGURE 9.2.1. Block diagram of state space representation.

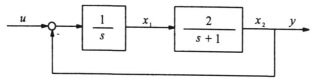

FIGURE 9.2.2.   A system in Example 9.2.1.

state variables.

State space models

The state space representation of a system can be obtained from the system differential equation, or from some other representation, such as a system transfer function, a block diagram and so on. If the block diagram of a system is available, the state space representation of the system may be obtained directly from it. We can mark the selected state variable on the block diagram, and write the state equation and output equation in a straightforward way.

Example 9.2.1

The block diagram of a system is given in Fig. 9.2.2, where the selected state variables x_1 and x_2 are marked. By inspection, we have

$$x_1 = \frac{1}{s}(u - x_2)$$

$$x_2 = \frac{2}{s+1}x_1$$

Transferred into the time domain, these equations become

$$\dot{x}_1 = u - x_2$$
$$\dot{x}_2 = 2x_1 - x_2$$

Then the state equation may be written as

$$\dot{\mathbf{x}} = \begin{pmatrix} 0 & -1 \\ 2 & -1 \end{pmatrix} \mathbf{x} + \begin{pmatrix} 1 \\ 0 \end{pmatrix} u$$

and the output equation can be found as

$$y = [0 \quad 1]\mathbf{x}$$

Example 9.2.2

Consider a third order system with a transfer function

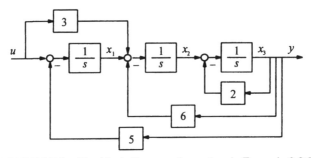

FIGURE 9.2.3. The block diagram of a system in Example 9.2.2.

$$G(s) = \frac{3s+1}{s^3 + 2s^2 + 6s + 5}$$

Dividing the numerator and denominator of the transfer function by s^3 yields

$$G(s) = \frac{Y(s)}{U(s)} = \frac{3(1/s)^2 + (1/s)^3}{1 + 2(1/s) + 6(1/s)^2 + 5(1/s)^3}$$

Therefore,

$$Y(s) = [-2(\tfrac{1}{s}) - 6(\tfrac{1}{s})^2 - 5(\tfrac{1}{s})^3]Y(s) + [3(\tfrac{1}{s})^2 + ((\tfrac{1}{s}))^3]U(s)$$

Then the block diagram of the system can be drawn as shown in Fig. 9.2.3, where state variables x_1, x_2 and x_3 are chosen. The variables in the block diagram may be related by the following equations:

$$\dot{x}_1 = -5x_3 + u$$

$$\dot{x}_2 = x_1 - 6x_3 + 3u$$

$$\dot{x}_3 = x_2 - x_3$$

$$y = x_3$$

The state space representation of the system is

$$\dot{\mathbf{x}} = \begin{pmatrix} 0 & 0 & -5 \\ 1 & 0 & -6 \\ 0 & 1 & -2 \end{pmatrix} \mathbf{x} + \begin{pmatrix} 1 \\ 3 \\ 0 \end{pmatrix} u$$

$$y = (0 \quad 0 \quad 1)\mathbf{x}$$

Controllability and observability canonical forms

Assume a *SISO* system has a transfer function

$$G(s) = \frac{b_{n-1}s^{n-1} + b_{n-2}s^{n-2} + \cdots + b_0}{s^n + a_{n-1}s^{n-1} + \cdots + a_0} \tag{9.2.12}$$

The state space representation of the system can be written in two standard forms: the controllability canonical form and observability canonical form.

If the state space representation of this *SISO* system is (9.2.4) and (9.2.5), the controllability canonical form is

$$\mathbf{x} = \begin{pmatrix} x_1 \\ x_2 \\ \vdots \\ x_{n-1} \\ x_n \end{pmatrix}, \quad \mathbf{A} = \begin{pmatrix} 0 & 1 & 0 & \cdots & 0 \\ 0 & 0 & 1 & \cdots & 0 \\ \vdots & \vdots & \vdots & & \vdots \\ 0 & 0 & 0 & \cdots & 1 \\ -a_0 & -a_1 & -a_2 & \cdots & -a_{n-1} \end{pmatrix}, \quad \mathbf{b} = \begin{pmatrix} 0 \\ 0 \\ \vdots \\ 0 \\ 1 \end{pmatrix} \tag{9.2.13}$$

and

$$\mathbf{c}^T = [b_0, b_1, \cdots, b_{n-1}] \tag{9.2.14}$$

The observability canonical form is

$$\mathbf{x} = \begin{pmatrix} x_1 \\ x_2 \\ \vdots \\ x_{n-1} \\ x_n \end{pmatrix}, \quad \mathbf{A} = \begin{pmatrix} 0 & 0 & \cdots & 0 & -a_0 \\ 1 & 0 & \cdots & 0 & -a_1 \\ 0 & 1 & \cdots & 0 & -a_2 \\ \vdots & \vdots & & \vdots & \vdots \\ 0 & 0 & \cdots & 1 & -a_{n-1} \end{pmatrix}, \quad \mathbf{b} = \begin{pmatrix} b_0 \\ b_1 \\ \vdots \\ b_{n-2} \\ b_{n-1} \end{pmatrix} \tag{9.2.15}$$

and

$$\mathbf{c}^T = [0, 0, \cdots, 1] \tag{9.2.16}$$

With these two canonical forms, we can easily write the state space representation of a *SISO* system from the system transfer function, or vice versa. It is obvious that the state space representation in (9.2.6) and (9.2.7) is a controllability canonical form and the result in Example 9.2.2 is an observability canonical form.

Transfer functions and transfer matrices

A system can be described by different models. The multivariable transfer function model can also be obtained from the state space representation. Taking the Laplace transformation of (9.2.8) and (9.2.9) yields

$$s\mathbf{X}(s) = \mathbf{A}\mathbf{X}(s) + \mathbf{B}\mathbf{U}(s) \tag{9.2.17}$$

$$\mathbf{Y}(s) = \mathbf{C}\mathbf{X}(s) + \mathbf{D}\mathbf{U}(s) \tag{9.2.18}$$

Rearranging, we can obtain

$$\mathbf{Y}(s) = [\mathbf{C}(s\mathbf{I} - \mathbf{A})^{-1}\mathbf{B} + \mathbf{D}]\mathbf{U}(s) \qquad (9.2.19)$$

The transfer function of the system will be

$$\mathbf{G}(s) = \frac{\mathbf{Y}(s)}{\mathbf{U}(s)} = \mathbf{C}(s\mathbf{I} - \mathbf{A})^{-1}\mathbf{B} + \mathbf{D} \qquad (9.2.20)$$

where $\mathbf{G}(s)$ is referred to as a transfer function matrix.

Example 9.2.3

For a system discussed in Example 9.2.1, we have the state space representation

$$\dot{\mathbf{x}} = \begin{pmatrix} 0 & -1 \\ 2 & -1 \end{pmatrix}\mathbf{x} + \begin{pmatrix} 1 \\ 0 \end{pmatrix}u$$

$$y = [0 \quad 1]\mathbf{x}$$

The system transfer function can be calculated as

$$G(s) = \frac{Y(s)}{U(s)} = \mathbf{c}^T(s\mathbf{I} - \mathbf{A})^{-1}\mathbf{b}$$

$$= [0 \quad 1]\begin{pmatrix} s & 1 \\ -2 & s+1 \end{pmatrix}^{-1}\begin{pmatrix} 1 \\ 0 \end{pmatrix}$$

$$= [0 \quad 1]\frac{1}{s(s+1)+2}\begin{pmatrix} s+1 & -1 \\ 2 & s \end{pmatrix}\begin{pmatrix} 1 \\ 0 \end{pmatrix}$$

$$= \frac{1}{s(s+1)+2}[0 \quad 1]\begin{pmatrix} s+1 \\ 2 \end{pmatrix}$$

$$= \frac{2}{s^2+s+2}$$

We can also obtain this result from the block diagram in Fig. 9.2.2.

9.3. SOLUTION TO THE STATE EQUATION

With the state space representation, we deal with the system in the time-domain. The system state equation is a differential matrix equation of first order, whose solution requires matrix manipulation.

Solution of homogeneous state equation

If the input of a system is zero, the state equation reduces to a homogeneous equation as

$$\dot{\mathbf{x}} = \mathbf{A}\mathbf{x} \qquad (9.3.1)$$

where $\mathbf{x}$ is an n vector, and $\mathbf{A}$ is an $n \times n$ system matrix. Taking Laplace transforms of both sides of (9.3.1) yields

$$s\mathbf{X}(s) - \mathbf{x}(0) = \mathbf{A}\mathbf{X}(s) \qquad (9.3.2)$$

where $\mathbf{X}(s) = \mathcal{L}[\mathbf{x}(t)]$. Rearrangement gives

$$(s\mathbf{I} - \mathbf{A})\mathbf{X}(s) = \mathbf{x}(0)$$

or

$$\mathbf{X}(s) = (s\mathbf{I} - \mathbf{A})^{-1}\mathbf{x}(0) \qquad (9.3.3)$$

Note that

$$(s\mathbf{I} - \mathbf{A})^{-1} = \frac{\mathbf{I}}{s} + \frac{\mathbf{A}}{s^2} + \frac{\mathbf{A}^2}{s^3} + \cdots + \frac{\mathbf{A}^k}{s^{k+1}} + \cdots \qquad (9.3.4)$$

which can be proved by

$$(s\mathbf{I} - \mathbf{A})(\frac{\mathbf{I}}{s} + \frac{\mathbf{A}}{s^2} + \frac{\mathbf{A}^2}{s^3} + \cdots + \frac{\mathbf{A}^k}{s^{k+1}} + \cdots)$$

$$= s(\frac{\mathbf{I}}{s} + \frac{\mathbf{A}}{s^2} + \frac{\mathbf{A}^2}{s^3} + \cdots + \frac{\mathbf{A}^k}{s^{k+1}} + \cdots) + (-\frac{\mathbf{A}}{s} - \frac{\mathbf{A}^2}{s^2} - \frac{\mathbf{A}^3}{s^3} - \cdots - \frac{\mathbf{A}^k}{s^k} - \cdots)$$

$$= \mathbf{I}$$

Taking inverse Laplace transform of (9.3.4), we have,

$$\mathcal{L}^{-1}[(s\mathbf{I} - \mathbf{A})^{-1}] = \mathbf{I} + \mathbf{A}t + \frac{\mathbf{A}^2}{2!}t^2 + \cdots + \frac{\mathbf{A}^k}{k!}t^k + \cdots \qquad (9.3.5)$$

Denote $e^{\mathbf{A}t} = \mathcal{L}^{-1}[(s\mathbf{I} - \mathbf{A})^{-1}]$, then from (9.3.3), the solution of (9.3.1) is

$$\mathbf{x}(t) = e^{\mathbf{A}t}\mathbf{x}(0) \qquad (9.3.6)$$

State transition matrix

It is noted that (9.3.6) gives the relationship between states at instants t and 0. Generally, the relationship between states at t and t_0 can be written as

$$\mathbf{x}(t) = \Phi(t, t_0)\mathbf{x}(t_0) \qquad (9.3.7)$$

where $\Phi(t, t_0)$ is an $n \times n$ matrix, and is referred to as a state transition matrix. If the system (9.3.1) is time-invariant, then

$$\Phi(t, t_0) = e^{A(t-t_0)} \qquad (9.3.8)$$

The state transition matrix is a unique solution of the equation

$$\dot{\Phi}(t, t_0) = A\Phi(t, t_0) \qquad \Phi(t_0, t_0) = I \qquad (9.3.9)$$

which can be proved from the fact that

$$x(t_0) = \Phi(t_0, t_0)x(t_0) = Ix(t_0)$$

and that

$$A\Phi(t, t_0)x(t_0) = A x(t) = \dot{x}(t) = \dot{\Phi}(t, t_0)x(t_0)$$

The state transition matrix has the following properties:

1. $\dot{\Phi}(t, t_0) = A\Phi(t, t_0)$,
2. $\Phi(t_0, t_0) = I$
3. $\Phi(t_2, t_1)\Phi(t_1, t_0) = \Phi(t_2, t_0)$
4. $\Phi(t_1, t_0) = \Phi^{-1}(t_0, t_1)$

Laplace transform method for calculating e^{At}

The calculation of e^{At} is essential for the solution of a time-invariant system. When the system matrix A is known, the state transition matrix e^{At} can be obtained in different ways, one of which is the Laplace transform method. The calculation formula is (9.3.5).

Example 9.3.1

Assume a system matrix A to be

$$A = \begin{pmatrix} 0 & 1 \\ -2 & -3 \end{pmatrix}$$

Then,

$$|sI - A| = (s + 1)(s + 2)$$

and the inverse of $(sI - A)$ is

$$(sI - A)^{-1} = \begin{pmatrix} s & -1 \\ 2 & s+3 \end{pmatrix}^{-1} = \frac{1}{(s+1)(s+2)} \begin{pmatrix} s+3 & 1 \\ -2 & s \end{pmatrix}$$

The state transition matrix will be

$$e^{At} = \mathcal{L}^{-1}[(sI - A)^{-1}]$$

$$= \begin{pmatrix} \mathcal{L}^{-1}\left(\dfrac{s+3}{(s+1)(s+2)} \right) & \mathcal{L}^{-1}\left(\dfrac{1}{(s+1)(s+2)} \right) \\ \mathcal{L}^{-1}\left(\dfrac{-2}{(s+1)(s+2)} \right) & \mathcal{L}^{-1}\left(\dfrac{s}{(s+1)(s+2)} \right) \end{pmatrix}$$

$$= \begin{pmatrix} 2e^{-t} - e^{-2t} & e^{-t} - e^{-2t} \\ -2e^{-t} + 2e^{-2t} & -e^{-t} + 2e^{-2t} \end{pmatrix}$$

Cayley-Hamilton Theorem

We can also use the Cayley-Hamilton theorem to calculate the state transition matrix.

The Cayley-Hamilton Theorem states that every square matrix satisfies its own characteristic equation, i.e., if the characteristic equation of matrix $\mathbf{A}$ is

$$f(s) = s^n + f_{n-1}s^{n-1} + \cdots + f_1 s + f_0 = 0 \qquad (9.3.10)$$

then

$$f(\mathbf{A}) = \mathbf{A}^n + f_{n-1}\mathbf{A}^{n-1} + \cdots + f_1\mathbf{A} + f_0\mathbf{I} = 0 \qquad (9.3.11)$$

The proof of the theorem holds for the assumption that all the eigenvalues of $\mathbf{A}$ are distinct. Denote $\mathbf{W}$ as the modal matrix of $\mathbf{A}$, then

$$\mathbf{A} = \mathbf{W}\Lambda\mathbf{W}^{-1}, \qquad \mathbf{A}^2 = \mathbf{W}\Lambda^2\mathbf{W}^{-1}, \qquad \cdots \qquad \mathbf{A}^n = \mathbf{W}\Lambda^n\mathbf{W}^{-1}$$

where

$$\Lambda = \mathrm{diag}(\lambda_1, \lambda_2, \cdots, \lambda_n)$$

It follows that

$$f(\mathbf{A}) = \mathbf{W}(\Lambda^n + f_{n-1}\Lambda^{n-1} + \cdots + f_1\Lambda + f_0\mathbf{I})\mathbf{W}^{-1} = 0 \qquad (9.3.12)$$

because Λ is a diagonal matrix whose (i, i) element satisfies

$$\lambda_i^n + f_{n-1}\lambda_i^{n-1} + \cdots + f_1\lambda_i + f_0 = 0 \qquad (9.3.13)$$

The proof is completed.

For an $n \times n$ square matrix $\mathbf{A}$ and its eigenvalues λ_i, denote the characteristic polynomial of $\mathbf{A}$ as $\Delta(\mathbf{A})$. Then, we have $\Delta(\mathbf{A}) = 0$ and $\Delta(\lambda_i) = 0$. For a polynomial $f(x)$ of order greater than n, dividing it by $\Delta(x)$ yields

$$f(x) = g(x)\Delta(x) + R(x) \qquad (9.3.14)$$

where $R(x)$ has an order less than n, and usually $n - 1$ is assumed. It follows that

$$f(\lambda_i) = g(\lambda_i)\Delta(\lambda_i) + R(\lambda_i) \qquad (9.3.15)$$

Similarly, with the Cayley-Hamilton theorem, we have

$$f(\mathbf{A}) = g(\mathbf{A})\Delta(\mathbf{A}) + R(\mathbf{A}) = R(\mathbf{A}) \qquad (9.3.16)$$

With (9.3.15) and (9.3.16), $e^{\mathbf{A}t}$ can be obtained.

Example 9.3.2

Recalling Example 9.3.1, the matrix $\mathbf{A}$ has eigenvalues $\lambda_1 = -1$ and $\lambda_2 = -2$. Since $\mathbf{A}$ is a 2×2 matrix, i.e., $n = 2$, we assume that $R(\mathbf{A})$ in (9.3.16) is of order 1, or

$$e^{\mathbf{A}t} = \alpha_1 \mathbf{A} + \alpha_0 \mathbf{I}$$

The coefficients of this equation may be obtained from the following equations:

$$e^{-t} = -\alpha_1 + \alpha_0$$
$$e^{-2t} = -2\alpha_1 + \alpha_0$$

Solving them yields

$$\alpha_0 = 2e^{-t} - e^{-2t}$$
$$\alpha_1 = e^{-t} - e^{-2t}$$

Therefore, the state transition matrix is

$$e^{\mathbf{A}t} = \alpha_1 \mathbf{A} + \alpha_0 \mathbf{I}$$
$$= (e^{-t} - e^{-2t})\begin{pmatrix} 0 & 1 \\ -2 & -3 \end{pmatrix} + (2e^{-t} - e^{-2t})\begin{pmatrix} 1 & 0 \\ 0 & 1 \end{pmatrix}$$
$$= \begin{pmatrix} 2e^{-t} - e^{-2t} & e^{-t} - e^{-2t} \\ -2e^{-t} + 2e^{-2t} & -e^{-t} + 2e^{-2t} \end{pmatrix}$$

Solution of state equation

The solution of the nonhomogeneous state equation

$$\dot{\mathbf{x}} = \mathbf{A}\mathbf{x} + \mathbf{b}u$$

can be obtained by the Laplace transform method. Taking Laplace transforms on both sides of the above equation yields

$$s\mathbf{X}(s) - \mathbf{x}(0) = \mathbf{A}\mathbf{X}(s) + \mathbf{b}U(s) \qquad (9.3.17)$$

where $\mathbf{X}(s) = \mathcal{L}[\mathbf{x}(t)]$. Rearrangement gives

$$(s\mathbf{I} - \mathbf{A})\mathbf{X}(s) = \mathbf{x}(0) + \mathbf{b}U(s)$$

or

TABLE 9.6. Solutions of systems.

Item	Homogeneous state equation	Nonhomogeneous state equation
State in s-domain	$\mathbf{X}(s) = (s\mathbf{I} - \mathbf{A})^{-1}\mathbf{x}(0)$	$\mathbf{X}(s) = (s\mathbf{I} - \mathbf{A})^{-1}\mathbf{x}(0) + (s\mathbf{I} - \mathbf{A})^{-1}\mathbf{b}U(s)$
State in t-domain	$\mathbf{x}(t) = e^{\mathbf{A}(t-t_0)}\mathbf{x}(t_0)$	$\mathbf{x}(t) = e^{\mathbf{A}(t-t_0)}\mathbf{x}(t_0) + \int_{t_0}^{t} e^{\mathbf{A}(t-\tau)}\mathbf{b}u(\tau)d\tau$
Output in s-domain	$\mathbf{Y}(s) = \mathbf{C}(s\mathbf{I} - \mathbf{A})^{-1}\mathbf{x}(0)$	$\mathbf{Y}(s) = \mathbf{C}(s\mathbf{I} - \mathbf{A})^{-1}\mathbf{x}(0) + \mathbf{C}(s\mathbf{I} - \mathbf{A})^{-1}\mathbf{b}U(s)$
Output in t-domain	$y(t) = \mathbf{C}e^{\mathbf{A}(t-t_0)}\mathbf{x}(t_0)$	$y(t) = \mathbf{C}e^{\mathbf{A}(t-t_0)}\mathbf{x}(t_0) + \mathbf{C}\int_{t_0}^{t} e^{\mathbf{A}(t-\tau)}\mathbf{b}u(\tau)d\tau$

$$\mathbf{X}(s) = (s\mathbf{I} - \mathbf{A})^{-1}\mathbf{x}(0) + (s\mathbf{I} - \mathbf{A})^{-1}\mathbf{b}U(s) \qquad (9.3.18)$$

Introducing the convolution theorem of Laplace transform studied in Chapter 2 yields

$$\mathbf{x}(t) = e^{\mathbf{A}t}\mathbf{x}(0) + \int_{0}^{t} e^{\mathbf{A}(t-\tau)}\mathbf{b}u(\tau)d\tau \qquad (9.3.19)$$

If the initial time is t_0, (9.3.19) takes the form

$$\mathbf{x}(t) = e^{\mathbf{A}(t-t_0)}\mathbf{x}(t_0) + \int_{t_0}^{t} e^{\mathbf{A}(t-\tau)}\mathbf{b}u(\tau)d\tau \qquad (9.3.20)$$

The solutions of the system equations are summarized in Table 9.6.

System Response
The response of a system to any inputs may be calculated by (9.3.19) or (9.3.20) as well as the output equation.

Example 9.3.3
　　Consider a linear system

$$\dot{\mathbf{x}} = \begin{pmatrix} -12 & 2/3 \\ -36 & -1 \end{pmatrix}\mathbf{x} + \begin{pmatrix} 1/3 \\ 1 \end{pmatrix}u$$

$$y = [1 \ \ 1]\mathbf{x}$$

with the initial condition $\mathbf{x}(0) = [2 \ \ 1]^T$. The response of the system to a unit step input can be obtained as follows:
　　The characteristic equation of the system is

$$|s\mathbf{I} - \mathbf{A}| = (s + 4)(s + 9) = 0$$

and the inverse of $(s\mathbf{I} - \mathbf{A})$ is

$$(s\mathbf{I} - \mathbf{A})^{-1} = \begin{pmatrix} s+12 & -2/3 \\ 36 & s+1 \end{pmatrix}^{-1} = \frac{1}{(s+4)(s+9)}\begin{pmatrix} s+1 & 2/3 \\ -36 & s+12 \end{pmatrix}$$

The state transition matrix will be

$$e^{\mathbf{A}t} = \mathcal{L}^{-1}[(s\mathbf{I} - \mathbf{A})^{-1}] = \begin{pmatrix} \mathcal{L}^{-1}\left(\dfrac{s+1}{(s+4)(s+9)}\right) & \mathcal{L}^{-1}\left(\dfrac{2}{(s+4)(s+9)}\right) \\ \mathcal{L}^{-1}\left(\dfrac{-36}{(s+4)(s+9)}\right) & \mathcal{L}^{-1}\left(\dfrac{s+12}{(s+4)(s+9)}\right) \end{pmatrix}$$

$$= \begin{pmatrix} \mathcal{L}^{-1}\left(\dfrac{-3/5}{(s+4)} + \dfrac{8/5}{(s+9)}\right) & \mathcal{L}^{-1}\left(\dfrac{2/15}{(s+4)} + \dfrac{-2/15}{(s+9)}\right) \\ \mathcal{L}^{-1}\left(\dfrac{-36/5}{(s+4)} + \dfrac{36/5}{(s+9)}\right) & \mathcal{L}^{-1}\left(\dfrac{8/5}{(s+4)} + \dfrac{-3/5}{(s+9)}\right) \end{pmatrix}$$

$$= \begin{pmatrix} -\dfrac{3}{5}e^{-4t} + \dfrac{8}{5}e^{-9t} & \dfrac{2}{15}(e^{-4t} - e^{-9t}) \\ \dfrac{36}{5}(-e^{-4t} + e^{-9t}) & \dfrac{8}{5}e^{-4t} - \dfrac{3}{5}e^{-9t} \end{pmatrix}$$

Then, the state $\mathbf{x}(t)$ will be

$$\mathbf{x}(t) = e^{\mathbf{A}t}\mathbf{x}(0) + \int_0^t e^{\mathbf{A}(t-\tau)}\mathbf{b}u(\tau)d\tau$$

$$= \begin{pmatrix} -\dfrac{3}{5}e^{-4t} + \dfrac{8}{5}e^{-9t} & \dfrac{2}{15}(e^{-4t} - e^{-9t}) \\ \dfrac{36}{5}(-e^{-4t} + e^{-9t}) & \dfrac{8}{5}e^{-4t} - \dfrac{3}{5}e^{-9t} \end{pmatrix}\begin{pmatrix} 2 \\ 1 \end{pmatrix}$$

$$+ \int_0^t \begin{pmatrix} -\dfrac{3}{5}e^{-4(t-\tau)} + \dfrac{8}{5}e^{-9(t-\tau)} & \dfrac{2}{15}(e^{-4(t-\tau)} - e^{-9(t-\tau)}) \\ \dfrac{36}{5}(-e^{-4(t-\tau)} + e^{-9(t-\tau)}) & \dfrac{8}{5}e^{-4(t-\tau)} - \dfrac{3}{5}e^{-9(t-\tau)} \end{pmatrix}\begin{pmatrix} \dfrac{1}{3} \\ 1 \end{pmatrix}d\tau$$

$$= \begin{pmatrix} -\dfrac{16}{15}e^{-4t} + \dfrac{46}{15}e^{-9t} \\ -\dfrac{64}{5}e^{-4t} + \dfrac{69}{5}e^{-9t} \end{pmatrix} + \int_0^t \begin{pmatrix} -\dfrac{1}{15}e^{-4(t-\tau)} + \dfrac{6}{15}e^{-9(t-\tau)} \\ -\dfrac{4}{5}e^{-4(t-\tau)} + \dfrac{9}{5}e^{-9(t-\tau)} \end{pmatrix}d\tau$$

$$= \begin{pmatrix} -\dfrac{16}{15}e^{-4t} + \dfrac{46}{15}e^{-9t} \\ -\dfrac{64}{5}e^{-4t} + \dfrac{69}{5}e^{-9t} \end{pmatrix} + \begin{pmatrix} \dfrac{1}{36} + \dfrac{1}{60}e^{-4t} - \dfrac{2}{45}e^{-9t} \\ \dfrac{1}{5}e^{-4t} - \dfrac{1}{5}e^{-9t} \end{pmatrix}$$

$$= \begin{pmatrix} \dfrac{1}{36} - \dfrac{21}{20}e^{-4t} + \dfrac{136}{45}e^{-9t} \\ -\dfrac{63}{5}e^{-4t} + \dfrac{68}{5}e^{-9t} \end{pmatrix}$$

Therefore, the system response will be

$$y(t) = [1 \quad 1]\mathbf{x}(t) = [1 \quad 1]\begin{pmatrix} \dfrac{1}{36} - \dfrac{21}{20}e^{-4t} + \dfrac{136}{45}e^{-9t} \\ -\dfrac{63}{5}e^{-4t} + \dfrac{68}{5}e^{-9t} \end{pmatrix}$$

$$= \dfrac{1}{36} - \dfrac{273}{20}e^{-4t} + \dfrac{748}{45}e^{-9t}$$

9.4. CONTROLLABILITY AND OBSERVABILITY

There are two basic problems we need to consider. The first one is the coupling between the input and the state: Can any state be controlled by the input? This is a controllability problem. Another is the relationship between the state and the output: Can all the information about the state be observed from the output? This is an observability problem.

Controllability and observability for linear systems

A system is said to be controllable at time t_0 if it is possible to find an unconstrained control vector to transfer any initial state to the origin in a finite time interval. Stated mathematically, the system is controllable at t_0 if for any $\mathbf{x}(t_0)$, there exists $\mathbf{u}_{[t_0,\ t_1]}$ that gives $\mathbf{x}(t_1) = 0$ ($t_1 > t_0$). If this is true for all initial time t_0 and all initial states $\mathbf{x}(t_0)$, the system is completely controllable.

A system is said to be observable at time t_0 if it is possible to determine the state $\mathbf{x}(t_0)$ from the output function over a finite time interval. In mathematical terms, the system is observable at t_0 if any $\mathbf{x}(t_0)$ can be estimated by the observation of $\mathbf{y}_{[t_0,\ t_1]}$ ($t_1 > t_0$). If this is true for all time t_0 and all states $\mathbf{x}(t_0)$, the system is completely observable.

Controllability criterion

Consider a linear system

$$\dot{\mathbf{x}} = \mathbf{A}\mathbf{x} + \mathbf{b}u \qquad (9.4.1)$$

$$y = \mathbf{c}^T\mathbf{x} \qquad (9.4.2)$$

The solution of (9.4.1) has been obtained as

$$\mathbf{x}(t) = e^{\mathbf{A}t}\mathbf{x}(0) + \int_0^t e^{\mathbf{A}(t-\tau)}\mathbf{b}u(\tau)d\tau \qquad (9.4.3)$$

From the definition of complete controllability, the system is completely controllable if and only if for any initial state $\mathbf{x}(0)$, there is an input $u(t)$ such that at time $t_1 > t_0$, we have

$$\mathbf{x}(t_1) = 0 = e^{\mathbf{A}t_1}\mathbf{x}(0) + \int_0^{t_1} e^{\mathbf{A}(t_1-\tau)}\mathbf{b}u(\tau)d\tau$$

or

$$\mathbf{x}(0) = -\int_0^{t_1} e^{-\mathbf{A}\tau}\mathbf{b}u(\tau)d\tau \qquad (9.4.4)$$

By the Cayley-Hamilton theorem, $e^{\mathbf{A}\tau}$ can be written as

$$e^{-\mathbf{A}\tau} = \alpha_0(\tau)\mathbf{I} + \alpha_1(\tau)\mathbf{A} + \cdots + \alpha_{n-1}(\tau)\mathbf{A}^{n-1} \qquad (9.4.5)$$

Therefore

$$\mathbf{x}(0) = -\sum_{k=0}^{n-1}\mathbf{A}^k\mathbf{b}\int_0^{t_1}\alpha_k(\tau)u(\tau)d\tau \qquad (9.4.6)$$

Denote

$$\int_0^{t_1}\alpha_k(\tau)u(\tau)d\tau = \beta_k \qquad (9.4.7)$$

Then, (9.4.6) becomes

$$\mathbf{x}(0) = -\sum_{k=0}^{n-1}\mathbf{A}^k\mathbf{b}\beta_k = [\mathbf{b}|\,\mathbf{Ab}|\,\mathbf{A}^2\mathbf{b}|\cdots|\,\mathbf{A}^{n-1}\mathbf{b}]\begin{pmatrix}\beta_0\\\beta_1\\\vdots\\\beta_{n-1}\end{pmatrix} \qquad (9.4.8)$$

This result shows that every state vector $\mathbf{x}(0)$ must be expressed as a linear combination of the columns of the matrix

$$\mathbf{P} = [\mathbf{b}|\mathbf{Ab}|\mathbf{A}^2\mathbf{b}|\cdots|\mathbf{A}^{n-1}\mathbf{b}] \qquad (9.4.9)$$

It implies that the matrix $\mathbf{P}$ is of full rank, i.e.,

$$\text{Rank } [\mathbf{b}|\mathbf{Ab}|\mathbf{A}^2\mathbf{b}|\cdots|\mathbf{A}^{n-1}\mathbf{b}] = n \qquad (9.4.10)$$

For a *MIMO* system (9.2.8), the necessary and sufficient condition for the system to have completely controllability is

$$\text{Rank } [\mathbf{B}|\mathbf{AB}|\mathbf{A^2B}| \cdots |\mathbf{A}^{n-1}\mathbf{B}] = n \qquad (9.4.11)$$

Matrix **P** is referred to as a controllability matrix.

Observability criterion

Since observability is the coupling between the state variables and the output of a system, when we discuss observability, we can assume the system input to be zero, i.e., we only consider a linear system

$$\dot{\mathbf{x}} = \mathbf{Ax} \qquad (9.4.12)$$

$$y = \mathbf{c}^T\mathbf{x} \qquad (9.4.13)$$

The solution of (9.4.12) and (9.4.13) can be obtained as

$$y(t) = \mathbf{c}^T e^{-\mathbf{A}t}\mathbf{x}(0) \qquad (9.4.14)$$

By the Cayley-Hamilton theorem, $e^{\mathbf{A}t}$ can be written as

$$e^{-\mathbf{A}\tau} = \alpha_0(\tau)\mathbf{I} + \alpha_1(\tau)\mathbf{A} + \cdots + \alpha_{n-1}(\tau)\mathbf{A}^{n-1}$$

It follows that

$$y(t) = -\sum_{k=0}^{n-1}\alpha_k(t)\mathbf{c}^T\mathbf{A}^k\mathbf{x}(0) = [\alpha_0(t)|\alpha_1(t)|\cdots|\alpha_{n-1}(t)]\begin{pmatrix} \mathbf{c}^T \\ \mathbf{c}^T\mathbf{A} \\ \vdots \\ \mathbf{c}^T\mathbf{A}^{n-1} \end{pmatrix}\mathbf{x}(0) \quad (9.4.15)$$

The determination of $\mathbf{x}(0)$ from $y(t)$ $0 \le t \le t_1$ requires that the matrix **Q**, where

$$\mathbf{Q} = \begin{pmatrix} \mathbf{c}^T \\ \mathbf{c}^T\mathbf{A} \\ \vdots \\ \mathbf{c}^T\mathbf{A}^{n-1} \end{pmatrix} = [\mathbf{c}|\mathbf{A}^T\mathbf{c}|(\mathbf{A}^T)^2\mathbf{c}|\cdots|(\mathbf{A}^T)^{n-1}\mathbf{c}]^T \qquad (9.4.16)$$

is of full rank, i.e.,

$$Rank[\mathbf{c}|\mathbf{A}^T\mathbf{c}|(\mathbf{A}^T)^2\mathbf{c}|\cdots|(\mathbf{A}^T)^{n-1}\mathbf{c}]^T = n \qquad (9.4.17)$$

For the general system (9.2.8), the necessary and sufficient condition of a linear system for complete observability is

$$Rank[\mathbf{C}^T|\mathbf{A}^T\mathbf{C}^T|(\mathbf{A}^T)^2\mathbf{C}^T|\cdots|(\mathbf{A}^T)^{n-1}\mathbf{C}^T] = n \qquad (9.4.18)$$

Matrix **Q** is referred to as an observability matrix.

Example 9.4.1

Determine the controllability and observability of a system if the state space representation of the system is

$$\dot{\mathbf{x}} = \begin{pmatrix} -2 & 2 & 1 \\ 0 & -2 & 0 \\ 1 & -4 & 0 \end{pmatrix} \mathbf{x} + \begin{pmatrix} 0 \\ 0 \\ 1 \end{pmatrix} u$$

$$\mathbf{y} = \begin{pmatrix} 1 & 0 & 1 \\ 0 & 1 & 0 \end{pmatrix} \mathbf{x}$$

Since

$$\mathbf{b} = \begin{pmatrix} 0 \\ 0 \\ 1 \end{pmatrix} \qquad \mathbf{Ab} = \begin{pmatrix} -2 & 2 & 1 \\ 0 & -2 & 0 \\ 1 & -4 & 0 \end{pmatrix} \begin{pmatrix} 0 \\ 0 \\ 1 \end{pmatrix} = \begin{pmatrix} 1 \\ 0 \\ 0 \end{pmatrix}$$

$$\mathbf{A}^2 \mathbf{b} = \mathbf{A}(\mathbf{Ab}) = \begin{pmatrix} -2 & 2 & 1 \\ 0 & -2 & 0 \\ 1 & -4 & 0 \end{pmatrix} \begin{pmatrix} 1 \\ 0 \\ 0 \end{pmatrix} = \begin{pmatrix} -2 \\ 0 \\ 1 \end{pmatrix}$$

then,

$$\mathbf{P} = \begin{pmatrix} 0 & 1 & -2 \\ 0 & 0 & 0 \\ 1 & 0 & 1 \end{pmatrix} \qquad \text{and} \qquad \text{Rank } \mathbf{P} = 2 < 3$$

The system is uncontrollable.

Since

$$\mathbf{C} = \begin{pmatrix} 1 & 0 & 1 \\ 0 & 1 & 0 \end{pmatrix}, \qquad \mathbf{CA} = \begin{pmatrix} 1 & 0 & 1 \\ 0 & 1 & 0 \end{pmatrix} \begin{pmatrix} -2 & 2 & 1 \\ 0 & -2 & 0 \\ 1 & -4 & 0 \end{pmatrix} = \begin{pmatrix} -1 & -2 & 1 \\ 0 & -2 & 0 \end{pmatrix}$$

$$\mathbf{CA}^2 = (\mathbf{CA})\mathbf{A} = \begin{pmatrix} -1 & -2 & 1 \\ 0 & -2 & 0 \end{pmatrix} \begin{pmatrix} -2 & 2 & 1 \\ 0 & -2 & 0 \\ 1 & -4 & 0 \end{pmatrix} = \begin{pmatrix} 3 & -2 & -1 \\ 0 & 4 & 0 \end{pmatrix}$$

we have

$$\mathbf{Q} = \begin{pmatrix} 1 & 0 & 1 \\ 0 & 1 & 0 \\ -1 & -2 & 1 \\ 0 & -2 & 0 \\ 3 & -2 & -1 \\ 0 & 4 & 0 \end{pmatrix}$$

Because of

$$\det \mathbf{Q}^T\mathbf{Q} = \det \begin{pmatrix} 1 & 0 & -1 & 0 & 3 & 0 \\ 0 & 1 & -2 & -2 & -2 & 4 \\ 1 & 0 & 1 & 0 & -1 & 0 \end{pmatrix} \begin{pmatrix} 1 & 0 & 1 \\ 0 & 1 & 0 \\ -1 & -2 & 1 \\ 0 & -2 & 0 \\ 3 & -2 & -1 \\ 0 & 4 & 0 \end{pmatrix}$$

$$= \det \begin{pmatrix} 11 & -4 & -3 \\ -4 & 29 & 0 \\ -3 & 0 & 3 \end{pmatrix} = 957 + 0 + 0 - 261 - 0 - 48 = 648 \neq 0$$

we have $\text{Rank}\,\mathbf{Q} = 3$. The system is completely observable.

The basic concepts of controllability and observability are summarized in Table 9.7.

Canonical forms for controllability and observability
It is easy to verify that a *SISO* system that has a controllable canonical form in (9.2.13) and (9.2.14) is completely controllable.

Also, a *SISO* system that has an observable canonical form in (9.2.15) and (9.2.16) is completely observable.

A general state space representation of a completely controllable (observable) *SISO* system can be transferred into a controllable (observable) canonical form. This can be done by writing the system transfer function, i.e., using (9.2.20) and (9.2.12 ~ 9.2.16) to obtain a controllable (observable) canonical form.

The similarity transformation can be used to transfer a general state space representation to a controllable canonical form. For a general form (9.4.1) and (9.4.2), if the characteristic equation of the system is known as

TABLE 9.7. Controllability and observability.

	Controllability	Observability
Reflection	Coupling between the input and the state	Coupling between the state and the output
Definition	For any $\mathbf{x}(t_0)$, $\mathbf{u}_{[t_0,\,t_1]}$ exists to give $\mathbf{x}(t_1) = 0$ $(t_1 > t_0)$	Any $\mathbf{x}(t_0)$ can be observed by output $\mathbf{y}_{[t_0,\,t_1]}$ $(t_1 > t_0)$.
Criterion matrix	$\mathbf{P} = [\mathbf{B}|\mathbf{AB}|\mathbf{A}^2\mathbf{B}| \cdots |\mathbf{A}^{n-1}\mathbf{B}]$	$\mathbf{Q} = [\mathbf{C}^T|\mathbf{A}^T\mathbf{C}^T|(\mathbf{A}^T)^2\mathbf{C}^T|\cdots|(\mathbf{A}^T)^{n-1}\mathbf{C}^T]$
General criterion	Rank $\mathbf{P} = n$	Rank $\mathbf{Q} = n$

$$s^n + a_{n-1}s^{n-1} + \cdots + a_1 s + a_0 = 0$$

the similarity transform matrix $\mathbf{M}$ is defined by

$$\mathbf{M} = \mathbf{PS} \qquad (9.4.19)$$

where $\mathbf{P}$ is the controllability matrix, and $\mathbf{S}$ has the form

$$\mathbf{S} = \begin{pmatrix} a_1 & a_2 & a_3 & \cdots & a_{n-1} & 1 \\ a_2 & a_3 & a_4 & \cdots & 1 & 0 \\ \vdots & \vdots & \vdots & & \vdots & \vdots \\ a_{n-1} & 1 & 0 & \cdots & 0 & 0 \\ 1 & 0 & 0 & \cdots & 0 & 0 \end{pmatrix} \qquad (9.4.20)$$

Other methods can be found in the literature [Kuo 1991].

Example 9.4.2

The state space representation of a system is

$$\dot{\mathbf{x}} = \mathbf{Ax} + \mathbf{b}u = \begin{pmatrix} 1 & 1 \\ 2 & 3 \end{pmatrix}\mathbf{x} + \begin{pmatrix} 1 \\ 1 \end{pmatrix}u$$

$$y = \mathbf{c}^T\mathbf{x} = [1 \ 0]\mathbf{x}$$

The characteristic equation of the system can be calculated as

$$\begin{vmatrix} s-1 & -1 \\ -2 & s-3 \end{vmatrix} = s^2 - 4s + 1 = 0$$

The related matrices are

$$\mathbf{P} = [\mathbf{b}|\mathbf{Ab}] = \begin{pmatrix} 1 & 2 \\ 1 & 5 \end{pmatrix}, \qquad \mathbf{S} = \begin{pmatrix} -4 & 1 \\ 1 & 0 \end{pmatrix}$$

$$\mathbf{M} = \mathbf{PS} = \begin{pmatrix} 1 & 2 \\ 1 & 5 \end{pmatrix}\begin{pmatrix} -4 & 1 \\ 1 & 0 \end{pmatrix} = \begin{pmatrix} -2 & 1 \\ 1 & 1 \end{pmatrix}$$

$$\mathbf{M}^{-1} = \frac{-1}{3}\begin{pmatrix} 1 & -1 \\ -1 & -2 \end{pmatrix}$$

With the similarity transform $\mathbf{x} = \mathbf{Mz}$, the controllable canonical form is obtained as

$$\dot{\mathbf{z}} = \hat{\mathbf{A}}\mathbf{z} + \hat{\mathbf{b}}u = \begin{pmatrix} 0 & 1 \\ -1 & 4 \end{pmatrix}\mathbf{z} + \begin{pmatrix} 0 \\ 1 \end{pmatrix}u$$

$$y = \hat{\mathbf{c}}^T\mathbf{z} = [-2 \ 1]\mathbf{z}$$

Controllability and observability criterion for a diagonal form

If the eigenvalues of system matrix A in (9.2.8) are distinct, the system (9.2.8) and (9.2.9) can be diagonalized by similarity transformation through the relations

$$\dot{z} = M^{-1}AMz + M^{-1}Bu \qquad (9.4.21)$$

$$y = CMz \qquad (9.4.22)$$

where $x = Mz$, M is the modal matrix of A, and $M^{-1}AM$ is a diagonal matrix. When the system matrix becomes diagonal, all the states are decoupled. If one row in $M^{-1}B$ is zero, the corresponding state variable cannot be changed by the input. In other words, for a linear system with distinct eigenvalues, the system is completely controllable if and only if no row of $M^{-1}B$ has all zero elements. If any column in CM is zero, then we cannot get any information about the corresponding state variable from the output. This means that for a linear system with distinct eigenvalues, the system is completely observable if and only if no column of CM has all zero elements.

Example 9.4.3

Consider a linear system

$$\dot{x} = \begin{pmatrix} -6 & -11 & -6 \\ 1 & 0 & 0 \\ 0 & 1 & 0 \end{pmatrix} x + \begin{pmatrix} 1 & 0 \\ 0 & 1 \\ 0 & 0 \end{pmatrix} u$$

$$y = [-1 \quad 0 \quad 1]x$$

Recalling Example 9.1.4, the modal matrix and its inverse have been found as

$$M = \begin{pmatrix} 1 & 4 & 9 \\ -1 & -2 & -3 \\ 1 & 1 & 1 \end{pmatrix} \qquad M^{-1} = \begin{pmatrix} 0.5 & 2.5 & 3 \\ -1 & -4 & -3 \\ 0.5 & 1.5 & 1 \end{pmatrix}$$

Then, the system can be transferred into

$$\dot{z} = M^{-1}AMz + M^{-1}Bu = \begin{pmatrix} -1 & 0 & 0 \\ 0 & -2 & 0 \\ 0 & 0 & -3 \end{pmatrix} z + \begin{pmatrix} 0.5 & 2.5 \\ -1 & -4 \\ 0.5 & 1.5 \end{pmatrix} u$$

$$y = CMz = [0 \quad -3 \quad -8]z$$

Thus, the system is completely controllable, but not completely observable.

If the eigenvalues of matrix A are not distinct, diagonalization may be impossible. By similarity transform, a Jordan canonical form can be obtained.

TABLE 9.8. Other criteria for controllability and observability.

System	Completely controllable	Completely observable
System with state matrix of diagonal form	No zero row in $\mathbf{M}^{-1}\mathbf{B}$. $\mathbf{M}$ is modal matrix of $\mathbf{A}$	No zero column in $\mathbf{CM}$. $\mathbf{M}$ is modal matrix of $\mathbf{A}$
System with state matrix of Jordan canonical form	(1) No two Jordan blocks are associated with the same eigenvalue. (2) The rows of $\mathbf{M}^{-1}\mathbf{B}$ corresponding to the last row (or the only row) of each Jordan block are nonzero.	(1) No two Jordan blocks are associated with the same eigenvalue. (2) The columns of $\mathbf{M}^{-1}\mathbf{B}$ corresponding to the first column (or the only column) of each Jordan block are nonzero.
SISO system with a controllable canonical form	Completely controllable	Unknown
SISO system with an observable canonical form	Unknown	Completely observable
SISO system with a transfer function	No pole-zero cancellation	

Criteria of controllability and observability for the Jordan canonical form can be found in the literature [Ogata 1990, Brogan 1991]. Table 9.8 provides a summary

Controllability, observability and pole-zero cancellation

For a *SISO* system, the pole-zero cancellation in the transfer function will result in an uncontrollable system, and/or an unobservable system. In other words, the transfer function has no cancellation if and only if the *SISO* system is completely controllable and completely observable.

Example 9.4.4

Consider a system as shown in Fig. 9.4.1, where the state variables are selected. It follows that

$$\frac{x_1}{u} = \frac{1}{s+a} \quad \text{and} \quad \frac{x_2}{x_1} = \frac{s+c}{s+b}$$

or

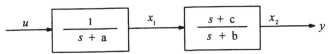

FIGURE 9.4.1. A system for Example 9.4.4.

$$\dot{x}_1 = -ax_1 + u$$

$$\dot{x}_2 = -bx_2 + \dot{x}_1 + cx_1 = (-a+c)x_1 - bx_2 + u$$

The state space representation of the system is

$$\begin{pmatrix} \dot{x}_1 \\ \dot{x}_2 \end{pmatrix} = \begin{pmatrix} -a & 0 \\ -a+c & -b \end{pmatrix} \begin{pmatrix} x_1 \\ x_2 \end{pmatrix} + \begin{pmatrix} 1 \\ 1 \end{pmatrix} u$$

$$y = \begin{bmatrix} 0 & 1 \end{bmatrix} \begin{pmatrix} x_1 \\ x_2 \end{pmatrix}$$

The controllability matrix is

$$\mathbf{P} = \begin{pmatrix} 1 & -a \\ 1 & -a-b+c \end{pmatrix}$$

and the observability matrix is

$$\mathbf{Q} = \begin{pmatrix} 0 & 1 \\ -a+c & -b \end{pmatrix}$$

If $b = c \neq a$, the pole-zero cancellation gives rise to RankP = 1 and RankQ = 2, i.e., the system is completely observable but uncontrollable. If $a = c \neq b$, then RankP = 2 and RankQ = 1, i.e., the system is completely controllable but unobservable. If $b = c = a$, RankP = 1 and RankQ = 1 are obtained, i.e., the system is both unobservable and uncontrollable.

Examination of controllability and observability with *PCET*
The *PCET* software package included with this textbook will enable the reader to examine the controllability and observability of a linear system with state space representation by using the subprogram *LSA* (Linear State Space Analysis). The procedure for examining the system controllability and observability is

    1. Type *PCET* to start.
    2. Select *LSA* to enter the sub-program.
    3. Select "1" to define system dimension.
    4. Select "2" to enter the system data.
    5. Select "6" to display the results of controllability and observability of the system on the screen.

The eight selections available after entering *LSA* are explained in detail in Table 9.9.

9.5. STATE FEEDBACK AND POLE ASSIGNMENT

Feedback is a common method in process control engineering to improve system performance. For a system with state space representation, feedback is commonly used to achieve suitable closed loop pole locations. This procedure is called pole assignment or eigenvalue assignment.

State feedback and output feedback
State variable feedback may be used to adjust the input to achieve satisfactory system performance. This arrangement, shown in Fig. 9.5.1a, is referred to as state feedback, whereas Fig. 9.5.1b represents output feedback. Consider a system with the state space representation

TABLE 9.9.   Function selection in *LSA* of *PCET*.

Step	Item	Function and results	Exit key
1	Define system dimension & option	Database file name Title of case study Number of state variables Number of input variables Number of output variables Field width for data input Initial guess for eigenvalue Tolerance for convergence	Esc
2	Enter data for matrices *A*, *B*, *C* & *D*	Input data of matrices **A**, **B**, **C** & **D**.	Esc
3	Load a case study from the database	The saved examples appear on screen for selection.	Return
4	Delete a case study from the database	The saved examples appear on screen for deletion.	Esc
5	Save current case study in database	Save the current data as a case study example with the name entered in selection 1 "Title of case study".	Return
6	Display results on screen	Controllability matrix Observability matrix Rank of controllability matrix Rank of observability matrix Conclusion Eigenvalues of **A** Modal matrix Inverse of modal matrix Transformed matrices **A**, **B** & **C**	Return or Esc
7	Print results	All items in '6' are printed on printer.	Return
8	Return to main menu	Back to main menu directly	Return

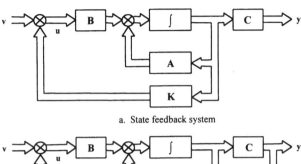

a. State feedback system

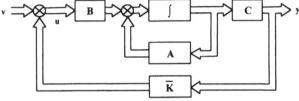

b. Output feedback system

FIGURE 9.5.1.   Feedback configurations.

$$\dot{x} = Ax + Bu \qquad (9.5.1)$$

$$y = Cx \qquad (9.5.2)$$

where $x \in \Re^n$, $y \in \Re^l$, $u \in \Re^m$ and A, B, C are compatible constant matrices. The state feedback can be represented by

$$u = Kx + v \qquad (9.5.3)$$

where K is an $m \times n$ constant feedback matrix, and v is a reference input vector that has the same dimension as the input. The output feedback can be described by

$$u = \overline{K}y + v \qquad (9.5.4)$$

where $\overline{K}$ is an $m \times l$ constant output feedback matrix.

Usually, only output variables rather than state variables are available. This does not mean that state feedback cannot be realized. The so-called observers can be used for the estimation or reconstruction of the state variables if inputs and outputs are measurable [Brogan 1991, Chen 1984]. The state x contains all pertinent information about the system. It is worthwhile to have an understanding of the effects of state feedback.

Closed loop matrix
For state feedback, substituting (9.5.3) into (9.5.1) and (9.5.2) yields the closed loop system equation and output equation

$$\dot{x} = (A + BK)x + Bv \qquad (9.5.5)$$

$$y = Cx \qquad (9.5.6)$$

The eigenvalues of the closed loop system matrix $A + BK$, i.e., the poles of the closed loop system, are determined by the characteristic equation

$$|sI - (A + BK)| = 0 \qquad (9.5.7)$$

Similarly, for the output feedback, we have

$$\dot{x} = (A + B\overline{K}C)x + Bv \qquad (9.5.8)$$

$$y = Cx \qquad (9.5.9)$$

The poles of the closed loop system are determined by

$$\left|sI - (A + B\overline{K}C)\right| = 0 \qquad (9.5.10)$$

The effect of feedback on controllability and observability
State feedback does not affect the controllability of the system, which can be proved directly from the definition of controllability. If the open loop system is controllable at t, then for any $x(t_0)$, there exists $u_{[t_0, t_1]}$ that gives $x(t_1) = 0$ $(t_1 > t_0)$. From (9.5.3), select the reference input v as

$$v_{[t_0, t_1]} = u_{[t_0, t_1]} - Kx$$

which also gives $x(t_1) = 0$ $(t_1 > t_0)$, i.e., the closed loop system is controllable at t_0. Similarly, if an open loop system is completely controllable, then the state feedback system is also completely controllable. Since the open loop system can be obtained by a feedback matrix $-K$ from the state feedback system, if the closed loop system is controllable, then the open loop system is controllable.

State feedback may affect the observability of the system, as illustrated by the following example.

Example 9.5.1
Consider a linear system with a state space representation

$$\dot{x} = \begin{pmatrix} a & b \\ c & d \end{pmatrix} x + \begin{pmatrix} e \\ f \end{pmatrix} u$$

$$y = [g \quad h]x$$

The observability matrix is

$$Q = \begin{pmatrix} g & h \\ ga + hc & gb + hd \end{pmatrix}$$

If the observability matrix $\mathbf{Q}$ is of full rank, the system is completely observable. If state feedback is implemented with a feedback matrix $\mathbf{K}$ where

$$\mathbf{K} = [\alpha \quad \beta]$$

then the closed loop system matrix is

$$\overline{\mathbf{A}} = \begin{pmatrix} a & b \\ c & d \end{pmatrix} + \begin{pmatrix} e \\ f \end{pmatrix}(\alpha \quad \beta) = \begin{pmatrix} a + \alpha e & b + \beta e \\ c + \alpha f & d + \beta f \end{pmatrix}$$

The observability matrix becomes

$$\overline{\mathbf{Q}} = \begin{pmatrix} g & h \\ ga + g\alpha e + hc + h\alpha f & gb + g\beta e + hd + h\beta f \end{pmatrix}$$

If α and β are chosen to satisfy

$$kg = ga + g\alpha e + hc + h\alpha f$$
$$kh = gb + g\beta e + hd + h\beta f$$

or

$$\alpha = \frac{kg - ga - hc}{ge + hf} \quad \text{and} \quad \beta = \frac{kh - gb - hd}{ge + hf}$$

where k is a constant, then $\mathrm{Rank}\,\mathbf{Q} = 1 < 2$, and the system is not completely observable.

The effects of feedback configurations as well as other items regarding the controllability and observability of systems are summarized in Table 9.10.

Pole assignment with state feedback
For a *SISO* system with the controllable canonical form of (9.2.13), the system poles are determined only by the entries in the last row of matrix A, which are the coefficients of the characteristic polynomial.
    For the system (9.2.13) and (9.2.14), if the state feedback matrix is

$$\mathbf{k}^T = [k_1, k_2, \cdots, k_n] \tag{9.5.11}$$

TABLE 9.10. Effects of some items on controllability and observability.

Item	Controllability	Observability
Similarity transformation	No change	No change
Pole-zero cancellation in a *SISO* system	Results in an uncontrollable and/or unobservable system	
State feedback	No change	May change
Output feedback	No change	No change

then the state space representation of the closed loop system will be

$$\dot{\mathbf{x}} = (\mathbf{A} + \mathbf{b}\mathbf{k}^T)\mathbf{x} + \mathbf{b}v$$

$$y = \mathbf{c}^T \mathbf{x}$$

where v is a reference input, and

$$\mathbf{A} + \mathbf{b}\mathbf{k}^T = \begin{pmatrix} 0 & 1 & 0 & \cdots & 0 \\ 0 & 0 & 1 & \cdots & 0 \\ \vdots & \vdots & \vdots & & \vdots \\ 0 & 0 & 0 & \cdots & 1 \\ -a_0 + k_1 & -a_1 + k_2 & -a_2 + k_3 & \cdots & -a_{n-1} + k_n \end{pmatrix}, \quad \mathbf{b} = \begin{pmatrix} 0 \\ 0 \\ \vdots \\ 0 \\ 1 \end{pmatrix}$$

$$\mathbf{c}^T = (b_0, b_1, \cdots, b_{n-1}) \tag{9.5.12}$$

The closed loop system still has a controllable canonical form, and the characteristic equation of the system will be

$$s^n + (a_{n-1} - k_n)s^{n-1} + \cdots + (a_1 - k_2)s + a_0 - k_1 = 0 \tag{9.5.13}$$

If the closed loop poles s_1, s_2, $\cdots$, s_n are assigned, the characteristic equation can also be written as

$$(s - s_1)(s - s_2) \cdots (s - s_n) = s^n + p_{n-1}s^{n-1} + \cdots + p_1 s + p_0 = 0 \tag{9.5.14}$$

Comparing (9.5.13) and (9.5.14) yields

$$k_n = a_{n-1} - p_{n-1}$$
$$k_{n-1} = a_{n-2} - p_{n-2}$$
$$\cdots\cdots\cdots\cdots \tag{9.5.15}$$
$$k_1 = a_0 - p_0$$

With the state feedback $\mathbf{k}$ in (9.5.15), the closed loop system poles can be assigned.

Example 9.5.2
   Consider the system in Example 9.4.2. With the similarity transform $\mathbf{x} = \mathbf{M}\mathbf{z}$, where $\mathbf{M}$ and its inverse are

$$\mathbf{M} = \begin{pmatrix} -2 & 1 \\ 1 & 1 \end{pmatrix}, \quad \mathbf{M}^{-1} = \frac{-1}{3}\begin{pmatrix} 1 & -1 \\ -1 & -2 \end{pmatrix}$$

the controllable canonical form has been found as

$$\dot{\mathbf{z}} = \hat{\mathbf{A}}\mathbf{z} + \hat{\mathbf{b}}u = \begin{pmatrix} 0 & 1 \\ -1 & 4 \end{pmatrix}\mathbf{z} + \begin{pmatrix} 0 \\ 1 \end{pmatrix}u$$

$$y = \hat{\mathbf{c}}^T \mathbf{z} = [-2 \ 1]\mathbf{z}$$

The system open loop characteristic equation was obtained as

$$s^2 - 4s + 1 = 0$$

If poles $s_1 = -1$ and $s_2 = -2$ are assigned, the closed loop characteristic equation is

$$s^2 + 3s + 2 = 0$$

From (9.5.15), the state feedback matrix $\mathbf{k}^T = [k_1, \ k_2]$ can be obtained as

$$k_1 = 1 - 2 = -1$$
$$k_2 = -4 - 3 = -7$$

or

$$\mathbf{k}^T = [k_1, \ k_2] = [-1 \ \ -7]$$

It should be noted that this solution is from the controllable canonical form. It means that the closed loop system is represented by

$$\dot{\mathbf{z}} = (\hat{\mathbf{A}} + \hat{\mathbf{b}}\mathbf{k}^T)\mathbf{z} + \hat{\mathbf{b}}v$$

$$= (\mathbf{M}^{-1}\mathbf{A}\mathbf{M} + \mathbf{M}^{-1}\mathbf{b}\mathbf{k}^T\mathbf{M}^{-1}\mathbf{M})\mathbf{z} + \mathbf{M}^{-1}\mathbf{b}v$$

$$= \mathbf{M}^{-1}(\mathbf{A} + \mathbf{b}\mathbf{k}^T\mathbf{M}^{-1})\mathbf{M}\mathbf{z} + \mathbf{M}^{-1}\mathbf{b}v$$

By the relationship $\mathbf{x} = \mathbf{M}\mathbf{z}$, the original system can be obtained as

$$\dot{\mathbf{x}} = (\mathbf{A} + \mathbf{b}\mathbf{k}^T\mathbf{M}^{-1})\mathbf{x} + \mathbf{b}v$$

$$y = \hat{\mathbf{c}}^T\mathbf{z} = \mathbf{c}^T\mathbf{x}$$

Obviously, the state feedback matrix for the original system should be

$$\mathbf{k}^T\mathbf{M}^{-1} = \frac{-1}{3}[-1 \ \ -7]\begin{pmatrix} 1 & -1 \\ -1 & -2 \end{pmatrix} = \frac{-1}{3}[6 \ \ 15] = [-2 \ \ -5]$$

Remarks
The pole assignment method presented in this section may be tedious. For a *SISO* system of lower order, the coefficient comparison method is a better choice. From the assigned closed loop poles, the assigned system characteristic equation can be easily obtained. The closed loop characteristic equation in the entries of state feedback matrix $\mathbf{K}$ can be obtained from Eq. (9.5.7). By comparing the coefficients of the two characteristic equations, the feedback matrix can be found.

9.6. STABILITY AND LYAPUNOV CRITERIA

In the previous chapters, system stability and many stability criteria have been studied. In this section, another important stability criterion, the Lyapunov second method, will be studied.

Definitions of Lyapunov stability
For a general system

$$\dot{\mathbf{x}} = \mathbf{f}(\mathbf{x}, t) \tag{9.6.1}$$

the state $\mathbf{x}_e$ is referred to as an equilibrium state, if

$$\mathbf{f}(\mathbf{x}_e, t) = 0 \tag{9.6.2}$$

The equilibrium state is said to be stable in the sense of Lyapunov if for any positive number ε, there is a positive number $\delta(\varepsilon, t_0)$ such that if

$$\|\mathbf{x}(t_0) - \mathbf{x}_e\| \le \delta \tag{9.6.3}$$

then for all time $t = t_0$ the trajectory satisfies

$$\|\mathbf{x} - \mathbf{x}_e\| \le \varepsilon \tag{9.6.4}$$

where $\|\cdot\|$ represents Euclidean norm. Otherwise, the system is unstable. The equilibrium state is said to be asymptotically stable if it is stable and there exists a positive number δ' such that whenever

$$\|\mathbf{x}(t_0) - \mathbf{x}_e\| \le \delta' \tag{9.6.5}$$

the trajectory satisfies

$$\lim_{t \to \infty} \|\mathbf{x} - \mathbf{x}_e\| = 0 \tag{9.6.6}$$

Fig. 9.6.1 shows three possible trajectories in a phase plane.

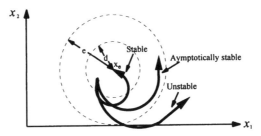

FIGURE 9.6.1.   Stable and unstable states.

The equilibrium state is said to be asymptotically stable in the large if δ' approaches infinity. In other words, if the system is stable and all the states approach the unique equilibrium state x_e. Usually, we assume $x_e = 0$, i.e., the equilibrium state is at the origin. For a linear system, $x_e = 0$ can always be true by a coordinate transformation.

Lyapunov stability theorem

Energy concepts are widely used in engineering problems. The energy of an object is always positive. If the derivative of the energy is always negative, the energy of the object will approach zero, i.e., the object will go into a static state. For a purely mathematical system, we construct a fictitious energy function, called the Lyapunov function, denoted by $V(x)$, or $V(x_1, x_2, \cdots, x_n)$. If $V(x)$ is always positive, and its derivative is always negative, then the system state must approach zero, which means that the system is asymptotically stable.

For a general system

$$\dot{x} = f(x, t)$$

and an equilibrium state at the origin $x_e = 0$, if there exists a scalar function $V(x, t)$ such that $V(x, t)$ is positive definite and $\dot{V}(x, t)$ is negative definite, then the equilibrium state $x_e = 0$ is asymptotically stable. If $V(x, t)$ also has the property $V(x, t) \to \infty$ as $x \to \infty$, then the equilibrium state is asymptotically stable in the large.

Example 9.6.1

Discuss the stability of a general system with the state equation

$$\dot{x}_1 = x_2 - 3x_1(x_1^2 + x_2^2)$$
$$\dot{x}_2 = -x_1 - 3x_2(x_1^2 + x_2^2)$$

Select a scalar function

$$V(x) = x_1^2 + x_2^2$$

Because $V(x) > 0$ for $x > 0$ and $V(x) = 0$ for $x = 0$, the scalar function $V(x)$ is positive definite. Obviously,

$$\dot{V}(x) = \frac{d}{dt}(x_1^2 + x_2^2) = 2x_1\dot{x}_1 + 2x_2\dot{x}_2$$

$$= 2x_1[x_2 - 3x_1(x_1^2 + x_2^2)] + 2x_2[-x_1 - 3x_2(x_1^2 + x_2^2)]$$

$$= 2x_1x_2 - 6x_1^2(x_1^2 + x_2^2) - 2x_2x_1 - 6x_2^2(x_1^2 + x_2^2)$$

$$= -6(x_1^2 + x_2^2)^2$$

Since $\dot{V}(\mathbf{x}) < 0$ for $\mathbf{x} \neq \mathbf{0}$ and $V(\mathbf{x}) = 0$ for $\mathbf{x} = 0$, $\dot{V}(\mathbf{x})$ is negative definite. Then the system at the origin is asymptotically stable. When $\mathbf{x} \to \infty$, i.e., $x_1 \to \infty$ or $x_2 \to \infty$, we have $V(\mathbf{x}) \to \infty$. Therefore, the system at the origin is also asymptotically stable in the large.

Stability analysis for a linear time-invariant system
For a linear time-invariant system

$$\dot{\mathbf{x}} = \mathbf{Ax} \qquad (9.6.7)$$

if the Lyapunov function $V(\mathbf{x})$ is chosen as the following quadratic form

$$V(\mathbf{x}) = \mathbf{x}^T \mathbf{Px} \qquad (9.6.8)$$

where $\mathbf{P}$ is a real symmetrical positive definite matrix, then its derivative is

$$\dot{V}(\mathbf{x}) = \dot{\mathbf{x}}^T \mathbf{Px} + \mathbf{x}^T \mathbf{P}\dot{\mathbf{x}}$$
$$= \mathbf{x}^T \mathbf{A}^T \mathbf{Px} + \mathbf{x}^T \mathbf{PAx}$$
$$= \mathbf{x}^T (\mathbf{A}^T \mathbf{P} + \mathbf{PA})\mathbf{x} \qquad (9.6.9)$$

We denote

$$\dot{V}(\mathbf{x}) = -\mathbf{x}^T \mathbf{Qx}$$

Asymptotic stability requires that $-\mathbf{Q}$ must be negative definite, i.e., $\mathbf{Q}$ must be positive definite, and

$$-\mathbf{Q} = \mathbf{A}^T \mathbf{P} + \mathbf{PA} \qquad (9.6.10)$$

In fact, we always go backward: we specify a positive definite symmetrical matrix $\mathbf{Q}$ first, then find matrix $\mathbf{P}$ to examine its positive definiteness. The positive definiteness of matrix $\mathbf{P}$ can be checked by the Sylvester criterion as mentioned in Section 9.1. For convenience, $\mathbf{Q} = \mathbf{I}$ is usually chosen.

Theorem For the linear time-invariant system (9.6.7), the necessary and sufficient condition for the equilibrium state $\mathbf{x} = \mathbf{0}$ being asymptotically stable in the large is that given any $n \times n$ positive definite real symmetrical matrix $\mathbf{Q}$, the matrix $\mathbf{P}$ obtained from (9.6.10) is positive definite.

Lyapunov stability and related criterion are summarized in Table 9.11.

Example 9.6.2
Consider the linear system

TABLE 9.11. Lyapunov stability.

	Stable in the sense of Lyapunov	Asymptotically stable	Asymptotically stable in the large
Definition	$\forall\, \varepsilon > 0,\ \exists\, \delta(\varepsilon, t_0)$ such that if $\|\mathbf{x}(t_0) - \mathbf{x}_e\| \le \delta$ then for $t \ge t_0$, $\|\mathbf{x} - \mathbf{x}_e\| \le \varepsilon$	Stable and whenever $\|\mathbf{x}(t_0) - \mathbf{x}_e\| \le \delta'$ then $\lim_{t\to\infty} \|\mathbf{x} - \mathbf{x}_e\| = 0$	Asymptotically stable and $\delta' \to \infty$
General criterion	$\exists V(\mathbf{x}, t) > 0$ $\dot{V}(\mathbf{x}, t) \le 0$	$\exists V(\mathbf{x}, t) > 0,$ $\dot{V}(\mathbf{x}, t) < 0$	$\exists V(\mathbf{x}, t) > 0,$ $\dot{V}(\mathbf{x}, t) < 0$ and $V(\mathbf{x}, t) \to \infty$ as $\mathbf{x} \to \infty$
Linear system criterion		$-\mathbf{I} = \mathbf{A}^T\mathbf{P} + \mathbf{P}\mathbf{A}$ $\mathbf{P}$ is positive definite	$-\mathbf{I} = \mathbf{A}^T\mathbf{P} + \mathbf{P}\mathbf{A}$ $\mathbf{P}$ is positive definite
System	System may not be asymptotically stable.	System must be stable.	System must be asymptotically stable.

$$\dot{\mathbf{x}} = \begin{pmatrix} -1 & -2 \\ 1 & -4 \end{pmatrix} \mathbf{x}$$

Choose $\mathbf{Q} = \mathbf{I}$. Eq. (9.6.10) yields

$$\begin{pmatrix} -1 & 1 \\ -2 & -4 \end{pmatrix}\begin{pmatrix} p_1 & p_2 \\ p_2 & p_3 \end{pmatrix} + \begin{pmatrix} p_1 & p_2 \\ p_2 & p_3 \end{pmatrix}\begin{pmatrix} -1 & -2 \\ 1 & -4 \end{pmatrix} = \begin{pmatrix} -1 & 0 \\ 0 & -1 \end{pmatrix}$$

or

$$-2p_1 + 2p_2 = -1$$
$$-2p_1 - 5p_2 + p_3 = 0$$
$$-4p_2 - 8p_3 = -1$$

Solving these equations yields

$$p_1 = 23/60, \qquad p_2 = -7/60, \qquad p_3 = 11/60$$

The matrix $\mathbf{P}$ is

$$\mathbf{P} = \frac{1}{60}\begin{pmatrix} 23 & -7 \\ -7 & 11 \end{pmatrix}$$

Because

$$\frac{23}{60} > 0 \qquad \text{and} \qquad \begin{vmatrix} 23/60 & -7/60 \\ -7/60 & 11/60 \end{vmatrix} = 204/60 > 0$$

the matrix $\mathbf{P}$ is positive definite, and thus the system is asymptotically stable. The Lyapunov function can be chosen as

$$V(\mathbf{x}) = \mathbf{x}^T \mathbf{P} \mathbf{x} = \begin{bmatrix} x_1 & x_2 \end{bmatrix} \frac{1}{60} \begin{pmatrix} 23 & -7 \\ -7 & 11 \end{pmatrix} \begin{pmatrix} x_1 \\ x_2 \end{pmatrix} = \frac{1}{60}(23x_1^2 - 14x_1 x_2 + 11x_2^2)$$

$\dot{V}(\mathbf{x})$ is negative definite, since

$$\dot{V}(\mathbf{x}) = \frac{46}{60}x_1\dot{x}_1 - \frac{14}{60}\dot{x}_1 x_2 - \frac{14}{60}x_1\dot{x}_2 + \frac{22}{60}x_2\dot{x}_2$$

$$= \frac{46}{60}x_1(-x_1 - 2x_2) - \frac{14}{60}(-x_1 - 2x_2)x_2 - \frac{14}{60}x_1(x_1 - 4x_2) + \frac{22}{60}x_2(x_1 - 4x_2)$$

$$= -x_1^2 - x_2^2 < 0$$

Problems

9.1. If $\mathbf{A}$ and $\mathbf{B}$ are of the same dimension and $\mathbf{XA} = \mathbf{XB}$, can we conclude that $\mathbf{A} = \mathbf{B}$?

9.2. If $\mathbf{A}$ is an $n \times n$ matrix, and $\mathbf{A}^k = \mathbf{0}$, prove that

$$(\mathbf{I} - \mathbf{A})^{-1} = \mathbf{I} + \mathbf{A} + \mathbf{A}^2 + \mathbf{A}^3 + \cdots + \mathbf{A}^{k-1}$$

9.3. Find the inverse of the $n \times n$ matrix $\mathbf{P}$ where

$$\mathbf{P} = \begin{pmatrix} 1 & 0 & \cdots & 0 & a_{n-1} \\ 0 & 1 & \cdots & 0 & a_{n-2} \\ \vdots & \vdots & & \vdots & \vdots \\ 0 & 0 & \cdots & 1 & a_1 \\ 0 & 0 & \cdots & 0 & 1 \end{pmatrix}$$

Hint: use the result of Problem 9.2.

9.4. Determine the rank of the following matrices:

a. $\mathbf{A} = \begin{pmatrix} 2 & 3 & -4 \\ 1 & 5 & 7 \\ 4 & 6 & -8 \end{pmatrix}$

b. $\mathbf{A} = \begin{pmatrix} 2 & 1 & 4 \\ 2 & -5 & 3 \\ 4 & 6 & 8 \end{pmatrix}$

c. $\mathbf{A} = \begin{pmatrix} 6 & -12 & 0 & 9 \\ 0 & 7 & 1 & 3 \\ 4 & -8 & 0 & 6 \end{pmatrix}$

9.5. Prove that two similar matrices have the same eigenvalues.

9.6. Diagonalize the following matrices:

$$\mathbf{A}_1 = \begin{pmatrix} 2 & -1 \\ -1 & 2 \end{pmatrix} \qquad \mathbf{A}_2 = \begin{pmatrix} 8 & -8 & 2 \\ 4 & -3 & -2 \\ 3 & -4 & 1 \end{pmatrix}$$

9.7. Find the definiteness of the following quadratic forms:

a. $V = 3x_1^2 + 8x_1 x_2 + 4x_2^2$

b. $V = x_1^2 + 2x_2^2 + 10x_3^2 + 2x_1 x_2 + 4x_2 x_3 + 6x_3 x_1$

c. $V = 2x_1^2 + 10x_2^2 + 18x_3^2 - 8x_1 x_2 + 12x_2 x_3$

d. $V = x_1^2 - 4x_1 x_2 + 6x_3 x_1 + 5x_2^2 - 10x_2 x_3 + 8x_3^2$

9.8. Write the controllability canonical form and the observability canonical form of the state space representation for the systems whose transfer functions are:

a. $\dfrac{s^2 + 2s + 1}{s^3 + 5s^2 + 6s + 9}$
b. $\dfrac{8s + 2}{s^4 + 3s^3 + 5s^2 + 4s + 9}$

9.9. Write down the state space representation for the system in Fig. 9.P.1.

9.10. Given the state space representation of a system as

$$\dot{\mathbf{x}} = \begin{pmatrix} 1 & -1 \\ 2 & -3 \end{pmatrix} \mathbf{x} + \begin{pmatrix} 1 \\ 1 \end{pmatrix} u$$

$$y = [0 \quad 1]\mathbf{x}$$

find the system transfer function.

9.11. Find $e^{\mathbf{A}t}$ where the system matrix $\mathbf{A}$ is

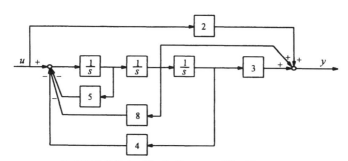

FIGURE 9.P.1.   Block diagram of Problem 9.9.

a. $A = \begin{pmatrix} -2 & -1 \\ -1 & -2 \end{pmatrix}$ b. $A = \begin{pmatrix} 0 & 1 & 0 \\ 0 & 0 & 1 \\ 0 & 1 & 0 \end{pmatrix}$

9.12. If the transition matrix of the system is

$$e^{At} = \begin{pmatrix} 2e^{-t} - e^{-2t} & 2(e^{-2t} - e^{-t}) \\ e^{-t} - e^{-2t} & 2e^{-2t} - e^{-t} \end{pmatrix}$$

what is the matrix A?

9.13. The state space representation of a system is

$$\dot{x} = \begin{pmatrix} 0 & 1 \\ -6 & -5 \end{pmatrix} x + \begin{pmatrix} 1 \\ 1 \end{pmatrix} u$$

$$y = [1 \quad 1]x$$

If the initial condition $x(0) = [0\ 1]^T$, what is the response of the system to a unit step input?

9.14. For the system with the state space presentation

$$\dot{x} = \begin{pmatrix} 0 & 3 \\ -3 & 0 \end{pmatrix} x + \begin{pmatrix} 0 \\ 1 \end{pmatrix} u$$

$$y = [1 \quad 2]x$$

if the initial state of the system at $t = 0$ is $x = [1\ 1]^T$, find the system response to a unit step input. What is the output value at $t = \pi/6$ and $t = \pi/3$?

9.15. Determine the controllability and observability of the following control systems:

a. $\dot{x} = \begin{pmatrix} -1 & 0 \\ 0 & -5 \end{pmatrix} x + \begin{pmatrix} 1 \\ 1 \end{pmatrix} u$

$y = [1 \quad 0]x$

b. $\dot{x} = \begin{pmatrix} -3 & 1 & 0 \\ 0 & 0 & 1 \\ -4 & -8 & -5 \end{pmatrix} x + \begin{pmatrix} 1 \\ 2 \\ 1 \end{pmatrix} u$

$y = (3 \quad 1 \quad 0)x + 2u$

c. $\dot{x} = \begin{pmatrix} -1 & 0 & 1 \\ 0 & -2 & 1 \\ 1 & -2 & 0 \end{pmatrix} x + \begin{pmatrix} 1 & 1 \\ 0 & 0 \\ 1 & 1 \end{pmatrix} u$

$y = \begin{pmatrix} 1 & 1 & 0 \\ 1 & 1 & 0 \end{pmatrix} x$

d. $\dot{\mathbf{x}} = \begin{pmatrix} -1 & 0 & 0 \\ 0 & -2 & 1 \\ 1 & -2 & 0 \end{pmatrix} \mathbf{x} + \begin{pmatrix} 0 & 0 \\ 0 & 0 \\ 1 & 1 \end{pmatrix} u$

$y = \begin{pmatrix} 1 & 0 & 0 \\ 1 & 0 & 0 \end{pmatrix} \mathbf{x}$

9.16. The state space representation of a system is

$$\dot{\mathbf{x}} = \begin{pmatrix} 2 & 1 \\ 2 & 3 \end{pmatrix} \mathbf{x} + \begin{pmatrix} 0 \\ 1 \end{pmatrix} u$$

$$y = [1 \quad 0]\mathbf{x}$$

Design a state feedback controller to assign system poles at –2 and –4.

9.17. For the system

$$\dot{\mathbf{x}} = \begin{pmatrix} 0 & 1 & 0 \\ 0 & 0 & 1 \\ 0 & -2 & -3 \end{pmatrix} \mathbf{x} + \begin{pmatrix} 0 \\ 0 \\ 1 \end{pmatrix} u$$

$$y = \begin{pmatrix} 0 & 1 & 1 \end{pmatrix} \mathbf{x}$$

design a state feedback matrix **k** to assign the poles of the closed loop system at $-1 \pm j1$ and -2.

9.18. Consider a control system with the system equations

$$\dot{x}_1 = x_2 - x_1$$

$$\dot{x}_2 = -x_2 - x_1^3$$

If we choose the Lyapunov function as

$$V(x) = \frac{1}{4}x_1^4 + \frac{1}{2}x_2^2$$

determine the system stability.

9.19. For the system

$$\dot{x}_1 = x_2 - x_1$$

$$\dot{x}_2 = -a(1 + x_2)^2 x_2 - x_1 \qquad (a > 0)$$

determine the system stability.

9.20. Find a Lyapunov function for the system

$$\dot{\mathbf{x}} = \begin{pmatrix} -1 & 1 \\ 2 & -3 \end{pmatrix} \mathbf{x}$$

and determine the system stability.

CHAPTER 10

OPTIMAL CONTROL

10.1. MODELS, CONSTRAINTS AND COST FUNCTION

Optimal control plays an important role in control engineering and systems science.

The word "optimal" intuitively means doing the best job with a defined criterion. The performance criterion, or performance index, or cost function, for judging a system is important because it is the goal we attempt to reach, and it will determine the nature of the optimal control. A system that is optimal for one criterion may be unsatisfactory for another criterion. Choosing an appropriate cost function for a given problem may be more difficult than solving the problem mathematically.

Optimal control problem
The objective of optimal control theory is to determine a control signal or a control law that enables a process to simultaneously satisfy the physical constraints of the system and minimize or maximize a given cost function.

To formulate an optimal control problem, the following information is required [Sage 1977]:

1. A mathematical description of the process to be controlled, e.g., a system state equation.

2. A description of physical constraints, such as constraints on the control variable.

3. A specification of a cost function.

Description of mathematical models
In general, the mathematical description of a continuous process can be described in a state space representation as

$$\dot{\mathbf{x}} = \mathbf{f}(\mathbf{x}(t), \mathbf{u}(t), t) \qquad (10.1.1)$$

300

and for a discrete process

$$\mathbf{x}(k+1) = \mathbf{f}(\mathbf{x}(k),\ \mathbf{u}(k),\ k) \qquad (10.1.2)$$

For a time invariant linear continuous system, the system may be described as

$$\dot{\mathbf{x}} = \mathbf{A}\mathbf{x} + \mathbf{B}\mathbf{u} \qquad (10.1.3)$$

For a discrete system, the state equation is

$$\mathbf{x}(k+1) = \overline{\mathbf{A}}\mathbf{x}(k) + \overline{\mathbf{B}}\mathbf{u}(k) \qquad (10.1.4)$$

In the above equations (10.1.1) through (10.1.4), $\mathbf{x}$ is the state vector and $\mathbf{u}$ is the input vector.

Description of constraints
For an optimal control problem, the state equation is a constraint on the relationship between the state and inputs. Sometimes, the state equation is the only constraint, whereas constraints may also exist on admissible values of the state variables and/or inputs.

Constraints on the state variables may be a fixed value, such as at the initial moment

$$\mathbf{x}(t_0) = \mathbf{x}_0 \qquad (10.1.5)$$

or at the terminal moment

$$\mathbf{x}(t_f) = \mathbf{x}_f \qquad (10.1.6)$$

It may be an initial constraint manifold

$$\mathbf{M}(\mathbf{x}_0,\ t_0) = \mathbf{0} \qquad (10.1.7)$$

or a terminal constraint manifold

$$\mathbf{N}(\mathbf{x}_f,\ t_f) = \mathbf{0} \qquad (10.1.8)$$

where $\mathbf{M}$ and $\mathbf{N}$ are vector functions, i.e., (10.1.7) and (10.1.8) may contain several scale constraint equations.

The constraint on inputs is a set of admissible controls. Usually, it is expressed as

$$\|\mathbf{u}(\mathrm{t})\| < M \qquad (10.1.9)$$

or

$$|u_i(\mathrm{t})| < M \qquad (i = 1,\ 2,\ \cdots,\ r) \qquad (10.1.10)$$

where M is a constant, $u_i(t)$ $(i = 1,\ 2,\ \cdots,\ r)$ are the components of input $\mathbf{u}$.

Cost function

A general cost function for a continuous system is assumed as

$$J = S(\mathbf{x}(t_f), t_f) + \int_{t_0}^{t_f} L(\mathbf{x}(t), \mathbf{u}(t), t)dt \qquad (10.1.11)$$

For a discrete system, the cost function is

$$J = S(\mathbf{x}(N)) + \sum_{k=0}^{N-1} L(\mathbf{x}(k), \mathbf{u}(k)) \qquad (10.1.12)$$

In Eq. (10.1.11) and (10.1.12), S and L are real scale functions. S is the cost associated with the error at the terminal moment, and L is the cost function associated with the transient state errors and control effort during the operating period of interest.

The optimization problem with the general cost function (10.1.11) is referred to as the Bolza problem. If the cost function is only associated with the terminal conditions, i.e.,

$$J = S(\mathbf{x}(t_f), t_f) \qquad (10.1.13)$$

then it is a Mayer problem. If the cost function only contains the integral part, i.e.,

$$J = \int_{t_0}^{t_f} L(\mathbf{x}(t), \mathbf{u}(t), t)dt \qquad (10.1.14)$$

then it is a Lagrange problem.

Different selections of S and L reflect different concerns about the system performance. For example, if $S = 0$ and $L = 1$, then

$$J = \int_{t_0}^{t_f} dt = t_f - t_0$$

Minimizing this cost function J means finishing the job during the shortest period. This is the minimum time problem. For another example, if $S = 0$ and $L = \mathbf{u}^T\mathbf{u}$, then

$$J = \int_{t_0}^{t_f} (u_1^2 + u_2^2 + \cdots + u_r^2)dt$$

which is often interpreted as control energy. This is a minimum energy problem.

Different cost functions are summarized in Table 10.1.

Mathematically, the general optimal control problem may be formulated as follows: From all admissible control functions $\mathbf{u}(t)$ (10.1.9) or (10.1.10), find an optimal $\mathbf{u}^*(t)$ that minimizes the cost function J of (10.1.11) subject to the

TABLE 10.1. Optimal control problems and their cost functions.

Problems		Cost function	Comments
Bolza problem	Continuous system	$J = S(\mathbf{x}(t_f), t_f) + \int_{t_0}^{t_f} L(\mathbf{x}(t), \mathbf{u}(t), t)dt$	A general case
	Discrete system	$J = S(\mathbf{x}(N)) + \sum_{k=0}^{N-1} L(\mathbf{x}(k), \mathbf{u}(k))$	
Mayer problem	Continuous system	$J = S(\mathbf{x}(t_f), t_f)$	A special case of the Bolza problem
	Discrete system	$J = S(\mathbf{x}(N))$	
Lagrange problem	Continuous system	$J = \int_{t_0}^{t_f} L(\mathbf{x}(t), \mathbf{u}(t), t)dt$	
	Discrete system	$J = \sum_{k=0}^{N-1} L(\mathbf{x}(k), \mathbf{u}(k))$	
Minimum time problem		$J = \int_{t_0}^{t_f} dt = t_f - t_0$	A special case of the Lagrange problem
Minimum energy problem		$J = \int_{t_0}^{t_f} (u_1^2 + u_2^2 + \cdots + u_r^2)dt$	

dynamic system constraints of (10.1.1) or (10.1.3) and all the initial and terminal boundary conditions (10.1.5), (10.1.6), (10.1.7) and (10.1.8).

In this chapter, we will only discuss the optimal control problem for continuous systems.

10.2. VARIATIONAL METHOD FOR OPTIMAL CONTROL

There are several different ways to solve optimal control problems. Variational calculus is one of the commonly used methods.

Functionals and their variations

As we discussed, the cost function $J(\mathbf{x})$ of (10.1.11) used in an optimal control problem is a function of the state vector $\mathbf{x}(t)$ and input vector $\mathbf{u}(t)$. $\mathbf{x}(t)$ and $\mathbf{u}(t)$ are functions of time. Therefore the cost function is a function of functions, and is referred to as a functional.

To simplify the problem at the beginning, we discuss the case where the functions are scale functions instead of vectors. A functional $J(x)$ assigns a unique real number to function x in the domain of the functional.

If x and Δx, the increment of x, are functions from which the functional J is defined, then the increment of J, denoted by ΔJ, is defined as

$$\Delta J = \Delta J(x, \Delta x) = J(x + \Delta x) - J(x) \qquad (10.2.1)$$

Clearly, the increment ΔJ depends on the functions x and Δx.
The derivative of a function $y(x)$ is defined as

$$dy = f_x'(x)dx \qquad (10.2.2)$$

where

$$dx = \lim_{(x+\Delta x)\to x} \Delta x$$

Similarly, the variation of a functional J may be defined as

$$\delta J = J_y'(y)\delta y \qquad (10.2.3)$$

where

$$\delta y = \lim_{(y(x)+\Delta y)\to y(x)} \Delta y$$

Obviously, the variation of functional $J(y)$ is a functional of y and its variation δy, i.e., $\delta J = \delta J(y, \delta y)$. It plays the same role in variational calculus as differentiation does in normal calculus. Variational calculus is a very useful methodology for minimizing or maximizing a functional. The parallels between functions and functionals are shown in Table 10.2.

Extremum of cost function
Let's discuss the Lagrange problem first. Assume a system described by the state space representation

$$\dot{\mathbf{x}} = \mathbf{f}(\mathbf{x}(t), \mathbf{u}(t), t) \qquad (10.2.4)$$

with the initial and terminal conditions

TABLE 10.2. Function and functional.

Items	Function $y = f(x)$	Functional $J[y(x)]$
Increment	$\Delta y = f(x + \Delta x) - f(x)$ $= f'(x)\Delta x + \varepsilon$	$\Delta J = J(y + \Delta y) - J(y)$ $= J'(y)\Delta y + \varepsilon$
Variation	$dx = \lim_{(x+\Delta x)\to x} \Delta x$ $dy = f_x'(x)dx$	$\delta y = \lim_{(y(x)+\Delta y)\to y(x)} \Delta y$ $\delta J = J_y'(y)\delta y$
Necessary condition for extremum	$dy/dx = 0$	$\delta J = 0$
Sufficient condition for extremum	$dy/dx = 0$ $(d^2 y/dx^2) > 0$ (minimum) $(d^2 y/dx^2) < 0$ (maximum)	$\delta J = 0$ $\delta^2 J > 0$ (minimum) $\delta^2 J < 0$ (maximum)

$$\mathbf{x}(t_0) = \mathbf{x}_0, \quad \mathbf{x}(t_f) = \mathbf{x}_f \tag{10.2.5}$$

Consider a cost function

$$J = \int_{t_0}^{t_f} L(\mathbf{x}(t), \mathbf{u}(t), t) dt \tag{10.2.6}$$

Eq.(10.2.4) may be rearranged as

$$\mathbf{u} = \mathbf{F}(\mathbf{x}, \dot{\mathbf{x}}, t) \tag{10.2.7}$$

Substituting (10.2.7) into (10.2.6) yields

$$J = \int_{t_0}^{t_f} L(\mathbf{x}(t), \dot{\mathbf{x}}(t), t) dt \tag{10.2.8}$$

Assume that $\mathbf{x}^*(t)$ is the optimal trajectory, defined as the solution of the state equation where $\mathbf{u}(t) = \mathbf{u}^*(t)$. Other trajectories around $\mathbf{x}^*$ may be denoted as

$$\mathbf{x}(t) = \mathbf{x}^*(t) + \varepsilon \eta(t) \quad (t_0 \le t \le t_f) \tag{10.2.9}$$

where ε is a small number, and $\eta(t)$ is a variation vector in $\mathbf{x}(t)$. From (10.2.5), $\eta(t)$ must satisfy the following conditions:

$$\eta(t_0) = \eta(t_f) = \mathbf{0} \tag{10.2.10}$$

Differentiating (10.2.9) yields

$$\dot{\mathbf{x}}(t) = \dot{\mathbf{x}}^*(t) + \varepsilon \dot{\eta}(t) \quad (t_0 \le t \le t_f) \tag{10.2.11}$$

It is obvious that

$$\mathbf{x}^*(t) = \mathbf{x}(t)\big|_{\varepsilon = 0} \quad (t_0 \le t \le t_f) \tag{10.2.12}$$

Thus, the cost function

$$J = \int_{t_0}^{t_f} L(\mathbf{x}^*(t) + \varepsilon \eta(t), \dot{\mathbf{x}}^*(t) + \varepsilon \dot{\eta}(t), t) dt \tag{10.2.13}$$

will be minimum or maximum at $\varepsilon = 0$, i.e.,

$$J^*(\mathbf{x}) = J(\mathbf{x}^*) = J(\mathbf{x})\big|_{\varepsilon = 0} \tag{10.2.14}$$

Therefore, the necessary condition for taking an extremum of $J(\mathbf{x})$ is

$$\frac{\partial J}{\partial \varepsilon}\bigg|_{\varepsilon = 0} = 0 \tag{10.2.15}$$

Euler-Lagrange equation

The derivative of (10.2.13) can be obtained:

$$\frac{\partial J}{\partial \varepsilon} = \int_{t_0}^{t_f} [\left(\frac{\partial L}{\partial \mathbf{x}}\right)^T \frac{\partial \mathbf{x}}{\partial \varepsilon} + \left(\frac{\partial L}{\partial \dot{\mathbf{x}}}\right)^T \frac{\partial \dot{\mathbf{x}}}{\partial \varepsilon}] dt$$

$$= \int_{t_0}^{t_f} [\left(\frac{\partial L(\mathbf{x}^* + \varepsilon\,\eta, \dot{\mathbf{x}}^* + \varepsilon\,\dot{\eta}, t)}{\partial \mathbf{x}}\right)^T \eta(t) + \left(\frac{\partial L(\mathbf{x}^* + \varepsilon\,\eta, \dot{\mathbf{x}}^* + \varepsilon\,\dot{\eta}, t)}{\partial \dot{\mathbf{x}}}\right)^T \dot{\eta}(t)] dt$$

For $\varepsilon = 0$, it follows that

$$\frac{\partial J}{\partial \varepsilon}\Big|_{\varepsilon=0} = \int_{t_0}^{t_f} [\left(\frac{\partial L(\mathbf{x}^*, \dot{\mathbf{x}}^*, t)}{\partial \mathbf{x}}\right)^T \eta(t) + \left(\frac{\partial L(\mathbf{x}^*, \dot{\mathbf{x}}^*, t)}{\partial \dot{\mathbf{x}}}\right)^T \dot{\eta}(t)] dt \qquad (10.2.16)$$

Using the formula of integration by parts and (10.2.10), the last term of (10.2.16) can be simplified as

$$\int_{t_0}^{t_f} \left(\frac{\partial L}{\partial \dot{\mathbf{x}}}\right)^T d\eta(t) = \left(\frac{\partial L}{\partial \dot{\mathbf{x}}}\right)^T \eta(t)\Big|_{t_0}^{t_f} - \int_{t_0}^{t_f} \eta^T(t)\frac{d}{dt}(\frac{\partial L}{\partial \dot{\mathbf{x}}}) dt$$

$$= -\int_{t_0}^{t_f} \eta^T(t)\frac{d}{dt}(\frac{\partial L}{\partial \dot{\mathbf{x}}}) dt$$

Substituting into (10.2.16) yields

$$\frac{\partial J}{\partial \varepsilon}\Big|_{\varepsilon=0} = \int_{t_0}^{t_f} [\frac{\partial L}{\partial \mathbf{x}} - \frac{d}{dt}(\frac{\partial L}{\partial \dot{\mathbf{x}}})]^T \eta(t) dt$$

Considering (10.2.15), the necessary condition for obtaining an extremum of cost function J is

$$\frac{\partial L}{\partial \mathbf{x}} - \frac{d}{dt}(\frac{\partial L}{\partial \dot{\mathbf{x}}}) = \mathbf{0} \qquad (10.2.17)$$

which is called the Euler-Lagrange equation.

The solution of the Euler-Lagrange equation, $\mathbf{x}^*(t)$, is the optimal trajectory. Substituting $\mathbf{x}^*(t)$ into (10.2.7), we can obtain the optimal control $\mathbf{u}^*(t)$. The minimum cost function may be obtained by (10.2.14).

In fact, the Euler-Lagrange equation can be obtained directly by variational calculus: The variation of the cost function (10.2.8) is

$$\delta J = \delta J(\mathbf{x}, \delta\mathbf{x}) = \int_{t_0}^{t_f} [\frac{\partial L(\mathbf{x}^*, \dot{\mathbf{x}}^*, t)}{\partial \mathbf{x}} \cdot \delta\mathbf{x} + \frac{\partial L(\mathbf{x}^*, \dot{\mathbf{x}}^*, t)}{\partial \dot{\mathbf{x}}} \cdot \delta\dot{\mathbf{x}}] dt \qquad (10.2.18)$$

Expressed entirely in terms containing dx, (10.2.18) becomes

$$\delta J(\mathbf{x},\delta\mathbf{x}) = \frac{\partial L(\mathbf{x},\dot{\mathbf{x}},t)}{\partial\dot{\mathbf{x}}}\cdot\delta\mathbf{x}\Big|_{t_0}^{t_f} + \int_{t_0}^{t_f}\{\frac{\partial L(\mathbf{x}^*,\dot{\mathbf{x}}^*,t)}{\partial\mathbf{x}} - \frac{d}{dt}[\frac{\partial L(\mathbf{x},\dot{\mathbf{x}},t)}{\partial\dot{\mathbf{x}}}]\}\delta\mathbf{x}(t)dt$$

$$(10.2.19)$$

Since $\mathbf{x}(t_0)$ and $\mathbf{x}(t_f)$ are specified, the terms outside the integral in (10.2.19) vanish. The necessary condition $\delta J = 0$ for minimizing cost function J yields the Euler-Lagrange equation (10.2.17).

Remarks

1. The derivative of a scale function J with respect to a vector $\mathbf{x}$, or $\dot{\mathbf{x}}$, is a vector. They are defined as

$$\frac{\partial L}{\partial\mathbf{x}} = \begin{bmatrix} \dfrac{\partial L}{\partial x_1} \\ \dfrac{\partial L}{\partial x_2} \\ \vdots \\ \dfrac{\partial L}{\partial x_n} \end{bmatrix}, \qquad \frac{\partial L}{\partial\dot{\mathbf{x}}} = \begin{bmatrix} \dfrac{\partial L}{\partial\dot{x}_1} \\ \dfrac{\partial L}{\partial\dot{x}_2} \\ \vdots \\ \dfrac{\partial L}{\partial\dot{x}_n} \end{bmatrix} \qquad (10.2.20)$$

2. Although the Euler-Lagrange equation is a necessary condition for minimizing the cost function, usually it is not difficult to find whether a solution $\mathbf{x}^*(t)$ is a minimizing curve or a maximizing curve.

3. Usually, the Euler-Lagrange equation cannot be easily solved analytically. Therefore, numerical algorithms may be used.

4. It is not necessary to substitute (10.2.7) into (10.2.6) to obtain (10.2.8) and eliminate $\mathbf{u}$ from the cost function. We can keep the variable $\mathbf{u}$ in the cost function to get another Euler-Lagrange equation, which will be discussed later. The substitution is made to describe how the Euler-Lagrange equation is related to the derivative terms.

Example 10.2.1

For a system of single state with state space representation

$$\dot{x} = u$$

find the optimal control $u^*(t)$ and the optimal trajectory $x^*(t)$ for the cost function

$$J = \int_0^{t_f}(x^2 + u^2)dt$$

and the boundary conditions $x(0) = 1$ and $x(t_f) = 0$. Substituting

$$L = x^2 + u^2 = x^2 + \dot{x}^2$$

into the Euler-Lagrange equation yields

$$\frac{\partial L}{\partial x} - \frac{d}{dt}(\frac{\partial L}{\partial \dot{x}}) = 2x - \frac{d}{dt}(2\dot{x}) = 2x - 2\ddot{x} = 0$$

The Laplace transformation is

$$s^2 X(s) - sx(0) - \dot{x}(0) - X(s) = 0$$

Therefore

$$X(s) = \frac{s + \dot{x}(0)}{s^2 - 1}$$

Taking the inverse Laplace transformation, we have

$$x(t) = \cosh t + \dot{x}(0)\sinh t$$

It follows that at $t = t_f$

$$\dot{x}(0) = \frac{-\cosh t_f}{\sinh t_f} = -\text{ctgh}(t_f)$$

Thus the optimal trajectory is

$$x^*(t) = \cosh t - \text{ctgh}(t_f)\sinh t$$

The optimal control is

$$u^*(t) = \dot{x}^*(t) = \sinh t - \text{ctgh}(t_f)\cosh t$$

Lagrange multiple
The state space representation (10.2.4) can be rewritten as

$$\mathbf{f}(\mathbf{x}(t), \mathbf{u}(t), t) - \dot{\mathbf{x}} = 0 \qquad (10.2.21)$$

We construct a cost function equivalent to (10.2.6) as

$$J_s = \int_{t_0}^{t_f} \{L(\mathbf{x}(t), \mathbf{u}(t), t) - \lambda^T[\mathbf{f}(\mathbf{x}(t), \mathbf{u}(t), t) - \dot{\mathbf{x}}]\}dt \qquad (10.2.22)$$

where

$$\lambda^T = [\lambda_1, \lambda_2, \cdots, \lambda_n] \qquad (10.2.23)$$

λ is a column vector, called the Lagrange multiple vector, or co-state vector, which has the same dimension as that of the state variable of the system. Obviously, the second term in the integrand of (10.2.22) is a scale function. From (10.2.21), the cost function J_s of (10.2.22) is equivalent to J in (10.2.6).

Denote the integrand in (10.2.22) as

$$L_s = L(\mathbf{x}(t), \mathbf{u}(t), t) - \lambda^T[\mathbf{f}(\mathbf{x}(t), \mathbf{u}(t), t) - \dot{\mathbf{x}}] \qquad (10.2.24)$$

where we do not eliminate the variable $\mathbf{u}$. It turns out that $\mathbf{x}$ and $\mathbf{u}$ must satisfy the Euler-Lagrange equations

$$\frac{\partial L_s}{\partial \mathbf{x}} - \frac{d}{dt}\left(\frac{\partial L_s}{\partial \dot{\mathbf{x}}}\right) = \mathbf{0}$$

$$\frac{\partial L_s}{\partial \mathbf{u}} - \frac{d}{dt}\left(\frac{\partial L_s}{\partial \dot{\mathbf{u}}}\right) = \mathbf{0}$$

Rearrangement gives

$$\frac{\partial L}{\partial \mathbf{x}} + \frac{\partial \mathbf{f}^T}{\partial \mathbf{x}}\lambda(t) = -\dot{\lambda}(t) \qquad (10.2.25)$$

$$\frac{\partial L}{\partial \mathbf{u}} + \frac{\partial \mathbf{f}^T}{\partial \mathbf{u}}\lambda(t) = \mathbf{0} \qquad (10.2.26)$$

where $\mathbf{x} \in \Re^n$, $\mathbf{u} \in \Re^r$ and

$$\frac{\partial L}{\partial \mathbf{x}} = \begin{bmatrix} \dfrac{\partial L}{\partial x_1} \\[6pt] \dfrac{\partial L}{\partial x_2} \\[2pt] \vdots \\[2pt] \dfrac{\partial L}{\partial x_n} \end{bmatrix}, \qquad \frac{\partial \mathbf{f}^T}{\partial \mathbf{x}} = \begin{bmatrix} \dfrac{\partial f_1}{\partial x_1} & \dfrac{\partial f_2}{\partial x_1} & \cdots & \dfrac{\partial f_n}{\partial x_1} \\[6pt] \dfrac{\partial f_1}{\partial x_2} & \dfrac{\partial f_2}{\partial x_2} & \cdots & \dfrac{\partial f_n}{\partial x_2} \\[2pt] \vdots & \vdots & & \vdots \\[2pt] \dfrac{\partial f_1}{\partial x_n} & \dfrac{\partial f_2}{\partial x_n} & \cdots & \dfrac{\partial f_n}{\partial x_n} \end{bmatrix} \qquad (10.2.27)$$

$$\frac{\partial L}{\partial \mathbf{u}} = \begin{bmatrix} \dfrac{\partial L}{\partial u_1} \\[6pt] \dfrac{\partial L}{\partial u_2} \\[2pt] \vdots \\[2pt] \dfrac{\partial L}{\partial u_r} \end{bmatrix}, \qquad \frac{\partial \mathbf{f}^T}{\partial \mathbf{u}} = \begin{bmatrix} \dfrac{\partial f_1}{\partial u_1} & \dfrac{\partial f_2}{\partial u_1} & \cdots & \dfrac{\partial f_n}{\partial u_1} \\[6pt] \dfrac{\partial f_1}{\partial u_2} & \dfrac{\partial f_2}{\partial u_2} & \cdots & \dfrac{\partial f_n}{\partial u_2} \\[2pt] \vdots & \vdots & & \vdots \\[2pt] \dfrac{\partial f_1}{\partial u_r} & \dfrac{\partial f_2}{\partial u_r} & \cdots & \dfrac{\partial f_n}{\partial u_r} \end{bmatrix} \qquad (10.2.28)$$

Example 10.2.2

Recalling Example 10.2.1, a new cost function with a Lagrange multiplier λ can be constructed as

$$J_s = \int_{t_0}^{t_f} [(x^2 + u^2) + \lambda(u - \dot{x})]dt$$

From (10.2.25) and (10.2.26), we have

$$\frac{\partial L}{\partial x} + \frac{\partial f^T}{\partial x}\lambda(t) = 2x = -\dot{\lambda}(t)$$

$$\frac{\partial L}{\partial u} + \frac{\partial f^T}{\partial u}\lambda(t) = 2u + \lambda = 0$$

Rearrangement gives

$$\dot{u} - x = 0$$

After substituting the state equation $\dot{x} = u$, we have

$$\ddot{x} - x = 0$$

This leads to the same result as in Example 10.2.1.

Bolza problem and Mayer problem
Following the steps of the Lagrange problem, we can solve the Bolza problem and Mayer problem. Since a universal variational formula will be discussed, we leave the Bolza problem and Mayer problem to the readers.

The Lagrange problem is summarized in Table 10.3.

10.3. A UNIVERSAL VARIATIONAL FORMULA

In this section, a universal variational formula for solving the general optimal

TABLE 10.3. Lagrange problem.

Items	Formula
System equation	$\mathbf{f}(\mathbf{x}(t), \mathbf{u}(t), t) - \dot{\mathbf{x}} = \mathbf{0}$
Initial condition	$\mathbf{x}(t_0) = \mathbf{x}_0$
Terminal condition	$\mathbf{x}(t_f) = \mathbf{x}_f$
Cost function	$J = \int_{t_0}^{t_f} L(\mathbf{x}(t), \mathbf{u}(t), t)dt$
Equivalent cost function	$J_s = \int_{t_0}^{t_f} \{L(\mathbf{x}(t), \mathbf{u}(t), t) - \lambda^T[\mathbf{f}(\mathbf{x}(t), \mathbf{u}(t), t) - \dot{\mathbf{x}}]\}dt$
Euler-Lagrange equations	$\dfrac{\partial L}{\partial \mathbf{x}} + \dfrac{\partial \mathbf{f}^T}{\partial \mathbf{x}}\lambda(t) = -\dot{\lambda}(t)$ $\qquad$ $\dfrac{\partial L}{\partial \mathbf{u}} + \dfrac{\partial \mathbf{f}^T}{\partial \mathbf{u}}\lambda(t) = \mathbf{0}$

control problem is introduced [Rao, Tsai and Jiang 1988].

Needed to develop a universal variational formula
For optimal control, most constrained optimization problems aimed at minimizing a general cost function can be represented by the Bolza problem [Donald 1970]:

$$J(\mathbf{u}) = S[\mathbf{x}(t),\ t]\Big|_{t=t_0}^{t=t_f} + \int_{t_0}^{t_f} L(\mathbf{x}, \mathbf{u}, t)dt \qquad (10.3.1)$$

where S and L are real scale functions. S is the cost associated with the error at the terminal time and the initial time, and L is the cost function associated with the transient state errors and control effort during the entire concerned period.

As we studied in section 10.2, the integrand in (10.3.1) leads to Euler-Lagrange equations. However, different terminal or initial manifolds will result in different boundary condition equations (transversality conditions), which are somewhat complicated and difficult to distinguish. There are many different transversality condition cases. For example, if the initial time t_0 and initial state x_0 are fixed, different differential equations and associated transversality conditions may be derived from eight cases that are listed below [Donald 1970]:

1. t_f is fixed, $\mathbf{x}(t_f) = \mathbf{x}_f$ is a specified final state;
2. t_f is fixed, $\mathbf{x}(t_f)$ is free;
3. t_f is fixed, $\mathbf{x}(t_f)$ is specified on the surface $\mathbf{N}(\mathbf{x}(t)) = \mathbf{0}$;
4. t_f is free, $\mathbf{x}(t_f)$ is a specified final state;
5. t_f is free, $\mathbf{x}(t_f)$ is free;
6. t_f is free, $\mathbf{x}(t_f)$ is specified on the moving point $\theta(t)$;
7. t_f is free, $\mathbf{x}(t_f)$ is specified on the surface $\mathbf{N}(\mathbf{x}(t)) = \mathbf{0}$;
8. t_f is free, $\mathbf{x}(t_f)$ is specified on the moving surface $\mathbf{N}(\mathbf{x}(t)) = \mathbf{0}$.

where t_f is the terminal time and $\mathbf{x}(t_f)$ is the terminal state vector. Besides these eight cases, mixed situations may arise. Similarly, initial conditions can also be varied. For each case, some boundary condition equations are involved. If we discuss these situations case by case, there are too many formulas to remember. We have to set up the boundary condition equations every time we need the formulas. Therefore, a universal variational formula has been developed to solve the general optimization problem.

Formulation of a general optimal control problem
A general optimal control problem can be formulated as follows: From all admissible control functions $\mathbf{u}(t)$, find an optimal $\mathbf{u}^*(t)$ that minimizes the general cost function (10.3.1) subject to the dynamic system constraint

$$\dot{\mathbf{x}} = \mathbf{f}(\mathbf{x}, \mathbf{u}, t) \qquad (10.3.2)$$

and the boundary conditions:

Initial constraint manifold: $\mathbf{M}(\mathbf{x}_0, t_0) = \mathbf{0}$ (10.3.3)

Terminal constraint manifold: $\mathbf{N}(\mathbf{x}_f, t_f) = \mathbf{0}$ (10.3.4)

First variation in cost function

A scalar function, the Hamiltonian, is defined as

$$H[\mathbf{x}(t), \mathbf{u}(t), \lambda(t), t] = L[\mathbf{x}(t), \mathbf{u}(t), t] + \lambda^T(t)\mathbf{f}(\mathbf{x}(t), \mathbf{u}(t), t) \quad (10.3.5)$$

where $\lambda(t)$ is a co-state vector.

With the method of Lagrange multipliers, we adjoin the system dynamic constraint, initial and terminal constraint manifolds by means of the Lagrange multipliers $\lambda(t)$, ξ and ν, respectively, where $\lambda(t)$, ξ and ν are vectors. Rewrite the cost function as:

$$J(\mathbf{u}) = S[\mathbf{x}(t), t]\big|_{t=t_0}^{t=t_f} + \xi^T \mathbf{M}[\mathbf{x}(t_0), t_0] + \nu^T \mathbf{N}[\mathbf{x}(t_f), t_f]$$

$$+ \int_{t_0}^{t_f} [H(\mathbf{x}, \mathbf{u}, \lambda, t) - \lambda^T(t)\dot{\mathbf{x}}]dt \quad (10.3.6)$$

To minimize cost function $J(\mathbf{u})$, the necessary condition is that the first variation in $J(\mathbf{u})$ vanishes. The corresponding control variable $\mathbf{u}(t)$ that makes the first variation in $J(\mathbf{u})$ vanish is the optimal control variable $\mathbf{u}^*(t)$.

We can form the first variation in equation (10.3.6), and rearrange it into the following form:

$$\delta J(\mathbf{u}) = \int_{t_0}^{t_f} \{(\frac{\partial H}{\partial \mathbf{x}} + \dot{\lambda})^T \delta\mathbf{x} + (\frac{\partial H}{\partial \lambda} - \dot{\mathbf{x}})^T \delta\lambda + (\frac{\partial H}{\partial \mathbf{u}})^T \delta\mathbf{u}\}dt$$

$$+ \delta t_0 \{-H - \frac{\partial S}{\partial t_0} + (\frac{\partial \mathbf{M}}{\partial t_0})^T \xi\} + \delta t_f \{H + \frac{\partial S}{\partial t_f} + (\frac{\partial \mathbf{N}}{\partial t_f})^T \nu\}$$

$$+ \delta\mathbf{x}^T(t_0)\{\lambda(t_0) - \frac{\partial S}{\partial \mathbf{x}(t_0)} + (\frac{\partial \mathbf{M}}{\partial \mathbf{x}(t_0)})^T \xi\}$$

$$+ \delta\mathbf{x}^T(t_f)\{-\lambda(t_f) + \frac{\partial S}{\partial \mathbf{x}(t_f)} + (\frac{\partial \mathbf{N}}{\partial \mathbf{x}(t_f)})^T \nu\} \quad (10.3.7)$$

To set equation (10.3.7) equal to zero so that the necessary condition for an extremum can be obtained, both the integral term and transversality condition must be equal to zero. The integral term leads to the well-known Euler-Lagrange equation that we encountered in last section. The transversality condition represents terms outside the integration symbol.

Euler-Lagrange equation

When the integral term is set equal to zero, the Euler-Lagrange equations for the Hamiltonian are formed:

$$\frac{\partial H}{\partial \mathbf{x}} + \dot{\lambda} = 0 \tag{10.3.8}$$

$$\frac{\partial H}{\partial \lambda} - \dot{\mathbf{x}} = 0 \tag{10.3.9}$$

$$\frac{\partial H}{\partial \mathbf{u}} = 0 \tag{10.3.10}$$

The Euler-Lagrange equations determine the structure of the optimal solution. If there exists a constraint on the control variable $\mathbf{u}$, $\partial H/\partial \mathbf{u}$ may not be zero for admissible $\mathbf{u}$. Then, the maximum principle that we will study later should be used to replace (10.3.10).

Transversality conditions

Setting the terms outside the integration symbol in (10.3.7) to zero, we can obtain the transversality conditions. For convenience, we define

$$\tilde{T}_0 = -H - \frac{\partial S}{\partial t_0} + (\frac{\partial \mathbf{M}}{\partial t_0})^T \xi \tag{10.3.11}$$

$$\tilde{T}_f = H + \frac{\partial S}{\partial t_f} + (\frac{\partial \mathbf{N}}{\partial t_f})^T \nu \tag{10.3.12}$$

$$\tilde{\mathbf{X}}_0 = \lambda(t_0) - \frac{\partial S}{\partial \mathbf{x}(t_0)} + (\frac{\partial \mathbf{M}}{\partial \mathbf{x}(t_0)})^T \xi \tag{10.3.13}$$

$$\tilde{\mathbf{X}}_f = -\lambda(t_f) + \frac{\partial S}{\partial \mathbf{x}(t_f)} + (\frac{\partial \mathbf{N}}{\partial \mathbf{x}(t_f)})^T \nu \tag{10.3.14}$$

Then, the transversality conditions are given by the following equations:

$$(\delta t_0)\tilde{T}_0 = 0 \tag{10.3.15}$$

$$(\delta t_0)\tilde{T}_f = 0 \tag{10.3.16}$$

$$(\delta \mathbf{x}^T(t_0))\tilde{\mathbf{X}}_0 = 0 \tag{10.3.17}$$

$$(\delta \mathbf{x}^T(t_f))\tilde{\mathbf{X}}_f = 0 \tag{10.3.18}$$

Note that $\tilde{T}_0$ and $\tilde{T}_f$ are scalars, while $\tilde{\mathbf{X}}_0$ and $\tilde{\mathbf{X}}_f$ are vectors.

Generally speaking, if a time variable (t_0 or t_f) or state variable ($\mathbf{x}(t_0)$ or $\mathbf{x}(t_f)$) is specified, its first variation will be equal to zero. For example, if t_0 is specified, say, $t_0 = 3$ sec., the first variation in t_0 is zero and (10.3.15) is automatically satisfied. The necessary condition for the first variation of J to be zero at $t = t_0$ is still guaranteed. When a variable is unspecified, its first variation is in general not equal to zero, and the corresponding term in Eqs. (10.3.15 $\sim$ 10.3.18) has to be assigned to zero so that an optimal solution can be guaranteed.

This universal formula (10.3.7), or its equivalent equations (10.3.8 $\sim$ 10.3.10) and (10.3.15 $\sim$ 10.3.18), contains all the cases for optimal control with the variational calculus technique, and is summarized in Table 10.4. Using this formula other than the case-by-case formulas, the calculus of the variation method in optimal control problems becomes easier. It also makes decision-

TABLE 10.4. Universal formula for optimal control.

Items	Formula
System equation	$\mathbf{f}(\mathbf{x}(t), \mathbf{u}(t), t) - \dot{\mathbf{x}} = 0$
Initial constraint manifold	$\mathbf{M}(\mathbf{x}_0, t_0) = 0$
Terminal constraint manifold	$\mathbf{N}(\mathbf{x}_f, t_f) = 0$
Cost function	$J(\mathbf{u}) = S[\mathbf{x}(t), t]\big|_{t=t_0}^{t=t_f} + \int_{t_0}^{t_f} L(\mathbf{x}, \mathbf{u}, t)dt$
Hamiltonian	$H[\mathbf{x}(t), \mathbf{u}(t), \lambda(t), t] = L[\mathbf{x}(t), \mathbf{u}(t), t] + \lambda^T \mathbf{f}(\mathbf{x}(t), \mathbf{u}(t), t)$
Equivalent cost function	$J(\mathbf{u}) = S[\mathbf{x}(t), t]\big|_{t=t_0}^{t=t_f} + \xi^T \mathbf{M}[\mathbf{x}(t_0), t_0] + v^T \mathbf{N}[\mathbf{x}(t_f), t_f]$ $+\int_{t_0}^{t_f}[H(\mathbf{x}, \mathbf{u}, \lambda, t) - \lambda^T(t)\mathbf{x}]dt$
Euler-Lagrange equation	$\dfrac{\partial H}{\partial \mathbf{x}} + \dot{\lambda} = 0 \qquad \dfrac{\partial H}{\partial \lambda} - \dot{\mathbf{x}} = 0 \qquad \dfrac{\partial H}{\partial \mathbf{u}} = 0$
Transversality conditions	$\tilde{T}_0 = -H - \dfrac{\partial S}{\partial t_0} + (\dfrac{\partial \mathbf{M}}{\partial t_0})^T \xi \qquad\qquad (\delta t_0)\tilde{T}_0 = 0$ $\tilde{T}_f = H + \dfrac{\partial S}{\partial t_f} + (\dfrac{\partial \mathbf{N}}{\partial t_f})^T v \qquad\qquad (\delta t_0)\tilde{T}_f = 0$ $\tilde{\mathbf{X}}_0 = \lambda(t_0) - \dfrac{\partial S}{\partial \mathbf{x}(t_0)} + (\dfrac{\partial \mathbf{M}}{\partial \mathbf{x}(t_0)})^T \xi \qquad (\delta \mathbf{x}^T(t_0))\tilde{\mathbf{X}}_0 = 0$ $\tilde{\mathbf{X}}_f = -\lambda(t_f) + \dfrac{\partial S}{\partial \mathbf{x}(t_f)} + (\dfrac{\partial \mathbf{N}}{\partial \mathbf{x}(t_f)})^T v \quad (\delta \mathbf{x}^T(t_f))\tilde{\mathbf{X}}_f = 0$

making easier for developing an expert system [Rao, Tsai and Jiang 1988] for optimal control.

Example 10.3.1

Suppose that t_0, $\mathbf{x}(t_0)$ and t_f are fixed and $\mathbf{x}(t_f)$ is free, i.e., $\delta t_0 = 0$, $\delta \mathbf{x}(t_0) = \mathbf{0}$ and $\delta t_f = 0$. Therefore, Eqs. (10.3.15 ~ 10.3.17) need not be considered, and the corresponding equations are Eqs. (10.3.8 ~ 10.3.10) and (10.3.18). Since $\mathbf{x}(t_f)$ is free, there is no terminal constraint manifold. Then Eq. (10.3.18) reduces to

$$-\lambda(t_f) + \frac{\partial S}{\partial \mathbf{x}(t_f)} = \mathbf{0} \tag{10.3.19}$$

Example 10.3.2

Consider another case where initial time t_0 is fixed, $\mathbf{x}(t_0)$ is free, t_f is free and $\mathbf{x}(t_f)$ is specified on the moving surface $\mathbf{N}(\mathbf{x}(t_f),t_f) = \mathbf{0}$. Fixed t_0 eliminates Eq. (10.3.15). From Eq. (10.3.17), free $\mathbf{x}(t_0)$ yields

$$\lambda(t_0) - \frac{\partial S}{\partial \mathbf{x}(t_0)} = \mathbf{0} \tag{10.3.20}$$

Free t_f and specified $\mathbf{x}(t_f)$ produce

$$H + \frac{\partial S}{\partial t_f} + (\frac{\partial \mathbf{N}}{\partial t_f})^T \mathbf{v} = 0 \tag{10.3.21}$$

$$-\lambda(t_f) + \frac{\partial S}{\partial \mathbf{x}(t_f)} + (\frac{\partial \mathbf{N}}{\partial \mathbf{x}(t_f)})^T \mathbf{v} = \mathbf{0} \tag{10.3.22}$$

By using the variational calculus method, the boundary condition equations for each case in optimal control can be obtained from the universal variational calculus formula (10.3.7) as we did for these two examples above.

10.4. LINEAR REGULATOR PROBLEMS

As we mentioned before, a feedback control system is less sensitive to disturbance than an open loop control system. The optimal control method covered in the last section does not use any on-line information from the state variables. Thus, the designed system may be sensitive to disturbances.

The linear quadratic regulation problem discussed in this section is probably the most important issue for optimal control applications. An optimal control law in a linear state feedback form will be obtained for the linear regulator problem.

Problem formulation
Consider a linear time invariant, multivariable system with the state space representation

$$\dot{x} = Ax + Bu \qquad (10.4.1)$$

where $x(t)$ is an $n \times 1$ state vector and $u(t)$ is an $m \times 1$ input vector. The initial time t_0 and initial condition $x(t_0)$ are fixed. The terminal time t_f is specified, but $x(t_f)$ is free. We will find a control law, i.e., an optimal input vector that is a function of state variables, to minimize the quadratic cost function

$$J(u) = \frac{1}{2}x^T(t_f)Hx(t_f) + \frac{1}{2}\int_{t_0}^{t_f}[x^T(t)Qx(t) + u^T(t)Ru(t)]dt \qquad (10.4.2)$$

where H, Q and R are symmetric square matrices, called weighting matrices, which represent the relative importance of the corresponding variables. It is assumed that the control function $u(t)$ is unconstrained. It is also assumed that H and Q are semi-positive definite, and R is positive definite; this condition is expressed by

$$H = H^T \geq 0, \quad Q = Q^T \geq 0, \quad R = R^T > 0 \qquad (10.4.3)$$

The quadratic cost function is used, since it results in an analytical control law with a clear physical meaning.

Control law of linear regulator problem
The Hamiltonian is

$$H[x(t), u(t), \lambda(t), t] = \frac{1}{2}x^T(t)Qx(t) + \frac{1}{2}u^T(t)Ru(t) + \lambda^T(t)(Ax + Bu) \qquad (10.4.4)$$

where λ is a Lagrange multiplier vector. Recalling Example 10.3.1, the following equations may be obtained:

$$\frac{\partial H}{\partial x} + \dot{\lambda} = Qx + A^T\lambda + \dot{\lambda} = 0 \qquad (10.4.5)$$

$$\frac{\partial H}{\partial \lambda} - \dot{x} = Ax + Bu - \dot{x} = 0 \qquad (10.4.6)$$

$$\frac{\partial H}{\partial u} = Ru + B^T\lambda = 0 \qquad (10.4.7)$$

From (10.4.7), we have

$$\mathbf{u} = -\mathbf{R}^{-1}\mathbf{B}^T\boldsymbol{\lambda} \tag{10.4.8}$$

Assume that

$$\boldsymbol{\lambda}(t) = \mathbf{K}(t)\mathbf{x}(t) \tag{10.4.9}$$

where $\mathbf{K}(t)$ is an $n \times n$ time variant square matrix. Substituting (10.4.9) into (10.4.8), we have

$$\mathbf{u}(t) = -\mathbf{R}^{-1}\mathbf{B}^T\mathbf{K}(t)\mathbf{x}(t) \tag{10.4.10}$$

Substituting (10.4.9) into (10.4.5) yields

$$\mathbf{Q}\mathbf{x}(t) + \mathbf{A}^T\mathbf{K}(t)\mathbf{x}(t) + \dot{\mathbf{K}}(t)\mathbf{x}(t) + \mathbf{K}(t)\dot{\mathbf{x}}(t) = 0 \tag{10.4.11}$$

Substituting (10.4.10) into (10.4.6), and then into (10.4.11), we get

$$\dot{\mathbf{x}} = \mathbf{A}\mathbf{x} - \mathbf{B}\mathbf{R}^{-1}\mathbf{B}^T\mathbf{K}\mathbf{x} \tag{10.4.12}$$

and

$$\mathbf{Q}\mathbf{x} + \mathbf{A}^T\mathbf{K}\mathbf{x} + \dot{\mathbf{K}}\mathbf{x} + \mathbf{K}\mathbf{A}\mathbf{x} - \mathbf{K}\mathbf{B}\mathbf{R}^{-1}\mathbf{B}^T\mathbf{K}\mathbf{x} = 0$$

or

$$\dot{\mathbf{K}} + \mathbf{K}\mathbf{A} + \mathbf{A}^T\mathbf{K} - \mathbf{K}\mathbf{B}\mathbf{R}^{-1}\mathbf{B}^T\mathbf{K} + \mathbf{Q} = 0 \tag{10.4.13}$$

This first order differential matrix equation is called the matrix Riccati equation. Solving this equation, we will get the control law (10.4.10). The initial condition for solving the Riccati equation can be obtained from substitution into (10.3.19):

$$-\mathbf{K}(t_f)\mathbf{x}(t_f) + \mathbf{H}\mathbf{x}(t_f) = 0$$

or

$$\mathbf{K}(t_f) = \mathbf{H} \tag{10.4.14}$$

Minimum value of the cost function

Using the control law (10.4.10) and the state space representation (10.4.12) of the closed loop system, the minimum value of the cost function can be found as follows:

The derivative of the quadratic form $\mathbf{x}^T\mathbf{K}\mathbf{x}$ can be found as

$$\frac{d(\mathbf{x}^T\mathbf{K}\mathbf{x})}{dt} = \dot{\mathbf{x}}^T\mathbf{K}\mathbf{x} + \mathbf{x}^T\dot{\mathbf{K}}\mathbf{x} + \mathbf{x}^T\mathbf{K}\dot{\mathbf{x}} \tag{10.4.15}$$

where

$$\dot{\mathbf{x}}^T\mathbf{K}\mathbf{x} = \mathbf{x}^T\mathbf{A}^T\mathbf{K}\mathbf{x} - \mathbf{x}^T\mathbf{K}\mathbf{B}\mathbf{R}^{-1}\mathbf{B}^T\mathbf{K}\mathbf{x} \tag{10.4.16}$$

$$\mathbf{x}^T \mathbf{K}\dot{\mathbf{x}} = \mathbf{x}^T \mathbf{K}\mathbf{A}\mathbf{x} - \mathbf{x}^T \mathbf{K}\mathbf{B}\mathbf{R}^{-1}\mathbf{B}^T \mathbf{K}\mathbf{x} \qquad (10.4.17)$$

Note that $\mathbf{K}$ and $\mathbf{R}^{-1}$ are symmetric. From (10.4.15), (10.4.16), (10.4.17), (10.4.10) and (10.4.12), it follows that

$$\mathbf{x}^T(t)\mathbf{Q}\mathbf{x}(t) + \mathbf{u}^T(t)\mathbf{R}\mathbf{u}(t)$$

$$= \mathbf{x}^T \mathbf{Q}\mathbf{x} + \mathbf{u}^T \mathbf{R}\mathbf{u} + \dot{\mathbf{x}}^T \mathbf{K}\mathbf{x} + \mathbf{x}^T \dot{\mathbf{K}}\mathbf{x} + \mathbf{x}^T \mathbf{K}\dot{\mathbf{x}} - \frac{d(\mathbf{x}^T \mathbf{K}\mathbf{x})}{dt}$$

$$= \mathbf{x}^T \mathbf{Q}\mathbf{x} + \mathbf{x}^T \mathbf{K}\mathbf{B}\mathbf{R}^{-1}\mathbf{R}\mathbf{R}^{-1}\mathbf{B}^T \mathbf{K}\mathbf{x} + \mathbf{x}^T \mathbf{A}^T \mathbf{K}\mathbf{x} + \mathbf{x}^T \mathbf{K}\mathbf{A}\mathbf{x}$$

$$\qquad -2\mathbf{x}^T \mathbf{K}\mathbf{B}\mathbf{R}^{-1}\mathbf{B}^T \mathbf{K}\mathbf{x} + \mathbf{x}^T \dot{\mathbf{K}}\mathbf{x} - \frac{d(\mathbf{x}^T \mathbf{K}\mathbf{x})}{dt}$$

$$= \mathbf{x}^T(\mathbf{Q} - \mathbf{K}\mathbf{B}\mathbf{R}^{-1}\mathbf{B}^T \mathbf{K} + \mathbf{A}^T \mathbf{K} + \mathbf{K}\mathbf{A} + \dot{\mathbf{K}})\mathbf{x} - \frac{d(\mathbf{x}^T \mathbf{K}\mathbf{x})}{dt}$$

Combining with (10.4.13) yields

$$\mathbf{x}^T(t)\mathbf{Q}\mathbf{x}(t) + \mathbf{u}^T(t)\mathbf{R}\mathbf{u}(t) = -\frac{d(\mathbf{x}^T \mathbf{K}\mathbf{x})}{dt}$$

From (10.4.2) and the above result, we have

$$J(\mathbf{u}) = \frac{1}{2}\mathbf{x}^T(t_f)\mathbf{H}\mathbf{x}(t_f) + \frac{1}{2}\int_{t_0}^{t_f}[\mathbf{x}^T(t)\mathbf{Q}\mathbf{x}(t) + \mathbf{u}^T(t)\mathbf{R}\mathbf{u}(t)]dt$$

$$= \frac{1}{2}\mathbf{x}^T(t_f)\mathbf{H}\mathbf{x}(t_f) - \frac{1}{2}\int_{t_0}^{t_f}\frac{d(\mathbf{x}^T \mathbf{K}\mathbf{x})}{dt}dt$$

$$= \frac{1}{2}\mathbf{x}^T(t_f)\mathbf{H}\mathbf{x}(t_f) - \frac{1}{2}\mathbf{x}^T(t_f)\mathbf{K}(t_f)\mathbf{x}(t_f) + \frac{1}{2}\mathbf{x}^T(t_0)\mathbf{K}(t_0)\mathbf{x}(t_0)$$

Substituting (10.4.14) into this expression, we have

$$J(\mathbf{u}) = \frac{1}{2}\mathbf{x}^T(t_0)\mathbf{K}(t_0)\mathbf{x}(t_0) \qquad (10.4.18)$$

Example 10.4.1

Given a system

$$\dot{\mathbf{x}} = \begin{bmatrix} 0 & 1 \\ 0 & 0 \end{bmatrix}\mathbf{x} + \begin{bmatrix} 0 \\ 1 \end{bmatrix}u$$

design a state feedback optimal control law to minimize the cost function

$$J(u) = \frac{1}{2}[x_1^2(t_f) + 2x_2^2(t_f)] + \frac{1}{2}\int_{t_0}^{t_f} [2x_1^2 + 4x_2^2 + 2x_1x_2 + \frac{1}{2}u^2]dt$$

where $t_0 = 0$ and $t_f = 3$.

It can be obtained that

$$Q = \begin{bmatrix} 2 & 1 \\ 1 & 4 \end{bmatrix}, \quad H = \begin{bmatrix} 1 & 0 \\ 0 & 2 \end{bmatrix}, \quad \text{and} \quad R = \frac{1}{2}$$

From (10.4.10), the control law is

$$u(t) = -R^{-1}B^T K(t)x(t) = -2[0 \quad 1]\begin{bmatrix} k_{11} & k_{12} \\ k_{12} & k_{22} \end{bmatrix}\begin{bmatrix} x_1 \\ x_2 \end{bmatrix}$$

The matrix $K(t)$ can be obtained from the Riccati equation, i.e.,

$$\begin{bmatrix} \dot{k}_{11} & \dot{k}_{12} \\ \dot{k}_{12} & \dot{k}_{22} \end{bmatrix} + \begin{bmatrix} k_{11} & k_{12} \\ k_{12} & k_{22} \end{bmatrix}\begin{bmatrix} 0 & 1 \\ 0 & 0 \end{bmatrix} + \begin{bmatrix} 0 & 0 \\ 1 & 0 \end{bmatrix}\begin{bmatrix} k_{11} & k_{12} \\ k_{12} & k_{22} \end{bmatrix}$$

$$- \begin{bmatrix} k_{11} & k_{12} \\ k_{12} & k_{22} \end{bmatrix}\begin{bmatrix} 0 \\ 1 \end{bmatrix}2[0 \quad 1]\begin{bmatrix} k_{11} & k_{12} \\ k_{12} & k_{22} \end{bmatrix} + \begin{bmatrix} 2 & 1 \\ 1 & 4 \end{bmatrix} = \begin{bmatrix} 0 & 0 \\ 0 & 0 \end{bmatrix}$$

with the terminal condition

$$\begin{bmatrix} k_{11}(3) & k_{12}(3) \\ k_{12}(3) & k_{22}(3) \end{bmatrix} = \begin{bmatrix} 1 & 0 \\ 0 & 2 \end{bmatrix}$$

The matrix Riccati equation may be divided into the following equations:

$$\dot{k}_{11}(t) = 2k_{12}^2(t) - 2$$

$$\dot{k}_{12}(t) = -k_{11}(t) + 2k_{12}(t)k_{22}(t) - 1$$

$$\dot{k}_{22}(t) = -2k_{12}(t) + 2k_{22}^2(t) - 4$$

The boundary condition is

$$k_{11}(3) = 1, \quad k_{12}(3) = 0, \quad k_{22}(3) = 2$$

The equations are nonlinear, and can be solved by numerical techniques.

Remarks

1. $K(t)$ is symmetric because the transpose of the Riccati equation reveals that $K(t)$ and $K^T(t)$ satisfy the same equation.

2. The solution of $K(t)$ is independent of the initial state $x(t_0)$. This means that we can design the controller without information about initial conditions. However, $K(t)$ is usually a time variant matrix with nonlinear entries.

Infinite time regulator problem
A commonly used simplification of the linear regulator problem is to let the terminal time approach infinity. Because the closed loop system is stable, we have $\mathbf{x}(\infty) = \mathbf{0}$. Therefore the cost at the terminal moment in the cost function (10.4.2) lacks significance. Consequently, it does not appear in the cost function, i.e., we have set $\mathbf{H} = \mathbf{0}$ in (10.4.2).

The cost function for an infinite time regulator problem is

$$J(\mathbf{u}) = \frac{1}{2}\int_{t_0}^{\infty}[\mathbf{x}^T(t)\mathbf{Q}\mathbf{x}(t) + \mathbf{u}^T(t)\mathbf{R}\mathbf{u}(t)]dt \qquad (10.4.18)$$

and the optimal control law will be

$$\mathbf{u}(t) = -\mathbf{R}^{-1}\mathbf{B}^T\mathbf{K}\mathbf{x}(t) \qquad (10.4.19)$$

where the constant symmetric square matrix may be obtained by the following algebraic Riccati equation:

$$\mathbf{K}\mathbf{A} + \mathbf{A}^T\mathbf{K} - \mathbf{K}\mathbf{B}\mathbf{R}^{-1}\mathbf{B}^T\mathbf{K} + \mathbf{Q} = \mathbf{0} \qquad (10.4.20)$$

It has been noted that the closed loop system is time invariant, and thus the implementation is easier.

The solution of the algebraic Riccati equation may not be unique. The desired unique answer is obtained by enforcing the requirement that $\mathbf{K}$ be positive definite [Nagrath and Gopal 1982].

Example 10.4.2
Recalling Example 10.4.1, we have the algebraic equations

$$2k_{12}^2 - 2 = 0$$

$$-k_{11} + 2k_{12}k_{22} - 1 = 0$$

$$-2k_{12} + 2k_{22}^2 - 4 = 0$$

The above equations generate four solutions:

$$k_{12} = -1, \quad k_{22} = -1, \quad k_{11} = 1$$

$$k_{12} = -1, \quad k_{22} = 1, \quad k_{11} = -3$$

$$k_{12} = 1, \quad k_{22} = -\sqrt{3}, \quad k_{11} = -2\sqrt{3} - 1$$

$$k_{12} = 1, \quad k_{22} = \sqrt{3}, \quad k_{11} = 2\sqrt{3} - 1$$

or

$$\mathbf{K}_1 = \begin{bmatrix} 1 & -1 \\ -1 & -1 \end{bmatrix}, \qquad \mathbf{K}_2 = \begin{bmatrix} -3 & -1 \\ -1 & 1 \end{bmatrix},$$

$$\mathbf{K}_3 = \begin{bmatrix} -2\sqrt{3}-1 & 1 \\ 1 & -\sqrt{3} \end{bmatrix}, \qquad \mathbf{K}_4 = \begin{bmatrix} 2\sqrt{3}-1 & 1 \\ 1 & \sqrt{3} \end{bmatrix}$$

Among the solutions above, only $\mathbf{K}_4$ is positive definite. Therefore, the optimal control law is

$$\mathbf{u}(t) = -\mathbf{R}^{-1}\mathbf{B}^T\mathbf{K}\mathbf{x}(t) = -2[0\ 1]\begin{bmatrix} 2\sqrt{3}-1 & 1 \\ 1 & \sqrt{3} \end{bmatrix}\mathbf{x}(t) = [-2\ -2\sqrt{3}]\mathbf{x}(t)$$

The linear quadratic linear regulation problem is summarized in Table 10.5.

10.5. PONTRYAGIN'S MINIMUM PRINCIPLE

So far, it has been assumed that the admissible controls are not constrained by any boundaries. However, constraints always occur for real systems. For example, because of structure restrictions a control valve only provides a limited range of flow rates. The current in an electric motor is also limited, as overloaded current will damage the windings. In this section, we will discuss the optimal control problem with an input constraint.

Problem formulation
As we studied in Section 10.3, for a system

TABLE 10.5.   Linear regulation problem.

System equation	$\dot{\mathbf{x}} = \mathbf{A}\mathbf{x} + \mathbf{B}\mathbf{u}$	
Quadratic cost function	Finite terminal time	$J(\mathbf{u}) = \dfrac{1}{2}\mathbf{x}^T(t_f)\mathbf{H}\mathbf{x}(t_f) + \dfrac{1}{2}\int_{t_0}^{t_f}[\mathbf{x}^T(t)\mathbf{Q}\mathbf{x}(t) + \mathbf{u}^T(t)\mathbf{R}\mathbf{u}(t)]dt$
	Infinite terminal time	$J(\mathbf{u}) = \dfrac{1}{2}\int_{t_0}^{\infty}[\mathbf{x}^T(t)\mathbf{Q}\mathbf{x}(t) + \mathbf{u}^T(t)\mathbf{R}\mathbf{u}(t)]dt$
Control law	Finite terminal time	$\mathbf{u}(t) = -\mathbf{R}^{-1}\mathbf{B}^T\mathbf{K}(t)\mathbf{x}(t)$
	Infinite terminal time	$\mathbf{u}(t) = -\mathbf{R}^{-1}\mathbf{B}^T\mathbf{K}\mathbf{x}(t)$
Riccati equation	Finite terminal time	$\dot{\mathbf{K}} + \mathbf{K}\mathbf{A} + \mathbf{A}^T\mathbf{K} - \mathbf{K}\mathbf{B}\mathbf{R}^{-1}\mathbf{B}^T\mathbf{K} + \mathbf{Q} = 0$
	Infinite terminal time	$\mathbf{K}\mathbf{A} + \mathbf{A}^T\mathbf{K} - \mathbf{K}\mathbf{B}\mathbf{R}^{-1}\mathbf{B}^T\mathbf{K} + \mathbf{Q} = 0$

$$\dot{\mathbf{x}} = \mathbf{f}(\mathbf{x}, \mathbf{u}, t) \tag{10.5 1}$$

and a cost function

$$J(u) = S[\mathbf{x}(t), t]\Big|_{t=t_0}^{t=t_f} + \int_{t_0}^{t_f} L[\mathbf{x}, \mathbf{u}, t]dt \tag{10.5.2}$$

the optimal control can be obtained by solving the following equations:

$$\frac{\partial H}{\partial \mathbf{x}} + \dot{\lambda} = \mathbf{0} \tag{10.5.3}$$

$$\frac{\partial H}{\partial \mathbf{u}} = \mathbf{0} \tag{10.5.5}$$

with the transversality conditions in (10.3.15 ~ 10.3.18). If the input $\mathbf{u}(t)$ is constrained to

$$a_i \le |u_i(t)| \le b_i \tag{10.5.6}$$

the input cannot vary arbitrarily, and (10.5.5) may be meaningless.

Pontryagin's minimum principle
The Pontryagin's minimum principle states that the optimal control for minimizing the cost function (10.5.2) must minimize the Hamiltonian, i.e.,

$$H[\mathbf{x}^*(t), \mathbf{u}^*(t), \lambda^*(t), t] \le H[\mathbf{x}^*(t), \mathbf{u}(t), \lambda^*(t), t] \tag{10.5.7}$$

where
   $H[\mathbf{x}^*(t), \mathbf{u}^*(t), \lambda^*(t), t]$ is the Hamiltonian with the optimal control;
   $H[\mathbf{x}^*(t), \mathbf{u}(t), \lambda^*(t), t]$ is the Hamiltonian with an admissible control;
   $\mathbf{x}^*(t)$ is the $n \times 1$ optimal trajectory;
   $\mathbf{u}^*(t)$ and $\mathbf{u}(t)$ are the $m \times 1$ optimal and admissible control functions;
   $\lambda^*(t)$ is the $n \times 1$ optimal Lagrange multiplier vector.
   The principle asserts that minimizing the cost function requires minimizing the Hamiltonian.

Example 10.5.1
   Assume a system satisfies the state equation

$$\dot{x} = ax + bu$$

We need to find an optimal control to minimize the cost function

$$J = \int_0^\infty x^2 dt$$

The admissible control u is constrained to

$$-1 \le u \le 1$$

The Hamiltonian is written as

$$H = x^2 + \lambda(ax + bu)$$

Obviously, $\partial H/\partial u = \lambda b = 0$ is meaningless. However, if λb is positive, then $u = -1$ (the minimum value of u) makes H a minimum, and if λb is negative, then $u = 1$ makes H a minimum. Therefore, in the range $-1 \leq u \leq 1$, the input u that minimizes the Hamiltonian H is located on the boundaries, i.e.,

$$u^* = -sign(\lambda b) = \begin{cases} -1 & \lambda b > 0 \\ 1 & \lambda b < 0 \end{cases}$$

Substituting into the state equation yields

$$\dot{x} = ax - bsign(\lambda b)$$

From (10.5.3), we have

$$\dot{\lambda} = -\frac{\partial H}{\partial x} = -2x - \lambda a$$

The numerical result can be obtained from these two nonlinear equations by using a computer program. The terminal conditions may be obtained from Eqs. (10.3.11 ~ 10.3.18).

Problems

10.1. For a single state variable system with the state space representation

$$\dot{x} = 2u$$

find the optimal control $u^*(t)$ and the optimal trajectory $x^*(t)$ with the Euler-Lagrange equation. The cost function is

$$J = \int_0^{t_f} (x^2 + u^2)dt$$

and the boundary conditions are $x(0)$ and $x(t_f)$.

10.2. Solve Problem 10.1 with the Lagrange multiple method.

10.3. Prove that the closed loop system in an infinite terminal time, linear regulator problem is asymptotically stable. (Hints: use the Lyapunov equation under the condition that the weighting matrix Q is positive definite and the solution matrix of the Riccati equation is positive definite.)

10.4. Given a system

$$\dot{\mathbf{x}} = \begin{bmatrix} 0 & 1 \\ 0 & 0 \end{bmatrix} \mathbf{x} + \begin{bmatrix} 0 \\ 1 \end{bmatrix} u$$

and the cost function

$$J(\mathbf{u}) = \frac{1}{2} \int_0^\infty [\mathbf{x}^T(t) \mathbf{Q} \mathbf{x}(t) + R u^2(t)] dt$$

where

$$\mathbf{Q} = \begin{bmatrix} 1 & 0 \\ 0 & p \end{bmatrix}, \quad (p \geq 0) \quad \text{and} \quad R = 1$$

find the optimal control law (optimal feedback matrix) that minimizes the cost function.

10.5. Given a system

$$\dot{x}_1 = x_2 + u$$
$$\dot{x}_2 = -u$$

and the cost function

$$J = \int_0^t x_1^2(t) dt$$

Find the optimal control $u^*(t)$ and the state trajectory that minimize the cost function. Under this control, does state x_1 go to zero? Does state x_2 go to zero?

10.6. In Example 10.5.1, if the cost function is

$$J = \frac{1}{2} x_f^2 + \int_{t_0}^{t_f} dt = \frac{1}{2} x_f^2 + t_f - t_0$$

find the optimal control law to transfer state $x_0 = 0$ to $x_f = 1$.

CHAPTER 11

SYSTEM IDENTIFICATION AND ADAPTIVE CONTROL

11.1. MATHEMATICAL BACKGROUND

In order to design a control system, a mathematical model is needed to describe the process. Based on physical and/or chemical principles, we may derive the models. However, it is not always possible to develop a model theoretically. System identification is a useful way to construct process models and estimate unknown model parameters. The goal of system identification is to determine system equations from a given input and output time history.

All the mathematical models we discussed previously are deterministic models that do not include the effect of disturbances. The presence of disturbances is of course one of the main reasons for using control. Until now, we have treated the disturbances as inputs, such as step functions, impulse functions, and so on. However, many disturbances are neither eliminated at the source nor measured, and can only be described by stochastic process theory.

In this section, we briefly review some mathematical background needed for further study of system identification and adaptive control.

Probability

To each 'event' A of a class of possible events in a simple experiment, a number $P[A]$ is assigned to represent the possibility that the event will happen. This number is called probability.

The probability of any event is greater than or equal to zero, and less than or equal to 1, i.e.,

$$1 \geq P[A] \geq 0 \tag{11.1.1}$$

For a "certain" event Ω, we have

$$P[\Omega] = 1 \tag{11.1.2}$$

325

If Φ is an impossible event, then

$$P[\Phi] = 0 \qquad (11.1.3)$$

The sign '$\cup$' is used to represent 'or', and $A \cup B$ means event A or B or both will happen. The sign '$\cap$' is used to represent 'and'; i.e., $A \cap B$ means both A and B will happen. If event A and B are mutually exclusive, i.e., $A \cap B = \Phi$, it is obvious that

$$P[A \cup B] = P[A] + P[B] \qquad (11.1.4)$$

Usually for two events, we have

$$P[A \cup B] = P[A] + P[B] - P[A \cap B] \qquad (11.1.5)$$

If a certain event Ω consists of n events B_1, B_2, $\cdots$, B_n, i.e., $\Omega = [B_1, B_2, \cdots, B_n]$, then we have

$$P[B_1 \cup B_2 \cup \cdots \cup B_n] = 1 \qquad (11.1.6)$$

If also $B_i \cap B_j = \Phi(i \neq j)$, it follows that

$$\sum_{j=1}^{n} P[A \cap B_j] = P[A] \qquad (11.1.7)$$

Some commonly used properties of probability are given in Table 11.1.

Random variables

A random variable X is one that can have different values, but for each given real number a, the probability $P[X \leq a]$ is defined. The distribution function is defined as

$$F(x) = P[X \leq x] \qquad (11.1.8)$$

TABLE 11.1. Basic probability formulas.

Event	Conditions	Properties
Single event	Any event	$1 \geq P[A] \geq 0$
	Certain event Ω	$P[\Omega] = 1$
	Impossible event Φ	$P[\Phi] = 0$
Combination events	$A \cap B = \Phi$	$P[A \cup B] = P[A] + P[B]$
	$\Omega = [B_1, B_2, \cdots, B_n]$	$P[B_1 \cup B_2 \cup \cdots \cup B_n] = 1$
	$\Omega = [B_1, B_2, \cdots, B_n]$ $B_i \cap B_j = \Phi(i \neq j)$	$\sum_{j=1}^{n} P[A \cap B_j] = P[A]$

The probability density function is defined as

$$p(x) = \frac{dF(x)}{dx} \tag{11.1.9}$$

With this definition, it is easy to see that

$$p(x)\Delta x = P[x \le X \le x + \Delta x] \tag{11.1.10}$$

and

$$\int_a^b p(\xi)d\xi = P[a \le X \le b] \tag{11.1.11}$$

Obviously,

$$\int_{-\infty}^{\infty} p(\xi)d\xi = 1 \tag{11.1.12}$$

A joint distribution function is defined as

$$F(x, y) = P[X \le x, Y \le y] \tag{11.1.13}$$

Its corresponding probability density function is

$$p(x, y) = \frac{\partial^2 F(x, y)}{\partial x \partial y} \tag{11.1.14}$$

It follows that

$$p(x, y)\Delta x \Delta y = P[x \le X \le x + \Delta x, y \le Y \le y + \Delta y] \tag{11.1.15}$$

and

$$\int_c^d \int_a^b p(\xi, \eta)d\xi\, d\eta = P[a \le X \le b, c \le Y \le d] \tag{11.1.16}$$

Also,

$$\int_{-\infty}^{\infty} \int_{-\infty}^{\infty} p(\xi, \eta)d\xi\, d\eta = 1 \tag{11.1.17}$$

The definition and basic properties of random variables are summarized in Table 11.2.

Normal density function
The most useful probability density function is the normal density function, given by

$$p(\xi) = \frac{1}{\sqrt{2\pi}\sigma}\exp(-\frac{1}{2}\frac{(\xi-\mu)^2}{\sigma^2}) \tag{11.1.18}$$

TABLE 11.2. Definition and properties of random variables.

Item	One dimension	Two dimensions
Distribution function	$F(x) = P[X \leq x]$	$F(x, y) = P[X \leq x,\ Y \leq y]$
Probability density function	$p(x) = \dfrac{dF(x)}{dx}$	$p(x, y) = \dfrac{\partial^2 F(x, y)}{\partial x \partial y}$
Probability	$p(x)\Delta x =$ $P[x \leq X \leq x + \Delta x]$ $\int_a^b p(\xi)d\xi =$ $P[a \leq X \leq b]$	$p(x, y)\Delta x \Delta y = P[x \leq X \leq x + \Delta x,$ $y \leq Y \leq y + \Delta y]$ $\int_c^d \int_a^b p(\xi, \eta)d\xi\,d\eta =$ $P[a \leq X \leq b,\ c \leq Y \leq d]$
Properties	$\int_{-\infty}^{\infty} p(\xi)d\xi = 1$ $F(x) = \int_{-\infty}^{x} p(\xi)d\xi$ $F(-\infty) = 0$ $F(\infty) = 1$	$\int_{-\infty}^{\infty}\int_{-\infty}^{\infty} p(\xi, \eta)d\xi\,d\eta = 1$ $F(x,y) = \int_{-\infty}^{x}\int_{-\infty}^{y} p(\xi, \eta)d\xi\,d\eta$ $F(-\infty, y) = 0,\quad F(x, -\infty) = 0$ $F(x, \infty) = P[X \leq x]$ $F(\infty, y) = P[Y \leq y],\quad F(\infty, \infty) = 1$

The corresponding distribution function is called the normal distribution function. If the distribution function of a random variable is based on events that consist of a sum of a large number of independent random events, the distribution function will be normal. For example, the thermal motions of a large number of particles in a container may be described by a normal density function. The normal density (distribution) function is also called the Gaussian density (distribution) function.

Mean, variance and standard deviation
The mean, variance and standard deviation are the numerical characteristics of the probability density function $p(x)$ of the random variable X.

Expectation is an expected value of a random variable, and is defined as

$$\mu = \mathcal{E}[X] = \int_{-\infty}^{\infty} x p(x)dx \qquad (11.1.19)$$

The expectation of a random variable is also called the mean, which is like an arithmetic mean and is a weighted average of the random variable value.

The variance of a random variable is defined as

$$\sigma^2 = \text{var}[X] = \mathcal{E}[(X - \mu)^2] = \int_{-\infty}^{\infty} (x - \mu)^2 p(x)dx \qquad (11.1.20)$$

The variance of a random variable reflects the concentration of the probability density function around its mean.

The square root of the variance is referred to as the standard deviation, i.e., σ in (11.1.20).

The expectation of a random function $f(X)$ is defined as

$$\mathcal{E}[f(X)] = \int_{-\infty}^{\infty} f(x)p(x)dx \qquad (11.1.21)$$

The covariance of two random variables X and Y is defined as

$$\mathrm{cov}[X,Y] = \mathcal{E}[(X - \mu_x)(Y - \mu_y)]$$

$$= \int_{-\infty}^{\infty}\int_{-\infty}^{\infty}(x - \mu_x)(y - \mu_y)p(x,y)dxdy \qquad (11.1.22)$$

where $p(x, y)$ is the joint probability density function of random variables X and Y. μ_x and μ_y are the means of X and Y, respectively. If X and Y are independent of each other, we have

$$\mathrm{cov}[X,Y] = \mathcal{E}[(X - \mu_x)(Y - \mu_y)] = 0 \qquad (11.1.23)$$

The definitions and properties of the mean and variance are summarized in Table 11.3.

Example 11.1.1

The mean and variance of a random variable with a normal distribution function can be found as

$$\mathcal{E}[x] = \int_{-\infty}^{\infty} \xi \frac{1}{\sqrt{2\pi}\sigma} \exp(-\frac{1}{2}\frac{(\xi - \mu)^2}{\sigma^2})d\xi = \mu$$

TABLE 11.3. Means and variances.

Item	One dimension	Two dimensions
Definition of mean	$\mu = \mathcal{E}[X] = \int_{-\infty}^{\infty} xp(x)dx$ $\mathcal{E}[f(X)] = \int_{-\infty}^{\infty} f(x)p(x)dx$	$\mu_x = \mathcal{E}[X] = \int_{-\infty}^{\infty}\int_{-\infty}^{\infty} xp(x,y)dxdy$ $\mu_y = \mathcal{E}[Y] = \int_{-\infty}^{\infty}\int_{-\infty}^{\infty} yp(x,y)dxdy$
Definition of variance	$\sigma^2 = \mathrm{var}[X] = \mathcal{E}[(X - \mu)^2]$ $= \int_{-\infty}^{\infty}(x - \mu)^2 p(x)dx$	$\sigma_x^2 = \mathrm{var}[X] = \mathcal{E}[(X - \mu_x)^2]$ $= \int_{-\infty}^{\infty}\int_{-\infty}^{\infty}(x - \mu_x)^2 p(x,y)dxdy$
Covariance		$\mathrm{cov}[X,Y] = \mathcal{E}[(X - \mu_x)(Y - \mu_y)]$ $= \int_{-\infty}^{\infty}\int_{-\infty}^{\infty}(x - \mu_x)(y - \mu_y)p(x,y)dxdy$
Properties	$\mathcal{E}[(X - \mu)] = 0$ $\mathcal{E}[(X - \mu)^2] = \mathcal{E}[X^2] - \mu^2$	$\mathcal{E}[(X - \mu_x)^2]$ $= \int_{-\infty}^{\infty}\int_{-\infty}^{\infty} x^2 p(x,y)dxdy - \mu_x^2$
	If X and Y are independent of each other, then $\mathcal{E}[XY] = \mathcal{E}[X]\mathcal{E}[Y]$	

$$\text{var}[x] = \int_{-\infty}^{\infty} (\xi - \mu)^2 \frac{1}{\sqrt{2\pi}\sigma} \exp(-\frac{1}{2}\frac{(\xi - \mu)^2}{\sigma^2}) d\xi = \sigma^2$$

and the standard deviation is σ.

Random vector
A random vector is one whose entries are random variables, i.e.,

$$\mathbf{x} = [X_1, X_2, \cdots, X_n]^T \qquad (11.1.24)$$

where X_1, X_2, $\cdots$, X_n are the random variables. The expectation, or mean, of a random vector $\mathbf{x}$ is

$$\mathcal{E}(\mathbf{x}) = [\mu_1, \mu_2, \cdots, \mu_n]^T \qquad (11.1.25)$$

where μ_1, μ_2, $\cdots$, μ_n are the means of random variables X_1, X_2, $\cdots$, X_n respectively.

The covariance matrix, used to measure the probable deviation of a random vector from the mean value vector, is defined as

$$R_{xx} = \text{cov}(\mathbf{x}) = \mathcal{E}\{[\mathbf{x} - \mathcal{E}(\mathbf{x})][\mathbf{x} - \mathcal{E}(\mathbf{x})]^T\} \qquad (11.1.26)$$

The covariance matrix is a square matrix with the element in row i and column j given by the covariance of X_i and X_j, i.e.,

$$\mathcal{E}[(X_i - \mu_i)(X_j - \mu_j)]$$

Similarly, the covariance matrix of two different random vectors is defined as

$$R_{xy} = \text{cov}(\mathbf{x}, \mathbf{y}) = \mathcal{E}\{[\mathbf{x} - \mathcal{E}(\mathbf{x})][\mathbf{y} - \mathcal{E}(\mathbf{y})]^T\} \qquad (11.1.27)$$

which may not be a square matrix, since the dimensions of $\mathbf{x}$ and $\mathbf{y}$ may be different.

Stochastic process
A stochastic process $X(t)$ is a random variable that evolves in time t. It can be regarded as a family of random variables $\{X(t), t \in T\}$, where T is a time set. For a special time t', $X(t')$ is a random variable that has its own statistical characteristics such as mean and variance.

The mean function of a stochastic process is defined by

$$\mu(t) = \mathcal{E}[X(t)] = \int_{-\infty}^{\infty} x dF(x, t) \qquad (11.1.28)$$

where $F(x, t)$ is the probability distribution function of X at time t. It is obvious that $\mu(t)$ is a deterministic time function. The covariance function of a stochastic process is defined by

$$\text{cov}[X(s), X(t)] = \mathcal{E}\{[X(s) - \mu(s)][X(t) - \mu(t)]\} \qquad (11.1.29)$$

For a stochastic process vector, the covariance function is a square matrix and

$$\text{cov}[\mathbf{x}(s), \mathbf{x}(t)] = \mathcal{E}\{[\mathbf{x}(s) - \mu(s)][\mathbf{x}(t) - \mu(t)]\} \qquad (11.1.30)$$

where μ is a time function vector and

$$\mu(t) = \mathcal{E}[\mathbf{x}(t)] \qquad (11.1.31)$$

The finite-dimensional distribution function of a stochastic process is defined by

$$F(x_1, x_2, \cdots, x_n; t_1, t_2, \cdots, t_n)$$
$$= P[X(t_1) \le x_1, X(t_2) \le x_2, \cdots, X(t_n) \le x_n] \qquad (11.1.32)$$

If all the finite-dimensional distribution functions of a stochastic process are normal, the stochastic process is Gaussian or normal, and can be completely characterized by its mean and covariance function.

A stochastic process is called stationary if the finite dimensional distribution of $X(t_1)$, $X(t_2)$, $\cdots$, $X(t_n)$ is identical to the distribution of $X(t_1+\tau)$, $X(t_2+\tau)$, $\cdots$, $X(t_n+\tau)$ for all τ, n, t_1, t_2, $\cdots$, t_n. A stationary stochastic process is independent of time.

If the statistical characteristics of a stationary process can be calculated with a realization, i.e., the expectation can also be defined by a time integral

$$\mathcal{E}[x] = \lim_{T \to \infty} \frac{1}{T} \int_0^T x(t) dt = \int_{-\infty}^{\infty} x p(x) dx = \mu \qquad (11.1.33)$$

then the process is called ergodic. Such stationary processes have other properties, such as:

$$\text{var}[x] = \lim_{T \to \infty} \frac{1}{T} \int_0^T [x(t) - \mu]^2 dt = \int_{-\infty}^{\infty} (x - \mu)^2 p(x) dx = \sigma^2 \qquad (11.1.34)$$

$$\psi_{\mathbf{xx}}(t_1, t_2) = \lim_{T \to \infty} \frac{1}{T} \int_0^T [x(t_1) - \mu][x(t_2) - \mu] dt$$

$$= \int_{-\infty}^{\infty} \int_{-\infty}^{\infty} [x(t_1) - \mu][x(t_2) - \mu] p[x(t_1), x(t_2)] dx(t_1) dx(t_2) \qquad (11.1.35)$$

where $\psi_{\mathbf{xx}}$ is the covariance of random variables $x(t_1)$ and $x(t_2)$.

Discrete-time stochastic sequence
For a stochastic process $\{X(t), t \in T\}$, if we assign the time set T to integers, the stochastic process is referred to as a discrete-time stochastic sequence.

A typical and very useful stochastic sequence is a white noise sequence. The white noise sequence $X(t)$ is a stationary discrete-time stochastic sequence with the characteristics that $X(t)$ and $X(s)$ are independent of each other for $t \neq s$. Its covariance function is given by

$$\text{cov}[x(t), x(s)] = \begin{cases} \sigma^2 & t = s \\ 0 & t \neq s \end{cases} \qquad (11.1.36)$$

11.2. PARAMETER IDENTIFICATION-LEAST SQUARES ESTIMATION

The data, model and algorithm are three important factors for system identification. To introduce the basic system identification concepts, the following strong assumptions on data and models are made:

1. The system is linear and the order of the system is known. In other words, the structure of the model is known;

2. The system is stationary, i.e., the parameters are constant or change slowly, and the stochastic processes involved are stationary;

3. Only discrete-time models are considered.

Since system identification is usually performed with a computer, a discrete model is more convenient to use. Because the system model structure has alrcady been assumed, only the parameters of the model must be estimated; this is referred to as a parameter identification problem. Parameter identification or parameter estimation is defined as the experimental determination of values of parameters that govern the systems dynamic behavior, assuming that the structure of the process model is known [Eykhoff 1974].

System model
Consider a discrete-time system with the difference equation:

$$y(k) + a_1 y(k-1) + \cdots + a_n y(k-n) = b_1 u(k-1) + \cdots + b_n u(k-n) \quad (11.2.1)$$

where $y(\ )$ is an output variable, $u(\ )$ is an input variable, and the parameters $a_1, \cdots, a_n, b_1, \cdots, b_n$ are unknown and to be estimated.

This model can be rewritten as

$$y(k) + a_1 y(k-1) + \cdots + a_n y(k-n) - b_1 u(k-1) - \cdots - b_n u(k-n) = e(k) \,(11.2.2)$$

where $e(k)$ is an equation error that is added into the equation to represent the stochastic disturbance. $e(k)$ will be used to construct a cost function for the parameter estimation.

We assume that the outputs $y(0), y(1), \cdots, y(N)$ and inputs $u(0), u(1), \cdots, u(N)$ have been observed. The estimated parameters will best fit the observed data.

Least squares estimation (*LSE*)

Denote

$$\theta = [a_1, \cdots, a_n, b_1, \cdots, b_n]^T \qquad (11.2.3)$$

Since the equation error $e(k)$ is related to the system parameters, we denote it as $e(k, \theta)$. Then (11.2.2) can be rewritten in a vector form at the different moments as

$$y(n) = \phi^T(n)\theta + e(n,\theta)$$
$$y(n+1) = \phi^T(n+1)\theta + e(n+1,\theta)$$
$$\vdots \qquad\qquad (11.2.4)$$
$$y(N) = \phi^T(N)\theta + e(N,\theta)$$

where

$$\phi(k) = [-y(k-1), \cdots, -y(k-n), u(k-1), \cdots, u(k-n)]^T \quad (11.2.5)$$

A more compact matrix form can be constructed as

$$\mathbf{y} = \Phi\theta + \varepsilon(N,\theta) \qquad (11.2.6)$$

where

$$\mathbf{y} = [y(n), y(n+1), \cdots, y(N)]^T \qquad (11.2.7)$$

$$\Phi = [\phi(n), \phi(n+1), \cdots, \phi(N)]^T$$

$$= \begin{bmatrix} -y(n-1) & \cdots & -y(0) & u(n-1) & \cdots & u(0) \\ -y(n) & \cdots & -y(1) & u(n) & \cdots & u(1) \\ \vdots & & \vdots & \vdots & & \vdots \\ -y(N-1) & \cdots & -y(N-n) & u(N-1) & \cdots & u(N-n) \end{bmatrix} \qquad (11.2.8)$$

and

$$\varepsilon(N, \theta) = [e(n,\theta), e(n+1,\theta), \cdots, e(N,\theta)]^T \qquad (11.2.9)$$

is an equation error vector. To find the best parameter to fit the observed data, we define a cost function

$$J(\theta) = \sum_{k=n}^{N} e^2(k,\theta) = \varepsilon^T(N,\theta)\varepsilon(N,\theta) \qquad (11.2.10)$$

which can be rewritten as

$$J(\theta) = (\mathbf{y} - \Phi\theta)^T(\mathbf{y} - \Phi\theta) = \mathbf{y}^T\mathbf{y} - \mathbf{y}^T\Phi\theta - \theta^T\Phi^T\mathbf{y} + \theta^T\Phi^T\Phi\theta \quad (11.2.11)$$

The derivative of $J(\theta)$ with respect to θ is

$$\frac{dJ(\theta)}{d\theta} = -2\Phi^T\mathbf{y} + 2\Phi^T\Phi\theta = 0$$

or

$$\hat{\theta} = (\Phi^T\Phi)^{-1}\Phi^T\mathbf{y} \qquad (11.2.12)$$

where $\hat{\theta}$, instead of θ, is used to represent the estimated parameters. Vector $\mathbf{y}$ in (11.2.7) and matrix Φ in (11.2.8) are composed of observed outputs and inputs, from which the parameters in θ are estimated using (11.2.12).

Weighting least square estimation
The cost function with a weighting matrix is defined as

$$J_W(\theta) = \sum_{k=n}^{N} w(k)e^2(k, \theta) = \varepsilon^T(N, \theta)\mathbf{W}\varepsilon(N, \theta) \qquad (11.2.13)$$

where $\mathbf{W}$ is a real diagonal matrix. The estimated parameter vector can be obtained as

$$\hat{\theta}_W = (\Phi^T\mathbf{W}\Phi)^{-1}\Phi^T\mathbf{W}\mathbf{y} \qquad (11.2.14)$$

This result reduces to the ordinary *LSE* in (11.2.12) when $\mathbf{W} = \mathbf{I}$, where $\mathbf{I}$ is the identity matrix.

Random errors
All the observation of outputs and inputs may be contaminated with noise, or in other words, may contain random errors. The equation error $e(k, \theta)$ of (11.2.1), and thus its vector form $\varepsilon(k, \theta)$ in (11.2.6) may be generated by random errors, in which case they are considered as a stochastic sequence and a stochastic vector sequence, respectively. It is assumed that the stochastic sequence $e(k, \theta)$ is stationary and Gaussian with zero means, i.e.,

$$\mathcal{E}[e(k,\theta)] = 0 \qquad (11.2.15)$$

and its variance is denoted as

$$\text{var}[e(k,\theta)] = \sigma^2 \qquad (11.2.16)$$

The stochastic vector sequence $\varepsilon(k, \theta)$ also has a zero mean

$$\mathcal{E}[\varepsilon(k,\theta)] = \mathbf{0} \qquad (11.2.17)$$

and the covariance matrix is written as

$$\mathbf{R} = \text{cov}[\varepsilon(k,\theta)] = \mathcal{E}[\varepsilon(k,\theta)\varepsilon(k,\theta)^T] \qquad (11.2.18)$$

If the stochastic vector sequence $\varepsilon(k, \theta)$ is white, i.e., there is no time correlation between different moments, then (11.2.18) simplifies to

$$\text{cov}[\varepsilon(k,\theta)] = \mathcal{E}[\varepsilon(k,\theta)\varepsilon^T(k,\theta)] = \sigma^2 \mathbf{I} \qquad (11.2.19)$$

Stochastic characteristics of *LSE*
To evaluate the estimated parameters, we need to find their statistical characteristics, i.e., their means and covariances. Substituting (11.2.6) into (11.2.12), we have

$$\hat{\theta} = (\Phi^T \Phi)^{-1} \Phi^T [\Phi\theta + \varepsilon(N,\theta)] = \theta + (\Phi^T \Phi)^{-1} \Phi^T \varepsilon(N,\theta) \quad (11.2.20)$$

From (11.2.17), the expectation of $\hat{\theta}$ is

$$\mathcal{E}(\hat{\theta}) = \theta + \mathcal{E}[(\Phi^T \Phi)^{-1} \Phi^T \varepsilon(N,\theta)] = \theta \qquad (11.2.21)$$

This means that the mean value of the estimated parameter $\hat{\theta}$ is the same as the true value of the parameters, i.e., the least squares parameter estimation for the system (11.2.1) is unbiased. Similarly, the weighting least squares estimation is also unbiased.

The estimation error vector $\hat{\theta} - \theta$, which is different from the equation error vector in (11.2.9), can be found in (11.2.20), i.e.,

$$\hat{\theta} - \theta = (\Phi^T \Phi)^{-1} \Phi^T \varepsilon(N,\theta) \qquad (11.2.22)$$

Thus, the covariance matrix of the estimated error vector is

$$\begin{aligned}
\text{cov}(\hat{\theta} - \theta) &= \mathcal{E}[(\hat{\theta} - \theta)(\hat{\theta} - \theta)^T] \\
&= \mathcal{E}[(\Phi^T \Phi)^{-1} \Phi^T \varepsilon(N,\theta)\varepsilon^T(N,\theta)\Phi(\Phi^T \Phi)^{-1}] \\
&= (\Phi^T \Phi)^{-1} \Phi^T \mathcal{E}[\varepsilon(N,\theta)\varepsilon^T(N,\theta)]\Phi(\Phi^T \Phi)^{-1} \\
&= (\Phi^T \Phi)^{-1} \Phi^T \mathbf{R}\Phi(\Phi^T \Phi)^{-1}
\end{aligned} \qquad (11.2.23)$$

where $\mathbf{R}$ is given in (11.2.18). Similarly, the covariance matrix of the weighting least squares estimation error vector is

$$\text{cov}(\hat{\theta} - \theta) = (\Phi^T \mathbf{W}\Phi)^{-1} \Phi^T \mathbf{W}\mathbf{R}\mathbf{W}^T \Phi(\Phi^T \mathbf{W}\Phi)^{-1} \qquad (11.2.24)$$

If we select the weighting matrix as

$$\mathbf{W} = \mathbf{R}^{-1} \qquad (11.2.25)$$

then because $\mathbf{W} = \mathbf{W}^T$, we have

$$\text{cov}(\hat{\theta} - \theta) = (\Phi^T R^{-1} \Phi)^{-1} \tag{11.2.26}$$

The corresponding parameter estimation is

$$\hat{\theta}_W = (\Phi^T R^{-1} \Phi)^{-1} \Phi^T R^{-1} y = \hat{\theta}_{mv} \tag{11.2.27}$$

where the subscript mv represents minimum variance, and $\hat{\theta}_{mv}$ is referred to as a linear minimum variance estimation. It has been proven that the weighting least squares parameter estimation θ_{mv} with the weighting matrix $W = R^{-1}$ has the minimum covariance matrix [Deutsch 1965]. It has also been proven that if the equation error random sequence $e(k, \theta)$ is white, i.e., (11.2.19) is satisfied, then

$$\hat{\theta} = \hat{\theta}_{mv} \tag{11.2.28}$$

and the covariance of the estimation will be

$$\text{cov}(\hat{\theta} - \theta) = (\Phi^T R^{-1} \Phi)^{-1} = \sigma^2 (\Phi^T \Phi)^{-1} \tag{11.2.29}$$

This means that the least squares estimation has a minimum variance. The estimation with a minimum variance is called an efficient estimation.

The least squares estimation is also consistent, i.e., the covariance of the estimation error approaches 0 as the time goes to infinity.

The least squares estimation method is summarized in Table 11.4.

Recursive least squares estimation (RLSE)
Using (11.2.12) to estimate parameters, we need to calculate the inverse of $(\Phi^T \Phi)$ every time another observation is taken. The recursive least squares estimation method can overcome this problem.

TABLE 11.4.   Least squares estimation method.

Item	Mathematical expression
Model	$y(k) + a_1 y(k-1) + \cdots + a_n y(k-n)$ $-b_1 u(k-1) - \cdots - b_n u(k-n) = e(k)$
Vector model	$y = \Phi\theta + \varepsilon(N, \theta)$
Cost function	$J(\theta) = \sum_{k=n}^{N} e^2(k, \theta) = \varepsilon^T(N, \theta)\varepsilon(N, \theta)$
Estimated parameter	$\hat{\theta} = (\Phi^T \Phi)^{-1} \Phi^T y$
Properties	1. Unbiased: $\mathcal{E}(\hat{\theta}) = \theta$ 2. Efficient: $\hat{\theta} = \hat{\theta}_{mv}$ 3. Consistent: $\text{cov}(\hat{\theta} - \theta) \rightarrow 0 \quad (t \rightarrow \infty)$

If the equation (11.2.6) in mth step is denoted as

$$\mathbf{y}_m = \Phi_m \theta + \varepsilon_m(m,\theta) \qquad (11.2.30)$$

where

$$\mathbf{y}_m = [y(n),\, y(n+1),\, \cdots,\, y(m)]^T \qquad (11.2.31)$$

$$\Phi_m = [\phi(n),\, \phi(n+1),\, \cdots,\, \phi(m)]^T$$

$$= \begin{bmatrix} -y(n-1) & \cdots & -y(0) & u(n-1) & \cdots & u(n-p) \\ -y(n) & \cdots & -y(1) & u(n) & \cdots & u(n+1-p) \\ \vdots & & \vdots & \vdots & & \vdots \\ -y(m-1) & \cdots & -y(m-n) & u(m-1) & \cdots & u(m-n) \end{bmatrix} \qquad (11.2.32)$$

and

$$\varepsilon_m(m,\theta) = [e(n,\theta),\, e(n+1,\theta),\, \cdots,\, e(m,\theta)]^T \qquad (11.2.33)$$

then the m-step least squares parameter estimation is

$$\hat{\theta}_m = (\Phi_m^T \Phi_m)^{-1} \Phi_m^T \mathbf{y}_m \qquad (11.2.34)$$

After taking one more observation, we have a new equation:

$$\mathbf{y}_{m+1} = \Phi_{m+1}\theta + \varepsilon_{m+1}(m,\theta) \qquad (11.2.35)$$

where

$$\mathbf{y}_{m+1} = [y(n),\, y(n+1),\, \cdots,\, y(m),\, y(m+1)]^T = [\mathbf{y}_m^T,\, y(m+1)]^T \quad (11.2.36)$$

$$\Phi_{m+1} = [\phi(n),\, \phi(n+1),\, \cdots,\, \phi(m),\, \phi(m+1)]^T$$

$$= \begin{bmatrix} -y(n-1) & \cdots & -y(0) & u(n-1) & \cdots & u(n-p) \\ -y(n) & \cdots & -y(1) & u(n) & \cdots & u(n+1-p) \\ \vdots & & \vdots & \vdots & & \vdots \\ -y(m-1) & \cdots & -y(m-n) & u(m-1) & \cdots & u(m-n) \\ -y(m) & \cdots & -y(m-n+1) & u(m) & \cdots & u(m-n+1) \end{bmatrix}$$

$$= \begin{bmatrix} \Phi_m \\ \phi^T(m+1) \end{bmatrix} \qquad (11.2.37)$$

$$\phi^T(m+1) = [-y(m),\, \cdots,\, -y(m-n+1),\, u(m),\, \cdots,\, u(m-n+1)] \qquad (11.2.38)$$

$$\varepsilon_{m+1}(m,\theta) = [e(n,\theta),\, e(n+1,\theta),\, \cdots,\, e(m,\theta),\, e(m+1,\theta)]^T$$

$$= [\varepsilon_m^T(m,\theta),\, e(m+1,\theta)]^T \qquad (11.2.39)$$

then the $(m+1)$-step least squares parameter estimation is

$$\hat{\theta}_{m+1} = (\Phi_{m+1}^T \Phi_{m+1})^{-1} \Phi_{m+1}^T \mathbf{y}_{m+1} \qquad (11.2.40)$$

Denote

$$\mathbf{P}_m = (\Phi_m^T \Phi_m)^{-1} \qquad (11.2.41)$$

$$\mathbf{P}_{m+1} = (\Phi_{m+1}^T \Phi_{m+1})^{-1} \qquad (11.2.42)$$

Substituting (11.2.37) into (11.2.42), we have

$$\begin{aligned}
\mathbf{P}_{m+1} &= (\Phi_{m+1}^T \Phi_{m+1})^{-1} \\
&= (\Phi_m^T \Phi_m + \phi(m+1)\phi^T(m+1))^{-1} \\
&= (\mathbf{P}_m^{-1} + \phi(m+1)\phi^T(m+1))^{-1}
\end{aligned} \qquad (11.2.43)$$

From the following formula (see Problem 11.1)

$$(\mathbf{A} + \mathbf{BCD})^{-1} = \mathbf{A}^{-1} - \mathbf{A}^{-1}\mathbf{B}(\mathbf{C}^{-1} + \mathbf{DA}^{-1}\mathbf{B})^{-1}\mathbf{DA}^{-1} \qquad (11.2.44)$$

it follows that

$$\mathbf{P}_{m+1} = \mathbf{P}_m - \mathbf{P}_m \phi(m+1)[1 + \phi^T(m+1)\mathbf{P}_m \phi(m+1)]^{-1} \phi^T(m+1)\mathbf{P}_m \quad (11.2.45)$$

Since $[1 + \phi^T(m+1)\mathbf{P}_m \phi(m+1)]$ is a number instead of a matrix, we have

$$\mathbf{P}_{m+1} = \mathbf{P}_m - \frac{\mathbf{P}_m \phi(m+1)\phi^T(m+1)\mathbf{P}_m}{1 + \phi^T(m+1)\mathbf{P}_m \phi(m+1)} \qquad (11.2.46)$$

Thus, the parameter estimation becomes

$$\begin{aligned}
\hat{\theta}_{m+1} &= (\Phi_{m+1}^T \Phi_{m+1})^{-1} \Phi_{m+1}^T \mathbf{y}_{m+1} \\
&= \mathbf{P}_{m+1}[\Phi_m^T, \ \phi(m+1)]^T \begin{bmatrix} \mathbf{y}_m \\ y(m+1) \end{bmatrix} \\
&= \mathbf{P}_{m+1}[\Phi_m^T \mathbf{y}_m + \phi(m+1)y(m+1)] \\
&= [\mathbf{P}_m - \frac{\mathbf{P}_m \phi(m+1)\phi^T(m+1)\mathbf{P}_m}{1 + \phi^T(m+1)\mathbf{P}_m \phi(m+1)}][\Phi_m^T \mathbf{y}_m + \phi(m+1)y(m+1)] \\
&= \mathbf{P}_m \Phi_m^T \mathbf{y}_m - \frac{\mathbf{P}_m \phi(m+1)\phi^T(m+1)\mathbf{P}_m}{1 + \phi^T(m+1)\mathbf{P}_m \phi(m+1)} \Phi_m^T \mathbf{y}_m + \mathbf{P}_m \phi(m+1)y(m+1) \\
&\quad - \frac{\mathbf{P}_m \phi(m+1)\phi^T(m+1)\mathbf{P}_m}{1 + \phi^T(m+1)\mathbf{P}_m \phi(m+1)} \phi(m+1)y(m+1) \qquad (11.2.47)
\end{aligned}$$

The last two terms can be rearranged as

$$\mathbf{P}_m\phi(m+1)\frac{1+\phi^T(m+1)\mathbf{P}_m\phi(m+1)-\phi^T(m+1)\mathbf{P}_m\phi(m+1)}{1+\phi^T(m+1)\mathbf{P}_m\phi(m+1)}y(m+1)$$

$$=\frac{\mathbf{P}_m\phi(m+1)y(m+1)}{1+\phi^T(m+1)\mathbf{P}_m\phi(m+1)} \tag{11.2.48}$$

Substituting (11.2.48) and (11.2.34) into (11.2.47), we have

$$\hat{\theta}_{m+1}=\hat{\theta}_m+\frac{\mathbf{P}_m\phi(m+1)[y(m+1)-\phi^T(m+1)\hat{\theta}_m]}{1+\phi^T(m+1)\mathbf{P}_m\phi(m+1)} \tag{11.2.49}$$

If the parameter estimation $\hat{\theta}_m$ and matrix $\mathbf{P}_m$ at m step are known, then $\hat{\theta}_{m+1}$, the parameter estimation at $m+1$ step, can be estimated by (11.2.49) and the matrix $\mathbf{P}_{m+1}$ can be calculated by (11.2.46).

Remarks

1. Equation (11.2.49) has an intuitive meaning. The estimation at $m+1$ step is the sum of an estimation at m step and a revision term that is proportional to the difference between the new observation $y(m+1)$ and the estimation of this observation.

2. From (11.2.41) and (11.2.29), we have

$$\mathbf{P}_m=(\Phi^T\Phi)^{-1}=\text{cov}(\hat{\theta}-\theta)/\sigma^2 \tag{11.2.50}$$

The matrix P is a measure of the covariance of the estimation. Because the least squares estimation is consistent, we have

$$\mathbf{P}_m=(\Phi^T\Phi)^{-1}\to 0 \qquad (m\to\infty) \tag{11.2.51}$$

TABLE 11.5.   Recursive Least Squares Method.

Step	Description	Formulas
1	Initial value	e.g., $\hat{\theta}(0)=\mathbf{0}$, $\mathbf{P}_0=\alpha\mathbf{I}$
2	Update observation	$\phi^T(m+1)=[-y(m),\cdots,-y(m-n+1),$ $u(m),\cdots,u(m-n+1)]$
3	Calculate new estimation	$\hat{\theta}_{m+1}=\hat{\theta}_m+\dfrac{\mathbf{P}_m\phi(m+1)[y(m+1)-\phi^T(m+1)\hat{\theta}_m]}{1+\phi^T(m+1)\mathbf{P}_m\phi(m+1)}$
4	Update covariance	$\mathbf{P}_{m+1}=\mathbf{P}_m-\dfrac{\mathbf{P}_m\phi(m+1)\phi^T(m+1)\mathbf{P}_m}{1+\phi^T(m+1)\mathbf{P}_m\phi(m+1)}$
5	Recurrence	Return to Step 2.

3. The initial parameter estimation $\hat{\theta}(0)$ can be selected as zero or nonzero. The initial matrix $\mathbf{P}_0$ can be selected as

$$\mathbf{P}_0 = \alpha \mathbf{I} \qquad (11.2.52)$$

where α is a large number, say, 10^6, and $\mathbf{I}$ is the identity matrix.

The steps for using the recursive least squares method is shown in Table 11.5.

11.3. KALMAN FILTER

State estimation is another important aspect of system identification. As we studied in Chapters 9 and 10, state feedback plays an important role in linear system control theory. However, the state variables are usually not available, and the observation data are contaminated with noise. The Kalman filter is a state estimator that provides an optimal state estimation from noise observation data [Watanabe 1991].

Stochastic discrete-time model

For a discrete time state space representation, additional terms that are dependent on the noise appear in the following state equation and output equation:

$$\mathbf{x}_{k+1} = \mathbf{\Phi}\mathbf{x}_k + \mathbf{B}\mathbf{u}_k + \mathbf{\Gamma}\mathbf{v}_k \qquad (11.3.1)$$

$$\mathbf{y}_k = \mathbf{H}\mathbf{x}_k + \mathbf{e}_k \qquad (11.3.2)$$

where $\mathbf{\Phi}$, $\mathbf{B}$, $\mathbf{\Gamma}$ and $\mathbf{H}$ are compatible constant matrices, $\{\mathbf{v}_k\}$ and $\{\mathbf{e}_k\}$ are discrete time Gaussian white noise sequences with zero means, and

$$\text{cov}[\mathbf{v}_k] = \mathcal{E}[\mathbf{v}_k\mathbf{v}_k^T] = \mathbf{R}_1 \qquad (11.3.3)$$

$$\text{cov}[\mathbf{e}_k] = \mathcal{E}[\mathbf{e}_k\mathbf{e}_k^T] = \mathbf{R}_2 \qquad (11.3.4)$$

$$\text{cov}[\mathbf{v}_k, \mathbf{e}_k] = \mathcal{E}[\mathbf{v}_k\mathbf{e}_k^T] = 0 \qquad (11.3.5)$$

For the initial state $\mathbf{x}(0) = \mathbf{x}_0$, the following conditions are assumed:

1. $\mathbf{x}_0$ is Gaussian.
2. Its mean and variance are μ_0 and $\mathbf{R}_0$, i.e.,

$$\mathcal{E}[\mathbf{x}_0] = \mu_0, \qquad \text{and} \qquad \text{cov}[\mathbf{x}_0] = \mathbf{R}_0 \qquad (11.3.6)$$

3. It is not correlated with $\{\mathbf{v}_k\}$ and $\{\mathbf{e}_k\}$, i.e.,

$$\text{cov}[\mathbf{x}_0, \mathbf{v}_k] = 0, \qquad \text{cov}[\mathbf{x}_0, \mathbf{e}_k] = 0 \qquad (11.3.7)$$

It is also assumed that $\mathbf{R}_0$, $\mathbf{R}_1$ and $\mathbf{R}_2$ are positive semi-definite.

Prediction, filtering and smoothing
There are three kinds of state estimation: prediction, filtering and smoothing. If the estimated state is at the same time as the update observation time, we call it filtering; if before the observation time, it is smoothing; and if after, it is prediction.

Mathematically, assume that the update observation time is t, i.e., we use all the observation data at and before t, and the estimated state is at time τ. Then if $t = \tau$, it is called filtering; if $t > \tau$, it is called smoothing; and it is prediction if $t < \tau$. This relationship is shown in Fig. 11.3.1.

One step prediction
For the system (11.3.1), assume that the state estimation at the moment $k{-}1$ is $\hat{\mathbf{x}}_{k-1}$. Since $\mathcal{E}[\mathbf{v}_k] = \mathbf{0}$, reasonably, the estimated state at moment k is

$$\hat{\mathbf{x}}_{k/k-1} = \Phi\hat{\mathbf{x}}_{k-1} + \mathbf{Bu}_{k-1} \qquad (11.3.8)$$

where the subscript $k/k{-}1$ denotes "from the moment $k{-}1$ to that at the moment k." Accordingly, the output estimation, due to $\mathcal{E}[\mathbf{e}_k] = \mathbf{0}$, is

$$\hat{\mathbf{y}}_{k/k-1} = \mathbf{H}\hat{\mathbf{x}}_{k/k-1} \qquad (11.3.9)$$

We can assume that

$$\hat{\mathbf{x}}_k = \hat{\mathbf{x}}_{k/k-1} + \mathbf{K}_k(\mathbf{y}_k - \mathbf{H}\hat{\mathbf{x}}_{k/k-1}) \qquad (11.3.10)$$

where $\mathbf{y}_k$ is a new observation. $\mathbf{y}_k - \mathbf{H}\hat{\mathbf{x}}_{k/k-1}$ is the difference between the output observation and its estimation, and is used to revise the estimation based on the observation at the previous moment. $\mathbf{K}_k$ is a gain matrix to be determined.

Estimation errors and their covariance matrices
Define two estimation errors:

$$\tilde{\mathbf{x}}_{k/k-1} = \hat{\mathbf{x}}_{k/k-1} - \mathbf{x}_k \qquad (11.3.11)$$

$$\tilde{\mathbf{x}}_k = \hat{\mathbf{x}}_k - \mathbf{x}_k \qquad (11.3.12)$$

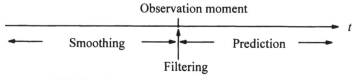

FIGURE 11.3.1. Filtering, prediction and smoothing.

They are the estimation errors before and after obtaining the observation y_k, respectively.

Substituting (11.3.2), (11.3.10) and (11.3.11) into (11.3.12), it follows that

$$\tilde{x}_k = \hat{x}_{k/k-1} + K_k(Hx_k + e_k - H\hat{x}_{k/k-1}) - x_k$$

$$= (I - K_k H)\tilde{x}_{k/k-1} + K_k e_k \tag{11.3.13}$$

From (11.3.6), it is easy to show (Problem 11.2) that

$$\mathcal{E}[\tilde{x}_k] = 0 \tag{11.3.14}$$

The covariance matrix of the estimation error will be

$$P_k = \mathcal{E}(\tilde{x}_k \tilde{x}_k^T) = \mathcal{E}\{(I - K_k H)\tilde{x}_{k/k-1}\tilde{x}_{k/k-1}^T (I - K_k H)^T\} + \mathcal{E}\{K_k e_k e_k^T K_k^T\}$$

$$+ \mathcal{E}\{(I - K_k H)\tilde{x}_{k/k-1}e_k^T\} + \mathcal{E}\{K_k e_k \tilde{x}_{k/k-1}^T(I - K_k H)^T\} \tag{11.3.15}$$

From (11.3.7), the last two terms of (11.3.15) are equal to zero, so that

$$P_k = (I - K_k H)P_{k/k-1}(I - K_k H)^T + K_k R_2 K_k^T \tag{11.3.16}$$

where

$$P_{k/k-1} = \mathcal{E}\{\tilde{x}_{k/k-1}\tilde{x}_{k/k-1}^T\} \tag{11.3.17}$$

Gain matrix

Eq. (11.3.16) can be rewritten as

$$P_k = P_{k/k-1} - K_k H P_{k/k-1} - P_{k/k-1}H^T K_k^T + K_k H P_{k/k-1}H^T K_k^T + K_k R_2 K_k^T \tag{11.3.18}$$

It is easy to be verified that

$$P_k = P_{k/k-1} - P_{k/k-1}H^T(HP_{k/k-1}H^T + R_2)^{-1}HP_{k/k-1}$$

$$+ [K_k - P_{k/k-1}H^T(HP_{k/k-1}H^T + R_2)^{-1}](HP_{k/k-1}H^T + R_2)[K_k - P_{k/k-1}H^T(HP_{k/k-1}H^T + R_2)^{-1}]^T \tag{11.3.19}$$

The last term of (11.3.19) includes K_k, and the others are non-negative. Thus, to minimize P_k, K_k can be selected as

$$K_k = P_{k/k-1}H^T(HP_{k/k-1}H^T + R_2)^{-1} \tag{11.3.20}$$

With this selection, the covariance of the estimation error becomes

$$P_k = P_{k/k-1} - P_{k/k-1}H^T(HP_{k/k-1}H^T + R_2)^{-1}HP_{k/k-1} = [I - K_k H]P_{k/k-1} \tag{11.3.21}$$

Recursive formula for covariance matrices

Substituting (11.3.8) and (11.3.1) into (11.3.11), we have

$$\tilde{\mathbf{x}}_{k/k-1} = \hat{\mathbf{x}}_{k/k-1} - \mathbf{x}_k = \Phi\hat{\mathbf{x}}_{k-1} - \Phi\hat{\mathbf{x}}_k - \Gamma\mathbf{v}_{k-1} = \Phi\tilde{\mathbf{x}}_{k-1} - \Gamma\mathbf{v}_{k-1} \qquad (11.3.22)$$

Therefore,

$$\begin{aligned}
\mathbf{P}_{k/k-1} &= \mathcal{E}\{\tilde{\mathbf{x}}_{k/k-1}\tilde{\mathbf{x}}_{k/k-1}^T\} \\
&= \Phi\mathcal{E}\{\tilde{\mathbf{x}}_{k-1}\tilde{\mathbf{x}}_{k-1}^T\}\Phi^T + \Gamma\mathcal{E}[\mathbf{v}_{k-1}\mathbf{v}_{k-1}^T]\Gamma^T \\
&= \Phi\mathbf{P}_{k-1}\Phi^T + \Gamma\mathbf{R}_1\Gamma^T \qquad (11.3.23)
\end{aligned}$$

where we have used the assumption that $\mathbf{v}$ is a white noise vector, i.e.,

$$\mathrm{cov}[\mathbf{x}_{k-1}, \mathbf{v}_{k-1}] = \mathbf{0} \qquad (11.3.24)$$

The discrete-time Kalman filter formulas are listed in Table 11.6.

State estimation steps

With the formulas given above, the estimated state can be recursively calculated by the following steps:

1. Determine the initial conditions $\hat{\mathbf{x}}_0 = \mathcal{E}[\mathbf{x}_0]$ and $\mathbf{P}_0 = \mathrm{cov}(\mathbf{x}_0)$.
2. Predict the state at the next step $\hat{\mathbf{x}}_{k/k-1}$ with (11.3.8).
3. Predict the covariance at the next step $\mathbf{P}_{k/k-1}$ with (11.3.23).

TABLE 11.6. Kalman filter.

Item	Formula
State equation	$\mathbf{x}_{k+1} = \Phi\mathbf{x}_k + \mathbf{B}\mathbf{u}_k + \Gamma\mathbf{v}_k$
Measurement equation	$\mathbf{y}_k = \mathbf{H}\mathbf{x}_k + \mathbf{e}_k$
Statistical characteristics	$\mathcal{E}[\mathbf{v}_k] = \mathbf{0}, \quad \mathcal{E}[\mathbf{e}_k] = \mathbf{0}, \quad \mathrm{cov}[\mathbf{v}_k] = \mathcal{E}[\mathbf{v}_k\mathbf{v}_k^T] = \mathbf{R}_1,$ $\mathrm{cov}[\mathbf{e}_k] = \mathcal{E}[\mathbf{e}_k\mathbf{e}_k^T] = \mathbf{R}_2, \quad \mathrm{cov}[\mathbf{v}_k, \mathbf{e}_k] = \mathcal{E}[\mathbf{v}_k\mathbf{e}_k^T] = \mathbf{0}$
Initial condition	$\mathcal{E}[\mathbf{x}_0] = \mu_0, \ \mathrm{cov}[\mathbf{x}_0] = \mathbf{R}_0, \ \mathrm{cov}[\mathbf{x}_0, \mathbf{v}_k] = \mathbf{0}, \ \mathrm{cov}[\mathbf{x}_0, \mathbf{e}_k] = \mathbf{0}$
State prediction estimation	$\hat{\mathbf{x}}_{k/k-1} = \Phi\hat{\mathbf{x}}_{k-1} + \mathbf{B}\mathbf{u}_{k-1}$
Covariance prediction	$\mathbf{P}_{k/k-1} = \Phi\mathbf{P}_{k-1}\Phi^T + \Gamma\mathbf{R}_1\Gamma^T$
State estimation	$\hat{\mathbf{x}}_k = \hat{\mathbf{x}}_{k/k-1} + \mathbf{K}_k(\mathbf{y}_k - \mathbf{H}\hat{\mathbf{x}}_{k/k-1})$
Gain matrix	$\mathbf{K}_k = \mathbf{P}_{k/k-1}\mathbf{H}^T(\mathbf{H}\mathbf{P}_{k/k-1}\mathbf{H}^T + \mathbf{R}_2)^{-1}$
Covariance iteration	$\mathbf{P}_k = [\mathbf{I} - \mathbf{K}_k\mathbf{H}]\mathbf{P}_{k/k-1}$

4. Calculate the gain matrix $\mathbf{K}_k$ with (11.3.20).
5. Estimate the state in the next step $\hat{\mathbf{x}}_k$ with (11.3.10).
6. Calculate the covariance $\mathbf{P}_k$ with (11.3.21).
7. Go to step 2.

Example 11.3.1

Consider a single state system with the state space representation

$$x_k = \phi x_{k-1} + v_{k-1}$$
$$y_k = x_k + e_k$$

where $\{v_k\}$ and $\{e_k\}$ are white noise sequences with zero means and variances of q and r, respectively. The Kalman filter formulas will be

$$\hat{x}_k = \hat{x}_{k/k-1} + K_k(y_k - \phi \hat{x}_{k-1})$$

$$K_k = P_{k/k-1} / (P_{k/k-1} + r)$$

$$P_{k/k-1} = \phi^2 P_{k-1} + q$$

$$P_k = [1 - K_k] P_{k/k-1}$$

If $\phi = 1$, $\hat{x}_0 = \mathcal{E}[x_0] = 0$, $P_0 = \text{cov}(x_0) = 100$, $q = 25$, $r = 15$ and the observed output data are 10, 20, 30, 40, $\cdots$, then the estimated state $\hat{x}_k$ can be determined. The calculation results of the first 6 steps are shown in Table 11.7.

11.4. ADAPTIVE CONTROL SYSTEMS

A process may vary with time or with applications so that the real process model may differ from the model used for designing the controller. The variation of a process may be caused by a change in the operating conditions or the

TABLE 11.7. The results of Example 11.3.1.

Step	Y_k	$P_{k/k-1}$	K_k	$\hat{x}_k$	P_k
0	0	--	--	0	100
1	10	125	0.893	8.93	13.39
2	20	38.39	0.719	16.89	10.79
3	30	35.79	0.705	26.13	10.57
4	40	35.57	0.703	35.89	10.55
5	50	35.55	0.703	45.81	10.55
6	60	35.55	0.703	55.79	10.55

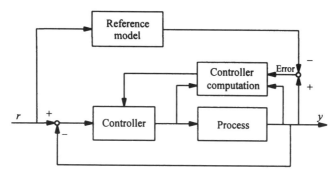

FIGURE 11.4.1. A Model reference adaptive control scheme.

environment. For such circumstances, an adaptive control strategy can be considered. An adaptive control system is one in which the controller parameters are adjusted automatically to compensate for the changing process conditions or environment. Adaptive control techniques provide a very useful approach to control system design. Model reference adaptive control and self-tuning control are the most popular adaptive control strategies. Here we only introduce some of their basic concepts.

Model-reference adaptive control
One typical adaptive control scheme is model-reference adaptive control (*MRAC*). Historically, this method is the first attempt in the design of stable adaptive systems.

A model-reference adaptive control scheme is shown in Fig. 11.4.1, where a reference model is incorporated in the system. Ideally, the response of the system to a specified input signal should be the same as the response of the reference model to the same input signal. The difference between the actual response of the system and the desired one of the reference model is used for the controller computation. The model reference adaptive control system consists of two loops. The inner loop is an ordinary feedback control loop, whereas the outer loop includes the reference model and controller computation device. The controller is compensated by the outer loop so that the error between the process output and reference model output is as small as possible.

A typical controller computation algorithm [Åström 1984] is

$$\frac{d\theta}{dt} = -\alpha\, e\, \text{grad}_\theta e \qquad (11.4.1)$$

where e is the error between process output and reference model output. θ is an adjustable parameter vector of the controller, while $\text{grad}_\theta e$ is the gradient of e

with respect to θ, which is of the same order as θ. The number α is a specified small coefficient. The gradient $\text{grad}_\theta e$ can be calculated from the data of process input and process output.

Constitution of a self-tuning control system

A natural way to compensate for the changing process parameters is to estimate the system parameters on-line, and to change the controller parameters accordingly. Conceptually, the design of an adaptive control system is a technique for combining a particular parameter estimation technique with any control law. This approach is often referred to as self-tuning control.

Fig. 11.4.2 shows a general block diagram of a self-tuning control system. The parameter identification may be carried out by the least squares estimation method described in Section 11.2. The estimated parameters of the system are used for the controller computations that determine the controller parameters. It can be seen in Fig. 11.4.2 that there are also two loops in the self-tuning system. As in a model reference adaptive control system, the inner loop is an ordinary feedback loop, which consists of the process and the controller. The outer loop is composed of a parameter estimator and controller computation device as well as the process and controller.

Self-tuning control approaches can be divided into two classes: explicit and implicit. In an explicit algorithm, the process model is explicitly estimated. In some cases, it is possible to directly parameterize the system in terms of controller parameters without knowing the process parameters explicitly. These algorithms are called implicit algorithms.

During the last two decades, research interest in self-tuning controllers was stimulated by the development of the self-tuning regulator [Åström and Wittenmark 1973], the self-tuning controller [Clarke and Gawthrop 1975, 1979] and the pole placement theory [Wellstead et al. 1979, Åström and Wittenmark 1980, Allidina and Hughes 1980]. The theoretical development of self-tuning control systems has been accompanied by a wide variety of applications in process control [Seborg et al. 1986].

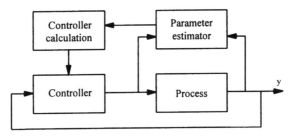

FIGURE 11.4.2. Self-tuning control scheme.

Generalized minimum variance self-tuning algorithm

There are different ways to calculate the controller setting. One is the minimum variance self-tuning algorithm [Clarke and Gawthrop 1975]. A pole assignment algorithm with a generalized minimum variance is introduced here as an example.

The model of a discrete time system is assumed to be

$$Ay(t) = q^{-k} Bu(t) + C\xi(t) \tag{11.4.2}$$

where A, B and C are polynomials in the backward shift operator q^{-1} with the form

$$A = a_0 + a_1 q^{-1} + \cdots + a_n q^{-n} \tag{11.4.3}$$

$$B = b_0 + b_1 q^{-1} + \cdots + b_m q^{-m} \tag{11.4.4}$$

$$C = c_0 + c_1 q^{-1} + \cdots + c_r q^{-r} \tag{11.4.5}$$

$y(t)$ and $u(t)$ are the output and input of the system, respectively. $\xi(t)$ is an uncorrelated zero-mean random sequence. k is the time delay in sampling instants associated with u. It is assumed that A and B are coprime (no common factors), and that $b_0 \neq 0$, $a_0 = c_0 = 1$, i.e., A and C are monic, and $k \geq 1$.

The cost function is defined as

$$J = E\left[\varphi^2(t+k)\right] \tag{11.4.6}$$

where $\varphi(t + k)$ is a generalized output, defined as

$$\varphi(t+k) = Py(t+k) + Qu(t) - Rw(t) \tag{11.4.7}$$

where $w(t)$ is a reference setpoint, and P, Q, and R are weighting polynomials in q^{-1}. For easy computation, we assume that P is monic.

Introduce the following identity

$$\frac{PC}{A} = F + q^{-k}\frac{G}{A} \tag{11.4.8}$$

where F is of order $k - 1$. F and G can be found by comparing coefficients on both sides of (11.4.8), or by the long division method. Substituting (11.4.8) and (11.4.2) into (11.4.7), we obtain

$$\varphi(t+k) = \frac{1}{C}[Hu(t) + Gy(t) + Ew(t)] + F\xi(t+k) \tag{11.4.9}$$

where H, G, and E are polynomials in q^{-1}, and

$$H = BF + QC \qquad (11.4.10)$$

$$E = -CR \qquad (11.4.11)$$

We can rewrite (11.4.9) into

$$\varphi(t+k) = \hat{\varphi}(t+k|t) + e(t+k) \qquad (11.4.12)$$

where

$$\hat{\varphi}(t+k|t) = \frac{1}{C}[Hu(t) + Gy(t) + Ew(t)] \qquad (11.4.13)$$

$$e(t+k) = F\xi(t+k) \qquad (11.4.14)$$

Since the two right hand terms of (11.4.12) are uncorrelated, according to the minimum variance control law proposed by Åström and Wittenmark (1973), we know

$$\hat{\varphi}(t+k|t) = 0 \qquad (11.4.15)$$

which indicates that the cost function of (11.4.6) is a minimum. It follows that

$$Hu(t) + Gy(t) + Ew(t) = 0 \qquad (11.4.16)$$

This is the control law. The controller contains feedback terms from $y(t)$ and feedforward terms from $w(t)$ as shown in Fig. 11.4.3.

The minimum cost function can be represented as

$$J_{min} = E[e^2(t+k)] = E[\{F\xi(t+k)\}^2] = [1 + f_1^2 + \cdots + f_{k-1}^2]\sigma^2 \quad (11.4.17)$$

where σ is the variance of $\xi(t)$, and F has the form

$$F = f_0 + f_1 q^{-1} + \cdots + f_{k-1}q^{1-k} \qquad (11.4.18)$$

where $f_0 = 1$, which comes from (11.4.8) with $a_0 = c_0 = p_0 = 1$.

The closed loop system with control law (11.4.16) can be presented as

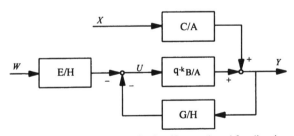

FIGURE 11.4.3. Configuration of a feedforward and feedback system.

TABLE 11.8. Procedure for minimum variance output self-tuning algorithm.

Steps	Contents	Parameters or formulas
1	Assign the closed loop transfer function	BL_z/L_p
2	Estimate system parameters	A, B, C
3	Calculate the polynomials P and Q	$BP + QA = L_p$
4	Calculate the polynomial R	$R = L_z$
5	Get the polynomials F and G	$\dfrac{PC}{A} = F + q^{-k}\dfrac{G}{A}$
6	Get the polynomials H and E	$H = BF + QC, \quad E = -CR$
7	Construct the control law	$Hu(t) + Gy(t) + EW(t) = 0$

$$y(t) = \frac{BR}{BP+QA}w(t-k) + \frac{H}{BP+QA}\xi(t) \tag{11.4.19}$$

The two transfer functions on the right hand side of (11.4.19) have the same denominator allowing for all closed loop poles to be assigned by choosing suitable weighting polynomials P and Q of the generalized output of the system, if the system parameters are known or have been estimated. The polynomials R, B and H determine the zeros of two transfer functions.

The parameters of the model (11.4.2) can be estimated by a parameter estimation method, e.g., the least squares technique. After the coefficients of the polynomial A, B and C are estimated, the minimum variance output self-tuning algorithm can be formatted by the procedure in Table 11.8.

Remarks

1. The numerator polynomial of the assigned closed loop transfer function has to include polynomial B, as was done in Step 1 in Table 11.8.

2. With the recursive least squares estimation method of estimating system parameters, Step 2 through Step 7 in Table 11.8 are repeated in every recursive cycle. The estimated system parameters are time variable. Therefore, the self-tuning controller is time variable.

3. Since the model parameters are directly estimated, this is an explicit self-tuning control algorithm.

Problems

11.1. Prove the matrix identity

$$(A + BCD)^{-1} = A^{-1} - A^{-1}B(C^{-1} + DA^{-1}B)^{-1}DA^{-1}$$

Hint: Multiply both sides of the formula by $(A + BCD)$.

11.2. Prove (11.3.14), i.e., that the state estimation is unbiased.

11.3. Consider a single state system with the state space representation

$$x_k = \phi x_{k-1} + v_{k-1}$$
$$y_k = x_k + e_k$$

where $\{v_k\}$ and $\{e_k\}$ are white noise sequences with zero means and variances of 25 and 15, respectively. If $\phi = 0.6$, $\hat{x}_0 = \mathcal{E}[x_0] = 0$, $P_0 = \text{cov}(x_0) = 10$ and the first 10 observed output data are 5.1, 6.0, 5.4, 4.8, 7.1, 6.3, 5.6, 4.4, 5.3, 5.8, $\cdots$, find the estimated state $\hat{x}_k$ in the first 10 steps with the Kalman filter. Give the variance of the estimation error.

11.4. Consider a second order system with the state space representation

$$\mathbf{x}_k = \mathbf{A}\mathbf{x}_{k-1} + \mathbf{b}v_{k-1} = \begin{bmatrix} -0.9 & 1 \\ 2 & -0.5 \end{bmatrix}\mathbf{x}_{k-1} + \begin{bmatrix} 1 \\ 0.3 \end{bmatrix}v_{k-1}$$
$$y_k = \mathbf{c}^T\mathbf{x}_k + e_k = (0.6 \ \ 1)\mathbf{x}_k + e_k$$

where $\{v_k\}$ and $\{e_k\}$ are white noise sequences with zero means and the variances $r_1 = 5$ and $r_2 = 2$, respectively. Assume $\hat{\mathbf{x}}_0 = \mathcal{E}[\mathbf{x}_0] = \mathbf{0}$ and

$$\mathbf{P}_0 = \text{cov}(\mathbf{x}_0) = \begin{bmatrix} 2 & 0 \\ 0 & 1 \end{bmatrix}$$

and the first 6 observed output data are 1.2, 2.4, 3.1, 2.1, 2.3, 0.4. Find the estimated state $\hat{\mathbf{x}}_k$ in the first 6 steps with the Kalman filter. Give the variance of the estimation error.

CHAPTER 12

INTELLIGENT PROCESS CONTROL

12.1. NEXT GENERATION OF PROCESS AUTOMATION

The growing complexity of industrial processes and the need for higher efficiency, greater flexibility, better product quality, and lower cost have changed the face of industrial practice. It has been widely recognized that quality control, process automation, management efficiency, marketing automation, and environmental protection are the keys to making modern industries stay competitive internationally. Recently, industry has begun to recognize the importance of intelligent process control in successfully implementing a multi-functional, flexible, company-wide manufacturing facility. The components of process control are not limited to system identification, process modeling, control scheme selection, and controller design. Intelligent process control will play vital role in company-wide integration, decision-making automation, operation support, maintenance automation, safe production, and environmental protection.

International marketplace competition
International competition has intensified the need for of high quality products that can compete in the global marketplace. As a result of this increased competition, the pace of product or system development has been quickened, thus forcing manufacturers into an era in which continuous quality improvement is a matter of survival, not simply competitive advantage. As the time scale of the product life cycle decreases and the demand for quality increases, more attention is focused on improving product quality and promoting the competitive ability of companies through better design, manufacturing, management and marketing. Obviously, increasing the manufacturing rate and improving product quality will become the most important factors for advancing in the international marketplace.

Global manufacturing industries are now undergoing a rapid structural change. As this process continues, the process industry will encounter difficulties as it confronts a changed and more competitive marketplace. Clearly, we cannot maintain our industrial base and high living standard without efficient competitive strategies.

To meet the global competition, manufacturers can use three strategies. The first strategy is to change the current strategies; that is, to improve our enterprise structure and to implement management automation. The second one is to develop marketing automation technology. The third strategy is using advanced manufacturing automation technology. The three automation implementations, namely M^3A (Management Automation, Marketing Automation and Manufacturing Automation), will change our enterprise structure and help us to increase the total factor productivity. A computer integrated process system (CIPS) is proposed to implement M^3A (as shown in Fig. 12.1.1) in process industry and to integrate all automation islands. In a CIPS environment, computer-aided market analysis decides the marketing strategies of a product; computer-aided design (CAD) creates the product on a video screen; computer-aided engineering analyzes the performance and productivity of designed product; computer-aided manufacturing (CAM) automates the shop floor process; an information management system (IMS) organizes the production process; and a knowledge integration system (KIS) coordinates the different activities among manufacturing, management and marketing. As a result, our companies will have the faster, cheaper, safer and better production and operation processes, and produce more competitive products.

In future economic competition, the competitive superiority of products in

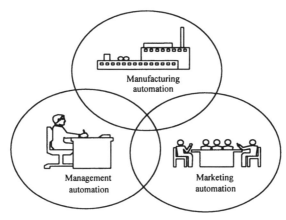

FIGURE 12.1.1. M^3A integration.

the international marketplace depends not only on the advantages of natural resources and manufacturing techniques but also on competitive marketing efforts. The success of industrial sectors relies on a global competition strategy, rather than a local one. The integration of M³A is an inevitable trend in the development of industrialization.

Development of industrial automation
It is widely recognized that industrialization is essentially a process of automation [Lu 1989; Rao 1991]. The development of industry automation can be divided into four stages, as shown in Table 12.1.

Stage 1: Labor-intensive industry. At this stage, there are no automatic controls and operations, and the efficiency and quality of the production mainly rests on the skills of human operators using simple tools. The equipment is not linked, and its maintenance depends heavily on the private experience of operators.

Stage 2: Equipment-intensive industry. Automatic equipment plays a dominant role in the increase of productivity. The equipment may consist of very complicated mechanical, electronic and computerized devices, and can be automatically operated only as stand alone machines. A typical example is a numerical control machine that is highly automatic in terms of a single machine. Meanwhile, sensor techniques and the advanced instruments have become the important means to obtain industrial data in equipment maintenance and automatic control. As a result of more powerful and affordable computing facilities on factory floors, our industry is now moving into the third stage.

Stage 3: Information-intensive industry. This environment stimulates the development of flexible manufacturing systems (FMS) in discrete manufactures, as well as distributed control systems (DCS) and information management

TABLE 12.1. Classification of industrial automation.

Automation level	Industrial stage	Technology	Manufacturing process	
			Discrete	Continuous
No automation	Labor intensive	Manual operation	Simple tools	
Data processing automation	Equipment intensive	Instrumentation control	Numeric control machines	Unit automation control
Information processing automation	Information intensive	Computer aided technology	CAD/CAM FMS	DCS and IMS
Decision making automation	Knowledge intensive	Computer integrated technology	CIMS	CIPS

systems (IMS) in continuous processes. At this stage, automation is realized at the level of information processing for groups of automatic machines. CAD/CAM technology is a typical manifestation of the characteristics of this stage. Based on mathematical models, computers assist human experts in numerical analysis, synthesis, simulation and graphics.

Stage 4: Knowledge-intensive industry. At this stage, computers help human experts in not only data and information processing, but also decision-making. Decision-making automation heavily relies on manufacture knowledge. In future industrial companies, intelligent computer aided design (ICAD) can automatically generate and analyze process (or product) models; computer integrated manufacturing systems (CIMS) and computer integrated process systems (CIPS) technology will be used to manage and implement the company-wide integration, and maintain the computerized automated production processes and equipment. This high-performance automation represents the next generation of manufacturing systems.

CIMS and CIPS

With the rapid growth of complexity in manufacturing processes, the industrial practice faces great challenges. The process design, control, operation, maintenance, as well as management involve many complex technical problems and economic issues that are not represented by mathematical models. Solving these ill-structured problems depends not only on mathematical models and calculation algorithms, but also on private knowledge and heuristic inference. Unfortunately, in the past decades, the primary objective of process control was focused only on finding new numerical processing techniques and algorithms, or establishing new mathematical control models. Nowadays, much attention has been drawn from both industry and academia to computer integrated manufacturing systems (CIMS) and computer integrated process systems (CIPS). CIMS and CIPS aim at integrating a product life cycle, including customer service, production planning, product design, program coding, operation quality testing, sales, service and production management, into a complete (or integrated) information processing system with minimum involvement in the operation process by human operators.

CIMS is a strategic thrust and an operating philosophy to achieve higher efficiencies within the whole cycle of product design, manufacturing, and marketing for discrete manufacturing. There are three flows in a continuous manufacturing process: information flow, material flow, and energy supply flow. The integration of CIMS emphasizes the information flow, whereas CIPS involves integrating all three flows. The decision-making among saving energy, increasing productivity, reducing manufacturing cost, as well as protecting the environment is a very important characteristic of CIPS. Discrete manufacturing

TABLE 12.2.   Distinctions between discrete and continuous manufacturing.

	Discrete manufacturing	Continuous manufacturing
Computer integration technology	CIMS	CIPS
Integration emphases	Information flows	Information flows Material flows Energy supply flows
Buffer area	Yes	No
Manufacturing environment	Common situation	Unusual situation
Current manufacturing technology	CAD/CAM, FMS	DCS, IMS

processes are operated in the common environment, while continuous manufacturing processes are very often operated under unusual situations such as high (low) temperature and/or high (low) pressure, high flammability, and high toxicity. The distinctions between discrete and continuous manufacturing are summarized in Table 12.2.

Many research papers associated with CIMS applications for the discrete manufacturing industry have been published. However, very few results are reported for CIPS in continuous processes.

Three levels in computer application in process industry
Data, information and knowledge are at different levels. Data are individual measurements from sensors. Information represents the relationship among the correlated data. Knowledge describes the connection among the structured information. Intelligence is the capability of utilizing knowledge [De Silva 1993]. These definitions are shown in Table 12.3. Brain activities to handle the information are based on domain-specific knowledge.

Data processing systems (DPS), information management systems (IMS) and knowledge integration systems (KIS) are three kinds of systems. They reflect three levels for computer applications in the manufacturing industry shown in Fig. 12.1.2. So far, most process companies are at the second level.

TABLE 12.3.   Data, information, knowledge and intelligence.

Item	Definition
Data	Individual measurements from sensors
Information	Relationship among the correlated data
Knowledge	Connection among the structured information
Intelligence	Capability of utilizing knowledge

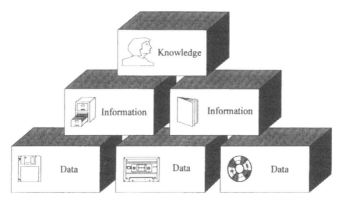

FIGURE 12.1.2.   Data, information and knowledge.

Philosophically, an information management system controls the data processing system, while a knowledge integration system can manage the information management systems.   For example, a distributed control system (DCS) in a process plant is a DPS. An IMS deals with all data from a DCS. Currently, DPS and IMS are being widely used in the process industry, and the implementation of KIS to improve quality and productivity is an important task. The three kinds of systems at the three levels and their functions are summarized in Table 12.4.

Why computer integrated process systems
As mentioned above, even though DPS and IMS are widely used in process control, the implementation of KIS to improve quality and productivity is an important issue. This is because the following problems exist in the process industry:

1. In a complicated computerized process, too much information needs to be processed, especially in an emergency situation. The coupling between system components is ever-accelerated, and the response time requirement is heightened. The stochastic occurrence of operational faults requires emergency handling. Especially in the chemical industry, this issue is more complicated due to the inefficiency and poor training of an operator in handling emergencies. DPS and IMS are not satisfactory for dealing with an emergency.

TABLE 12.4.   Functions of KIS, IMS and DPS.

Level	Computer system	Function
Knowledge	Knowledge integration system (KIS)	Managing information
Information	Information management system (IMS)	Managing data
Data	Data processing system (DPS)	Controlling equipment

2. Many problems and knowledge involved in industrial processes are often ill-structured, and difficult to formulate. In fact, mathematical modeling is not the only means to describe real manufacturing problems [Rao 1991]. The knowledge and experience we have about the real world are not well captured by numbers [Davis 1987]. Many manufacturing problems are ill-structured. They deal with non-numerical information and non-algorithmic procedures, and are suitable for the application of artificial intelligence (AI) techniques. Some important process variables cannot be measured on-line directly. They affect process operation, product quality, and productivity, and are only partially understood. Meanwhile, process control very often involves uncertain and fuzzy information. In such a process, mathematical modeling is not amenable, and purely algorithmic methods are difficult to use. DPS and IMS cannot deal with such information or data.

3. In industrial processes, too much information must be represented. Some may be expressed explicitly by graphics or neural networks, rather than only by numbers or symbols. So far, a few intelligent control systems have been successful in coupling symbolic reasoning with numerical computation. However, such integration cannot satisfy the requirements of continuous processes. The empirical data from process operations cannot be effectively processed with the existing expert systems or numerical models. In fact, many complicated industrial problems cannot be solved by a single technique such as symbolic reasoning, numerical computation, or neural networks. A new methodology to integrate neural networks, symbolic reasoning, numerical computation, and computer graphics has to be developed to handle these problems. Modern industrial processes are so complicated that no single tool can handle every situation. For example, in an integrated operation environment, the different existing software packages coded in computer languages such as C™, FORTRAN™ and Pascal™, as well as written with commercial AI tools such as KEE™, OPS5™, G2™ and M.1™, could be required to use together.

Of course, these problems cannot be handled by conventional process control techniques. To solve the problems above, new techniques and methodologies should be developed for the real-world industrial processes.

Implementation of computer integrated process systems
As a new advanced technology frontier, artificial intelligence (AI) has been widely applied to various disciplines, including process control. This study aims at processing non-numerical information, using heuristics and simulating the human capability for problem-solving. In fact, a much better terminology "Complex Information Processing System" was suggested for this field of "Artificial Intelligence."

Technological advances in automatic control have addressed the research interests of intelligent control, which encompasses the theory and applications of both artificial intelligence (AI) and automatic control [Fu 1971]. Intelligent process control is developed for implementing process automation, improving product quality, enhancing industrial productivity, preventing negative environmental effects, and ensuring operational safety.

In the development of computer integrated process systems, new problems are generated, and thus new technologies need to be developed:

1. With the development of process automation, the relationship between human operators and industrial processes has been fully changed. In many industrial companies, the processes are modern, highly computerized human-machine systems. Therefore, it is becoming increasingly difficult for operators to understand various signals and information from display boards, video, and computer screens. In this case, operators are fully isolated from modern process control systems (as shown in Fig. 12.1.3) because of the applications of DCS and IMS. Owing to the importance of the operator's role in these systems, keeping operators in a process control loop is necessary, and enhancing the quality of interaction between human operators and industrial processes is important. The

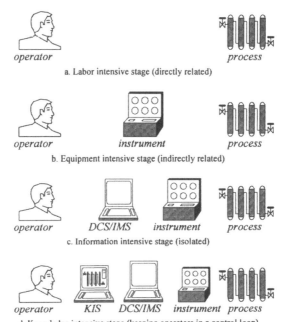

FIGURE 12.1.3. Relationship between operators and industrial processes.

development of an intelligent multimedia interface can better facilitate human interaction with the complex real-time monitoring and control systems. The multimedia interface can communicate with operators via multimedia and modes, such as natural language, graphics, animation, video presentation, and so on.

2. In an integrated process control environment, conflicting conclusions among unit control strategies, as well as among different domain engineers, usually arise because different intelligent process control systems can make different decisions based on different criteria, even though the data that fire the rules are nearly the same. For example, when a disturbance occurs, the "operation expert" will change the operation state in a manufacturing process, but the "control expert" may wish to keep the operation state unchanged (due to the set-point control strategy). The lack of conflict reasoning strategies has become a bottleneck in applications of integration techniques. In addition, company-wide optimization requires coordination among different production sections. Therefore, handling conflict conclusions is essential to integrate manufacturing, management, and marketing together.

3. The amount of manufacturing, management and marketing is rapidly increasing. In industrial processes, a large amount of operational data from different sensors and instruments have to be collected and processed in time with the useful knowledge and information acquired. The difficulties come from (1) how to process so much data in time, and (2) how to extract effective knowledge from new information because all data or information cannot be stored in a computer due to limited capacity. The existing database technology and numerical computing methods may be used for operational data processing. However, the problem of effectively acquiring knowledge from the data still remains difficult.

4. So far, most process plants have installed highly automatic facilities, where various computer hardware and software from different vendors have been used. To stimulate industrial companies to utilize the newest manufacturing technology, we need to develop a new system architecture to integrate the existing hardware and software from different vendors, rather than give up the existing process control computer facilities, operation environments, and information management systems in industrial plants. To get the maximum value from process control systems, various computers should be easily accessible to production personnel, engineers, and managers, so they can monitor process operation and make better decision.

The next generation of process automation is to implement computer integrated process systems to integrate manufacturing facilities and activities, i.e., integration of process design and control; integration of knowledge from multidisciplinary domains; integration of both empirical and analytical knowledge; integration of different symbolic processing systems; integration of numerical computation packages; integration of symbolic processing systems, numerical computation programs, neural networks, computer graphics packages and computer vision systems; integration of multimedia information and integration of multiple computer systems, and commercial software packages made by different vendors. Decision-making automation is actually a core in these integration activities.

Intelligence engineering

Traditionally, expert system development relies mainly on the knowledge engineer who, by definition, is a computer scientist and has the knowledge of artificial intelligence and computer science. Knowledge engineers interview domain experts to acquire the knowledge, then build expert systems. With increasingly powerful hardware platforms, a more user-friendly software environment, as well as the increased computational capability of engineers, a new era of AI applications is coming that will enable engineers to program intelligent systems to handle the problems at hand.

As we know, during the development of expert systems, knowledge acquisition is the most important but also most difficult task. Even with the help of knowledge engineers, some private knowledge (such as heuristics and personal experience as well as rules of thumb) may still be difficult to transfer. A new generation of engineers, namely intelligence engineers [Rao 1991], is to be trained to meet the needs in this subject. The name of intelligence engineer was set to distinguish from the knowledge engineer. As shown in Fig. 12.1.4, an intelligence engineer is a domain engineer, (for instance, a control engineer),

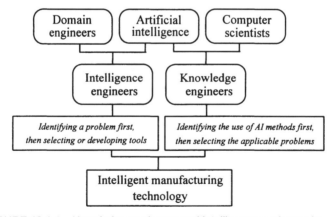

FIGURE 12.1.4.   Knowledge engineers and intelligence engineers in intelligent manufacturing

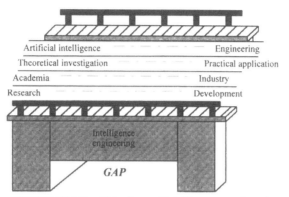

Artificial intelligence	Engineering
Theoretical investigation	Practical application
Academia	Industry
Research	Development

FIGURE 12.1.5. Objectives of intelligence engineering.

who has the certain domain knowledge in the related application area. Through a comparatively short training process, she/he learns the basic AI techniques, and gets the hands-on experience of programming expert systems. Thus, the intelligence engineer can produce much better expert systems to solve her/his domain problems, and extend AI applications successfully. In intelligence engineering, problems to be solved are identified first. Then, the problem-solving methods or tools are selected or developed. AI methodology will not be chosen unless there is a real need. On the other hand, AI methodology has been selected first in the knowledge engineering approach, which results in a very common failure in developing expert systems: the selected tools are not suitable for solving the problem at hand.

Intelligence engineering involves applying artificial intelligence techniques to solve engineering problems, and investigating theoretical fundamentals of AI techniques based on engineering methodology [Rao 1993]. Fig. 12.1.5 demonstrates the objectives of intelligence engineering. Different from knowledge engineering, intelligence engineering focuses on integrating knowledge from different disciplines and domains to solve the real engineering problems.

12.2. SYSTEM ARCHITECTURE FOR INTELLIGENT CONTROL

AI development and applications in process control

Computers have been successfully used for control and monitoring in process industries. The existing computer hardware and programming technologies are powerful, having allowed us to implement sophisticated control strategies, and to introduce intelligent control systems. However, the increasing demand for more effective data processing methods and control strategies for a wider range of industrial applications cannot be matched by current computing techniques.

Most existing intelligent systems for industrial control are implemented for design purposes [Pang et al. 1990]. Control system design is a very complex and not a completely understood process, particularly since it is abstract and requires creativity. Such design is largely a trial-and-error process, in which human design experts often employ heuristics or rules of thumb. An intelligent system can introduce a user-computer interface into such a process to improve the efficiency of design.

The primary function of controllers can be improved by introducing AI technology into control systems. The experience gained from building expert systems also shows that the power of expert systems is most apparent when the problem considered is sufficiently complex [Åström et al. 1986].

Fuzzy set theory was proposed by Zadeh [1978] to represent inherent vagueness. Recently, it has been applied to expert-aided control systems to deal with the imperfect knowledge of the control system. For example, a fuzzy controller has been integrated within a knowledge-based control system to tune servo controllers [De Silva and MacFarlane 1989].

In the analysis and synthesis of control systems, simulation is a major technique. The traditional simulation techniques are based on mathematical models and algorithms. These techniques cannot clearly simulate the dynamic behavior of the ill-conditioned processes. Knowledge-based simulation was suggested to solve these problems [Shannon 1986].

A functional approach to designing expert simulation systems for control, which performs model generations and simulations, was proposed by Lirov and his colleague [1988]. They chose differential game simulator designs to build this simulation system. The knowledge representation of the differential game models is described using semantic networks. The model generation methodology is a blend of several problem-solving paradigms, and the hierarchical dynamic goal system construction serves as the basis for model generation. This discrete event approach, based on the geometry of the games, can generally obtain the solution in much shorter time.

Classification of intelligent control systems
The research and development of intelligent control systems has continued for several years, and the efforts have produced four types of intelligent control systems in terms of system architecture. They are the symbolic reasoning system, the coupling inference system, the artificial neural network and the integrated distributed intelligent system. A brief description of each can be found in Table 12.5.

TABLE 12.5. Classification of intelligent control systems.

System	Description
Symbolic reasoning system	It processes symbolic information, and provides assistance to control engineers in decision-making for design and operation monitoring.
Coupling inference system	It links numerical computation programs with a symbolic reasoning system such that it can be used to solve engineering problems.
Artificial neural network	It models human brain and deals with empirical data, often used in system identification and fault diagnosis.
Integrated distributed intelligent system	It is a large intelligence integration environment, and can integrate different expert systems, neural networks, database, numerical computing packages and graphics programs to solve complex problems.

Symbolic reasoning systems

A symbolic reasoning system is a specific computer program, implemented by a specific programming technique, which is used to solve a specific problem in a specific domain, one that is difficult to handle by purely algorithmic methods. Many expert systems are symbolic reasoning systems that are composed of a database, knowledge base, inference engine and interface. Their functions are shown in Table 12.6.

The segregation of the database, the knowledge base and the inference engine in the expert system allows us to organize the different models and domain expertise efficiently because each of these components can be designed and modified separately.

Symbolic reasoning systems have been widely applied in intelligent manufacturing systems. Among successful AI applications, most intelligent systems are production systems. Production systems facilitate the representation of heuristic reasoning in such a way that intelligent systems can be built incrementally as the expertise knowledge increases.

Programming tools may be used to build expert systems. The tools usually

TABLE 12.6. The main components of an expert system.

Component	Function
Database	Storing the facts and data used in reasoning
Knowledge base	Storing knowledge about a specific domain
Inference engine	Determining where to start the reasoning process and how to resolve conflicts
Interface	Implementing the communication between the expert system and the user or other program

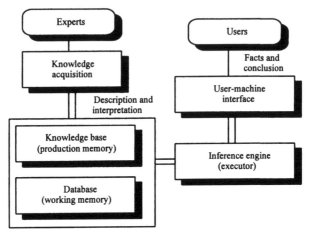

FIGURE 12.2.1. Structure of a symbolic reasoning system.

consist of three components: a data base (working memory), a knowledge base (production memory) and an inference engine as shown in Fig. 12.2.1. The working memory has a special buffer-like data structure and holds the knowledge that is accessible to the entire system. The Production memory of a symbolic reasoning system contains two kinds of knowledge: public knowledge (general knowledge) and private knowledge (expertise knowledge). The expertise knowledge is described by a set of production rules. The typical production rule is described as "IF (condition), THEN (action)". Every production rule consists of a condition-action pair. The inference engine is an executor. It determines which rules are relevant to a given data memory configuration and selects one to apply. Usually, this control strategy is called conflict resolution. The inference engine can be described as a finite-state machine with a three step cycle: matching rules, selecting rules and executing rules.

OPS5™ [Brownston et al. 1985], which served as one of the programming tools in our early projects, was the most widely used language for developing symbolic reasoning systems based on production rules. In OPS5™, each unit of working memory is an attribute-value element. Any attribute that is not assigned a value for a particular instance is given the default value designated as 'nil'. Its condition part of the condition-action pair is called the "LEFT-HAND-SIDE (LHS)", and the action part is called the "RIGHT-HAND-SIDE (RHS)". Each condition element specifies a pattern that is to be matched against working memory. The matching process is described as we define the syntax of LHS. The actions forming RHS are imperative statements that are executed in sequence when the rule is fired. Most permissible actions alter the working memory.

The syntax of a production rule in OPS5™ looks like

> (p rulename
>   condition
> →
>   action)

where the LHS and RHS are separated by the symbol '→', and 'p' represents the production rule.

Coupling inference systems

Many existing symbolic reasoning systems were developed for specific purposes. These systems can only process symbolic information and make heuristic inferences. The lack of numerical computation and coordination between applications limits their capability to solve real engineering problems.

In a manufacturing environment, we require not only a qualitative description of system behavior, but also a quantitative analysis. The former can predict the trend of the change of an operating variable, while the latter may provide us with a means to identify the change range of the variable. We often use qualitative and quantitative analyses together in solving engineering problems (Fig. 12.2.2). Usually, qualitative decisions are mainly based on symbolic and graphical information, while quantitative analysis is more conveniently performed using numerical information. These methods often complement each other.

Moreover, as a part of the accumulated knowledge of human expertise, many practical and successful numerical computation packages have already been

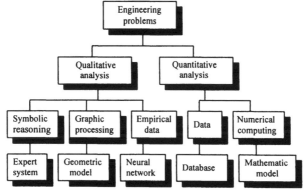

FIGURE 12.2.2. Qualitative and quantitative analyses.

developed. Although artificial intelligence should emphasize symbolic processing and non-algorithmic inference [Buchanan 1985], the utilization of numerical computation will make intelligent systems more able to deal with engineering problems. Like many modern developments, artificial intelligence and its applications should be viewed as a welcome addition to the technology rather than a substitute for numerical computation.

The coordination of symbolic reasoning and numerical computation is essential to develop intelligent manufacturing systems. More and more, the importance of coordinating symbolic reasoning and numerical computing in knowledge-based systems is being recognized. It has been realized that if applied separately, neither symbolic reasoning nor numeric computing can successfully address all problems in manufacturing. Complicated problems cannot be solved by purely symbolic or numerical techniques [Kowalik and Kitzmiller 1988].

Fig. 12.2.3 distinguishes the coupling intelligent system from the symbolic reasoning system from the viewpoint of software architecture. So far, many coupling intelligent systems have been developed in various engineering fields to enhance the problem solving capacity of the existing symbolic reasoning systems [Kitzmiller and Kowalik 1987]. In IDSOC (Intelligent Decisionmaker for the problem solving Strategy of Optimal Control) [Rao, Tsai and Jiang 1988], a set of numerical algorithms to compute certainty factors is coupled in the process of symbolic reasoning. Another coupling intelligent system, SFPACK [Pang et al. 1990], incorporates expert system techniques in a design package, then supports more functions to designers. Written in Franz Lisp™, CACE-III [James et al. 1985] can control the startup of several numerical routines programmed in FORTRAN™. Similar consideration was taken into account by Åström's group [Åström et al. 1986].

Methods of integrating symbolic reasoning systems and numerical

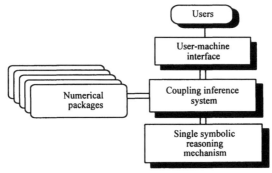

FIGURE 12.2.3. Structure of coupling intelligent systems.

computation packages have been proposed [Kitzmiller and Kowalik 1987]. A few developers tried to couple intelligent systems with conventional languages, such as FORTRAN™, so that these symbolic reasoning systems could be used as subroutines in a FORTRAN™ main program. Others suggested developing coupling intelligent systems in conventional languages in order to achieve the integration of numerical algorithms and symbolic inference. However, these methods preclude developing and using the individual programs separately, and are not cost-effective. It is very difficult to integrate new programs with these methods. Another disadvantage is that the procedural language environment cannot provide many of the good features that AI tools provide, such as easy debugging and allowance for interruption by human experts.

Numerical languages often have a procedural flavor, in which the program control is command-driven. They are very inefficient when dealing with processing strings. Symbolic languages are more declarative and data-driven. However, symbolic languages are very slow to execute numerical computations. Coupling symbolic processing with numerical computing is desirable to use numerical and symbolic languages in different portions of a software system.

Currently, not all of the expert system tools or environments provide the programming techniques for developing coupling intelligent systems. For example, it is very difficult to carry out numerical computation in OPS5™ [Brownston et al. 1985; Rao, Jiang and Tsai. 1988]. However, many software engineers are now building new tools for coupling inference systems that will be beneficial to the artificial intelligence applications in engineering domains.

Artificial neural networks

The human brain is an example of a natural neural network. Neurons are the living nerve cells, and neural networks are the networks of these cells. Such a network of neurons can think, feel, learn and remember. Artificial neural networks (ANNs), modeled after the known capabilities of the human brain, are copied into computer hardware and software [Vemuri 1988]. ANNs exhibit the following characteristics:

1. Learning: ANNs can modify their own behaviors in response to their environment.

2. Generalization: Once trained, an ANN's response can be insensitive to minor variations in their inputs.

3. Abstraction: Some ANNs are capable of abstracting the essence of a set of inputs.

4. Applicability: ANNs become the preferred technique for a large class of pattern recognition tasks that conventional computers do poorly.

Artificial neural networks consist of connected artificial neurons that are arranged in layers. The artificial neurons are designed to mimic the first-order

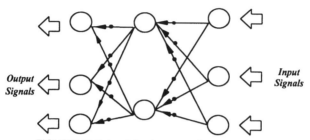

FIGURE 12.2.4.   A backpropagation neural network.

characteristic of biological neurons. A set of inputs is applied to a neuron. Each input, representing the output of another neuron in the previous layer, is multiplied by an associated weight, and all of the weighted inputs are then summed to produce a signal. This signal is usually further processed by an activation function to produce the neuron's output. This activation function simulates the nonlinear transfer characteristics of a biological neuron, and is often chosen to be a sigmoidal logistic function or a hyperbolic tangent function. The output of the neuron is then distributed to other neurons in the next layer.

Neural network structures can be divided into two classes based upon their operations during the recall phase: feedforward and feedback. In a feedforward network, the output of any given neuron cannot be fed back to itself directly or indirectly, and so its present output does not influence future output. Grandmother cells, ADALIN, MADALINE and backpropagation networks (Fig. 12.2.4) are feedforward networks. In a feedback network, any given neuron is allowed to influence its own output directly through self feedback or indirectly through the influence it has on other neurons from which it receives inputs. Kohonen self-organizing networks and Hopfield neural networks (Fig. 12.2.5) are feedback networks. In addition, ANNs can be classified based on net layers and structure, as well as the relationship among elements (or neurons).

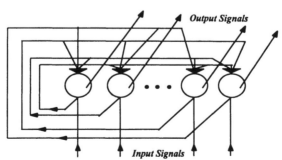

FIGURE 12.2.5.   Single layer Hopfield neural network.

Determining an ANN's structure in real applications is a difficult but very important task. Generally speaking, each of these ANN's architectures has a number of variations. For example, the backpropagation ANN is a typical feedforward structure, as shown in Fig. 12.2.4. The number of neurons in the input and output layers is determined based on applications. However, the number of hidden layers and the number of neurons within each hidden layer must be chosen by the engineers and is often determined with a trial-and-error method. Coupling ANNs with expert systems may be an effective method of handling the issue.

Inputs and outputs of a neural network can be viewed as vectors. A network is trained so that an input vector produces a desired (or at least consistent) output vector. Training is accomplished by sequentially applying input vectors, while adjusting network weights according to a predetermined procedure. During training, the network weights gradually converge to constants such that each input vector produces the corresponding desired output vector. Training algorithms are categorized into supervised and unsupervised. Supervised training requires the pairing of each input vector with a target vector representing the desired output. An unsupervised training algorithm requires no target vector for the output, and hence modifies network weights to produce output vectors that are consistent.

Recently, the interest in ANN has been growing rapidly since neural networks hold the promise of solving problems that have been proven to be extremely difficult for traditional numerical computation. There are a variety of ANN algorithms available. Each of them has specific advantages for some kinds of problems. Neural network problems can be categorized into the following groups based on what specific information we want to give the network as input and what we expect the network's output to be.

1. Mapping: In a mapping problem, an input pattern is associated with a particular output pattern.

2. Associative memory: An associative memory stores information by associating it with other information. The recall is performed by providing the association and having the network produce the stored information.

3. Categorization: The network is provided with an input pattern and responds to the category to which the pattern belongs.

4. Temporary mapping: In such a mapping, network weights vary with time. Many process control patterns have a temporal mapping characteristic.

Recently, ANNs have been applied to many pattern recognition problems, including image processing and speech recognition. Investigators have proposed using ANNs for modeling nonlinear process dynamics, filtering noisy signals, modeling the actions of human operators, interpreting advanced sensor data, detecting and diagnosing faults, and controlling process.

Integrated distributed intelligent systems

The concepts of the integrated distributed intelligent system were proposed by Rao, Tsai and Jiang [1987]. An integrated distributed intelligent system (IDIS) is a large-scale knowledge integration environment, which consists of several symbolic reasoning systems, numerical computation packages, neural networks, a database management subsystem, computer graphics programs, an intelligent multimedia interface and a meta-system. The integrated software environment allows for running programs written in different languages and communication among the programs as well as the exchange of data between programs and database. These isolated intelligent systems, numerical packages and models are under the control of a supervising intelligent system, namely, the meta-system. The meta-system manages the selection, coordination, operation and communication of these programs. In the next section, the configuration and functions of IDIS and the meta-system will be discussed in detail.

Development of intelligent control systems

The development of an intelligent system has five stages: identification, conceptualization, formalization, implementation and evaluation. The procedure for developing an intelligent control system is given in Table 12.7.

Two successful strategies in developing intelligent systems have been summarized as: "Plan big, start small" and "Learning by doing". Thus, the development of the real time intelligent control systems should be accomplished

TABLE 12.7. Procedure for development of intelligent control systems.

Stage	Goal	Content	Outcome
Identification	Defining problem characteristics	Clarify the problem. Organize participants. Prepare facilities and resources.	Objectives and requirement
Conceptualization	Finding concepts to represent domain knowledge	Decompose the problem. Acquire knowledge. Analyze input and output.	Concepts
Formalization	Designing data structures to organize knowledge	Represent knowledge. Analyze knowledge. Design interface.	Data structure
Implementation	Programming knowledge structure into a computer program	Choose or develop tool. Transfer knowledge into presentation framework	Software program
Evaluation	Testing and modifying the system	Test prototype system. Revise prototype system.	Applicable system

TABLE 12.8. Three phases for developing intelligent control systems.

Phase	Intelligent system	Objectives
1	Off-line system	Organizing the intelligent system Codifying the knowledge Evaluating intelligent system
2	On-line supervisory system	Evaluating the intelligent system interface with human operator, process computer systems and hardware instrumentation
3	On-line closed loop system	Reducing the operators' routine tasks Improving operation safety and efficiency

in three phases as follows:

Phase 1: Off-line system: The objectives of developing an off-line intelligent system are to codify the knowledge, and design and evaluate the system in an off-line environment, for training or off-line supervisory control [Rao, Jiang and Tsai 1988]. A system at this phase is basically a simulation where the entries are performed by the user over a keyboard. Such a system is excellent for training purposes because it is a highly interactive system. Most academic intelligent systems are developed mainly at this phase. Their main function is to provide decision support to human operators based on process operation information.

Phase 2: On-line supervisory system: The objective of developing an on-line supervisory system is to evaluate the intelligent system interfaces with human operators, process computer systems and hardware instrumentation. The intelligent system is physically connected to the actual process. The process information is fully or partially processed by the intelligent system, and reports and suggestions are generated.

Phase 3: On-line closed loop system: A closed loop intelligent system directly reads the inputs from the process and sends its outputs (actions) back to the process. However, human operators will be kept in the control loop. In most cases, the objective of these systems is to reduce the operators' routine tasks and improve operation safety and efficiency.

The three stages for development of intelligent control systems are summarized in Table 12.8.

12.3. INTEGRATED DISTRIBUTED INTELLIGENT SYSTEM FOR PROCESS CONTROL

Integration of existing expert systems and software packages

As discussed above, the existing intelligent systems can only be used alone and inflexibly. We cannot integrate the isolated intelligent systems that have been available, even though each of them is well developed for a specific task. In general, the best way to solve a complicated problem by intelligent system techniques is to distribute knowledge and to separate domain expertise. In such a case, several intelligent systems may be used together. Each of them should be developed for solving a sub-domain problem. However, the current technology prohibits us from integrating several knowledge-based systems that have been successfully developed. Here, we face the problem of heterogeneous knowledge integration and management.

The following drawbacks often arise in the existing intelligent systems:

1. lack of efficient search methods to process different knowledge in a large decision-making space,

2. lack of coordination of symbolic reasoning, neural networks, numeric computation, graphics representation as well as computer vision,

3. lack of integration of different intelligent systems, software packages and commercial tools,

4. lack of efficient management of intelligent systems,

5. lack of the capability of handling conflicts among various intelligent systems,

6. lack of a parallel configuration to deal with multiplicity of knowledge representation and problem solving strategies, and

7. difficulty in modifying knowledge bases by end users rather than the original developers.

Owing to these drawbacks, many industrial process control problems cannot be solved by the existing expert systems or numerical computation techniques. Many process variables that affect process operation and product quality and quantity can be only partially understood, and sometimes may not be directly measurable. For example, some empirical data from chemical processes cannot be effectively processed using the existing expert systems or numerical computation packages. A new methodology integrating neural networks, symbolic reasoning, numerical computation as well as graphic simulation may be suitable to process such data. However, how to acquire knowledge effectively from the data source still remains a difficult task.

Conflicts usually arise because different domain expert systems may make different decisions on the same problem based on different criteria. Most intelligent systems for industrial processes work in an interdisciplinary field, and often deal with conflicting requirements. These intelligent systems may produce conflicting decisions even when based on the same information.

Distributed intelligent system technology is a very practical method to improve the capability of knowledge base management and maintenance. Since solving most process control problems requires the knowledge and experience from different areas, the integration of distributed intelligent systems is critical to solving these complicated problems.

The modern industrial manufacturing process is becoming increasingly difficult for operators to understand and operate effectively. Due to the importance of the operator's role in these systems, the quality of interaction between human operators and computers is crucial. Thus, developing an intelligent multimedia interface, which communicates with operators via multiple media (language, graphics, animation and video) will facilitate human operator interaction with the complex real-time computer systems.

An integrated distributed intelligent system (IDIS) is a large knowledge integration environment. It can integrate isolated expert systems, numerical programs and commercial packages, by bringing them under the control of a meta-system.

The development of IDIS is a new direction for intelligent systems. The implementation of IDIS is a key issue in developing a CIPS environment. As shown in Fig. 12.3.1, the software platform should have the integration functions in four aspects: the integration of basic software techniques, the integration of different programming languages, the integration of commercial packages and

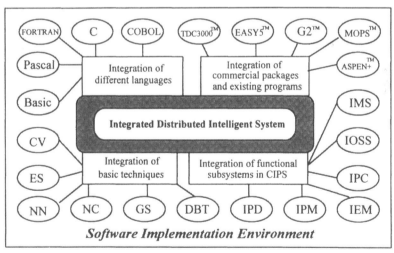

Note: CV: Computer Vision   ASPEN+: a commercial process simulation tool   DBT: Database Techniques
EASY5: a commercial package for control system design   ES: Expert Systems   GS: Graphics Simulation
G2: a commerical AI tool   IEM: Intelligent Equipment Maintenance   IMS: Information Management System
IOSS: Intelligent Operation Support System   IPC: Intelligent Process Control   IPD: Intelligent Process Design
IPM: Intelligent Process Monitoring   MOPS: a commercial information management system
NC: Numerical Computing   NN: Neural Networks   TDC3000: a commerical distributed control system

FIGURE 12.3.1.   Integrated distributed intelligent system for process control.

existing available programs, and the integration of functional subsystems in CIPS.

Meta-system

The key issue for implementing IDIS is to develop a meta-system. The meta-system can be viewed as a "control mechanism of meta-level knowledge." Knowledge can be categorized into "domain knowledge" and "meta-knowledge." The former is usually defined as facts, laws, formulae, heuristics and rules in a particular domain of knowledge about the specific problems, whereas the latter is defined as the knowledge about domain knowledge [Davis and Lenat 1982] and can be used to manage, control and utilize domain knowledge. Meta-knowledge possesses diversity, covering a broader area in content and varying considerably in nature. Also, it has a fuzzy property.

Presently, meta-level knowledge is a very broad concept. However, its application concentrates on implementing meta-rules that have been successfully used to control the selection and application of object-rules or a rule-base [Davis and Lenat 1982; Rao, Jiang and Tsai 1988]. As of yet, we cannot find any reported meta-level techniques that can solve the problems encountered while developing a large-scale integrated distributed intelligent system.

In the software engineering field, the terminology meta system is also widely used. For example, a meta system in software specification environments was also developed [Dedourek et al. 1989]. This meta system is designed to support the production of an information processing system throughout its life cycle. It greatly helps programmers reduce the time and cost of software development, and maintain and improve the existing software environments. Needless to say, the meta-systems have contributed a lot to database systems and software environments. In this book, we will discuss another meta-system that is developed to handle integration problems in distributed intelligent systems. Such a meta-system differs from that discussed by Dedourek and his colleagues [1989] in both its concepts and implementations.

Not just extending the concept of meta-level knowledge, but rather proposing an innovative idea, we develop a new high-level supervisory system, that is, a meta-system, to control IDIS. The main functions of a meta-system, shown in Fig. 12.3.2, are described below:

1. It coordinates all symbolic reasoning systems, numeric computing routines, neural networks as well as computer graphics programs in an integrated intelligent environment.

2. It distributes knowledge into separate expert systems, neural networks and numerical routines to achieve management effectiveness.

3. It helps us to acquire new knowledge and add new programs easily.

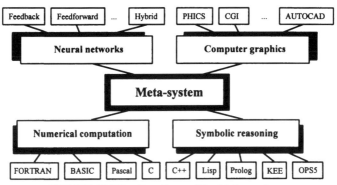

FIGURE 12.3.2. Functions of the Meta-system.

4. It can find a near-optimal solution for conflicting facts and events among different systems.

5. It allows for parallel processing in the integrated distributed intelligent system.

6. It communicates with the measuring devices and final control elements in a control system and transforms various input and output signals into standard communication signals.

Coordination function

The meta-system is a coordinator which manages all symbolic reasoning systems, numeric computation routines, neural networks, and computer graphics programs in an integrated distributed intelligent system. In the IDIS, different types of information (symbolic, graphic and numerical) are utilized and processed together. Even in the branch of symbolic processing, many different languages and tools (such as Lisp™, Prolog™, OPS5™, KEE™, C⁺⁺™, and so on) may be applied to build individual symbolic reasoning systems. In each language family, there also exist different expert systems. For example, among the OPS5™ family, there are IDSOC and IDSCA [Rao, Tsai and Jiang 1988; Rao, Jiang and Tsai 1988]. The meta-system controls the selection and operation of all programs, and executes the translation between different data or programs in an IDIS. For example, it may translate a set of numerical data into some symbolic sentences, and then send the symbolic information to an expert system that is to be invoked next.

Distribution function

The meta-system distributes knowledge into separate expert systems, numeric routines, neural networks, and computer graphics programs so that the integrated distributed intelligent system can be managed effectively. Such a modularity makes the knowledge bases of these intelligent systems easier to change by users rather than their original developers. Expanding the investigation of CACE-III [James et al. 1985], a distributed architecture is proposed to offer flexibility and efficiency in the decision-making process. With the meta-system, the knowledge sources can be separated into individual expert systems and numeric programs, which can be developed and applied for specific purposes at different times. Such a configuration of distributed knowledge allows users to write, debug and modify each program separately so that the overall integrated distributed intelligent system can be managed efficiently. Knowledge distribution can reduce the scope of rule search, thus minimizing the system running time. This function can also help reduce the number of rules or system interactions.

The basic approach is to divide the domain knowledge for a complex problem into a group of expert systems, numeric routines, neural networks, as well as computer graphics packages, then attempt to limit and specify the information flow among these programs. This approach makes the integrated intelligent environment much easier to modify.

Integration function
The meta-system is an integrator that can help us easily acquire new knowledge. It provides us with a free hand in integrating and utilizing new knowledge. Programs in an integrated distributed intelligent system are separated from each other, and strictly ordered only by the meta-system. Any communication between two programs must rely on translation by the meta-system. This configuration enables us to add or delete programs much more easily. When a new expert system or numerical package is integrated into the system, we just need to modify the interface and knowledge-base of the meta-system, while the other programs are kept unchanged. Therefore, any successfully developed software packages can be applied when needed. The time and money required for doing "repeated work" can be saved, thus making the commercial application of intelligent systems more feasible.

Conflicting reasoning function
The meta-system can provide a near optimal solution for conflicting results and facts among different intelligent systems. Knowledge sources are distributed into various distinctive programs corresponding either to different tasks or to different procedures. However, conflict usually arises because different domain expert systems can make different decisions based on different criteria, even though the data that fire the rules are nearly the same. In particular, intelligent process control is interdisciplinary in nature. It allows for the applications of the

knowledge from control engineering, artificial intelligence and computer science to be extensively applied to manufacturing processes. Intelligent process control systems often deals with conflicting requirements, because they work in an interdisciplinary field, and within which various domain knowledges are utilized and different expert systems are developed. These expert systems may produce conflicting decisions even based on the same information. In this case, the meta-system can pick a near optimal solution from the conflicting facts when the requirements from different domains contradict each other. In the future this function will play an even more important role in knowledge integration and management.

Currently, a method has been developed to search a near optimal solution in a conflicting decision making process, based on a priority ranking. That is, different objectives (such as criteria, facts, and methods) are assigned different priority factors; thus each solution will result in an overall priority factor. This procedure is similar to that used in certainty factor calculations. However, this method greatly relies on the expertise of the person who ranks the priorities in terms of various objectives. It is not easy to use in the general cases.

Parallel processing function
The meta-system provides the possibility of parallel processing in an integrated distributed intelligent system. In a production system, all rules and data are effectively scanned in parallel at each step to determine which rules are fireable on which data. This inherently parallel operation can perform very well in the integrated distributed intelligent system.

The one-to-one connection between a program and the meta-system allows for the execution of two or more programs at the same time, since in a parallel computer, several inference engines can deal with simultaneous bi-directional resolution. For example, forward-chaining can be employed on one processor and backward-chaining on another. This allows for the use of dynamic resolution strategies. The execution of a rule in the meta-system may result in several different actions that can invoke two or more programs to operate separately at the same time. This function greatly enhances the quality of real-time computation. Also, it may be applied to other computational processes.

Communication function
The meta-system can communicate with the measuring devices and the final control elements in the control systems and transform various non-standard signals into standard communication signals. A real-time manufacturing environment always deals with the transformation of communication signals, such as electronic (current, voltage, resistance), pneumatic, acoustic or even optical signals. The meta-system provides a user-friendly interface to

manufacturing systems so that these signals can be exchanged, depending on the needs of the user. For example, a rise in the output water temperature of a boiler results in an increase of electronic current from a thermocouple. Through the interface transformation of a meta-system, the signal can be expressed by a numerical value or symbolic information, and be used for computing or reasoning. The corresponding signal will be produced once a final result is obtained in an IDIS, and can be used to manipulate a final control element, such as a control valve.

In order to illustrate the function of a meta-system, let us visualize an integrated distributed intelligent system as a business structure. The meta-system acts as a manager who assigns different jobs to the employees according to their expertise, and monitors the progress of their work, as indicated by the coordination and distribution functions. When a newly hired employee joins the company, the manager is responsible for providing the new job assignment, and possibly for modifying the work plan. Meanwhile, the other workers still do essentially the same jobs they did before. This simulates the integration function. If the suggestions from two employees are in conflict, the manager must choose a near optimal solution between them or generate a new one, similar to the conflicting reasoning function of meta-system. The parallel processing function can be easily visualized as a situation where the manager can let two employees work on their own projects at the same time but share the same tools and resources. Of course, the function of parallel processing is more sophisticated than the simple management role described here. It requires further investigation and study. Finally, the transformation between the production plan and marketing information demonstrates the communication function.

Configuration of the meta-system
Like a common expert system, the meta-system has its own database, knowledge base and inference engine, but it distributes its activities into the separated, strictly ordered phases of information gathering and processing. The meta-system configuration, shown in Fig. 12.3.3, includes six main components: an interface to the external environment, an interface to internal subsystems, a meta-knowledge base, a global database, a static blackboard, and an inference mechanism.

1. *Interface to external environment:* The interface to the external environment is the communication bridge between the users and internal software systems as well as among the external software systems. The interface includes an icon-structured menu that consists of windows, a data structure, a security module, as well as an editor module. Windows display information through different media. The data structure is used to receive information. The

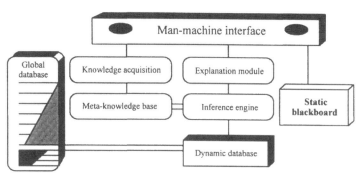

FIGURE 12.3.3. Configuration of the meta-system.

security module identifies the classes of users. The editor module helps users access all knowledge bases and databases at both the meta-level and subsystem-level. The interface can provide multimedia intelligent functionalities such as natural language recognition for oral inputs and computer vision for handwritten inputs. The interface can codify human expertise into the computer system such that it can adopt the most creative intelligence and knowledge in decision making, and can communicate with other intelligent software systems to extend the system to a much larger scale for solving more complicated tasks.

2. *Meta-knowledge base:* The meta-knowledge base is the intelligence resource of a meta-system. It serves as the foundation for the meta-system to carry out managerial tasks. The meta-knowledge base consists of a compiler and a structured frame knowledge representation facility. The compiler converts the external knowledge representation, which is obtained through the editor and is easy to understand by users, into internal representation forms available in the inference mechanism. The structure of knowledge representation can be a production rule or frame or their combination. Characterized by diversity, the meta-knowledge is better represented in object-oriented frame structures.

There are several modules in a frame to represent different components of meta-knowledge. These components are functioned for specific purposes. For example, the communication standardization module for heterogeneous subsystems and the conflicting resolution module are formed for general management purposes at the supervisory level.   The module containing knowledge about subsystems and the task assignment module have to be built according to each specific problem. The meta-knowledge base employs an open organization structure. It allows new intelligent functionalities to enter the meta-knowledge base to engage and assume more duties in decision-making.

3. *Global database:* The database of the meta-system is a global database for the IDIS, while databases in the subsystems are only attached to the individual

subsystems to which they belong. The interface converts the external data representation form into an internal form. The data flow in the global database is controlled either by the inference engine, depending on the corresponding module in the meta-knowledge base, or by users at certain security classes.

4. *Inference mechanism:* Due to the diversity of the meta-knowledge and the variety of its representation forms, the inference mechanism in the meta-system adopts various inference methods, such as forward chaining, backward chaining, uncertain reasoning, conflict reasoning, and so on. The inference mechanism conducts operation-to-process meta-knowledge. Additionally, it carries out various actions based on the reasoning results, such as passing data between any two subsystems, or storing new data in the database. Therefore, there are some functional modules in the mechanism, which further extends the functionality of the inference mechanism.

5. *Static blackboard:* The static blackboard is an external memory for temporary storage of needed data or information when the system is running. Limited by the on-board memory space, the subsystems in IDIS cannot execute at the same time. In fact, it is unnecessary to run the entire system simultaneously. Very often, the meta-system and all subsystems are run on the distributed hardware environment so that a buffer area in the external memory is needed for any two subsystems to exchange data or information. Besides the storage capability, the conversion of data in heterogeneous languages into an exchangeable standard form is also completed on the static blackboard.

6. *Interface to internal subsystems:* This component of the meta-system is established based on each specific application. The internal interface connects any individual subsystems that are used in problem-solving and under the control and management of the meta-system. Each module of the interface converts a nonstandard data form from a specific subsystem into a standard form in the integrated distributed intelligence environment. The conversion among the standard forms of different languages is carried out by the meta-system.

Implementation of the meta-system
Expert system developing tools now commercially available make the development of new expert systems much easier and faster. When selecting a tool to develop an expert system, commercial tools are the first choice, since a sophisticated, well-endowed development and run-time environment can save us a great deal of work when generating basic facilities such as reporting, debugging, graphics, databases, statistical packages, and other specialized functions on features. On the other hand, the commercial expert system development tool industry is still in its infancy [Bowerman and Glover 1988]. Most notably, a commercial tool for building integrated distributed intelligent environment has not been reported at this time. To face this challenge, several

meta-systems have been implemented in OPS5™, C™ language and Prolog™ language.

A meta-system was first implemented with OPS5™ on a VAX 11/780 computer, running the UNIX operating system [Rao et al. 1987]. OPS5™ is a Lisp-based tool. In the Lisp language, there is no difference between numerical data and symbolic attributes. A message can thus be an arbitrary Lisp expression. A Lisp programming environment supports good debugging features. OPS5™ supports a forward chaining inference process. Its pattern matching methodology permits variable bindings. However, it does not provide facilities for sophisticated object representation, and has difficulty with numerical computation. It is not an easy tool for a nonprogrammer to use.

A different meta-system is implemented in the C™ language environment. The C™ language is versatile in both numeric computation and symbolic manipulation. Its capability to handle numerical operation is much more powerful than Lisp™, Prolog™, OPS5™ and other expert system developing tools. It is also superior to FORTRAN™ and BASIC™ in terms of symbolic operations. This advantage makes C™ better equipped to integrate different forms of knowledge. Besides, the C™ language possesses merits of both high level and low level languages such that it is very flexible and convenient for program coding and control on hardware, especially under the UNIX operating system that is developed in C™. Furthermore, the C™ language can easily access other language environments by interfaces written in a mixture of C™ and the assembly language. C™ has its extension C^{++}™, which is an object-oriented programming language.

The advantage of implementing such a meta-system is that the symbolic process can follow the progress of the numerical computation by receiving posted information from the numerical procedure at several steps during number crunching execution. The symbolic process then has the options to let the numerical procedure continue, change some parameters, or abort the procedure altogether. This contrasts with shallow-coupled processes in which the heuristic process invokes a numerical routine via a procedure call, supplies the necessary input data, and then passively waits for the numerical process to finish its execution and provide the required output.

This new meta-system facilitates deep-coupled integration reasoning and algorithm-based numerical processes, by taking C™ as a fundamental knowledge representation language and serving as a functional supplement to the C™ language. Its important features are: (1) flexible reasoning, (2) a high-level representation language, which incorporates the extended C™ data structures to define knowledge bases explicitly, and (3) the collaboration of different types of programs by supplying a flexible interface based on a news posting and delivering model.

12.4.   NEW FRONTIERS AND CHALLENGES IN INTELLIGENT PROCESS CONTROL

Real time discrete event systems

Discrete event systems (DES) are used to analyze systems whose status has logical or symbolic as well as numerical values, and whose dynamics are governed by the occurrence of events that may also be described in non-numerical terms. A DES can be used in manufacturing processes to execute supervisory tasks. Real-time discrete event systems deal with not only the logical results of system behaviors, but also the time at which the results are generated. In general, real-time discrete event systems share the following characteristics [Ostroff and Wonham 1990]:

(1) Events occur at discrete times and states have discrete values;

(2) Processes are event driven rather than clock driven;

(3) Processes are typically non-deterministic;

(4) Interaction between processes may be synchronous or asynchronous;

(5) "Hard" real-time deadlines must often be met.

Currently, many methods have been developed to analyze real-time discrete event systems by using programming languages, state machines, Petri nets, predicate logic and real-time temporal logic.

Industrial processes can be influenced by stochastic, environmental intended operational commands. Small events slightly change the value of the process parameters that can be compensated by continuous process control. However, large events cannot fully be compensated for due to the limited control range. Thus, the processes have to be switched to other operation modes by protection functions or operator commands which are summarized by term event-related control.

DES has been used to analyze process safety and fault-tolerance. By combining hardware, software and human operators within one model, we can determine the effects that failure or fault in a component will have on another one to help prevent accident [Levenson and Stolzy 1987]. Real-time discrete event systems provide us with a new approach to designing the most economical redundancy of system components to achieve the required performance [Zhou and DiCesare 1993]. Distributed computer process control systems improve flexibility and reliability, and reduce cost. However, they also bring some drawbacks, such as communication delay and failures. A real-time DES can distribute tasks and handle the components to cope with communication delay and failures.

Machine vision for process monitoring

As an emerging technology frontier, machine vision applies computer techniques to image analysis and interpretation. Machine vision uses optical non-contact sensors to automatically receive and interpret an image of a real scene to obtain information and control processes [Rodd and Wu 1990]. In a manufacturing environment, a sensor (typically a television camera) acquires electromagnetic energy (such as visible spectrum, i.e., light) from a scene and converts the energy to an image that computers can use. The computer extracts data from the image, compares them with previously developed standards, and outputs the final results in the form of a response.

A basic machine vision system includes the following components: lighting, to dedicate illumination; optics, to couple the image to a sensor; a sensor, to convert optical image into electronic signal; an A/D converter, to sample and quantify the analog signals; an image processor, to reduce noise and analyze the image; a computer, to process information and make decisions; an operator interface; input/output communication channels; and a television monitor.

Machine vision applications can be classified as: gauging to deal with a quantitative correlation to design data; inspection to qualitatively detect unwanted defects and artifacts of an unknown shape at an unknown position; verification to conduct the correct fabrication assembly operation; recognition to identify an object based on descriptions associated with the object; identification to identify an object through symbols on the object; location analysis to assess the position of an object; guidance to direct an object by adaptively providing position information for feedback.

In an information-intensive manufacturing environment, vision machines are already valuable tools. It has been found that approximately 60% of the applications relate to process inspection (such as gauging, cosmetic and verification), 25% to robot guidance, and 15% to part identification and recognition [Sanz 1989].

Soft sensor for estimating non-measurable variables

Chemical processes are highly computerized human-machine systems. Most chemical process operations are usually run under unusual situations, such as high temperature, high pressure, low temperature, low pressure, high flammability, and high toxicity. In these cases, some valuable control variables associated with product quality cannot be measured directly or easily even though special measurement techniques are being used. On the other hand, it is becoming increasingly difficult for operators to understand various signals and information from the display board, video, and computer screen. Owing to the importance of the operator's role in process operation, the quality of interaction between human operators and computers is important. Thus, the development of soft sensors will better facilitate human interaction with the complex real-time monitoring and control systems.

Sometimes, even though a few important process variables associated with product quality cannot be measured, a number of non-specific measurements are still available. It is possible to use these measurements to implement symbolic inferences to predict these important, non-measurable quality variables. Currently, however, there does not seem to be a systematic way to develop soft sensors and integrate them into an overall system scheme. To implement a soft sensor, an inferential measurement based on a suitable model can be used. Data collecting, model building and system design are needed. The model can be updated from off-line laboratory testing.

There are two reasons for developing soft sensors: (1) to provide the estimated values for the non-measurable quality variables associated with product quality to improve real-time control systems; (2) to provide operators with the interpretation of the important variables to enhance the interaction between chemical processes and human operators, and thus control the product quality.

A soft sensor system is an intelligent software that uses inferential measurements based on operators' skills, numerical models and laboratory data, to obtain the approximate values of non-measurable process variables. These variables have the following features: (1) they are quantitative, (2) they cannot be measured by existing devices, (3) they are not sampled in real-time operating processes, and (4) they are related to other measurable variables directly or indirectly.

An intelligent soft sensor system for digester quality control was implemented to support a batch sulphite pulping process operation [Rao and Corbin 1992]. By predicting the Kappa number and cooking time, this expert system can quickly process the knowledge and provide operation support. A mathematical model for the cooking process using the S-factor has also been implemented. It can assist operators to better manage the operation and thus produce good quality pulp. This system can also be used to facilitate on-line supervisory control, detect unexpected events and faults during process operation, and train new operators and engineers.

Another soft sensor using a neural network method was developed at a refinery to predict the volatility of bottom atmosphere gas oil (BAGO). In processing oil products in a petroleum refinery, the prediction of the volatility of BAGO is an essential requirement for better operation. However, the traditional numerical and measurement techniques failed to predict it, since the process is too complex and many variables cannot be measured. A soft sensor based on neural networks was used to predict the volatility of BAGO and diagnose operation faults [Kim et al. 1992].

Intelligent on-line monitoring and control

Chemical production consists of a complex and long sequence of operations. Knowledge used in the operation includes complex technologies from different areas. In this case, operators have to monitor an increasingly large amount of raw data and supplemental information during a potentially critical situation. It would be hard for process operators, even experts to possess all operation knowledge. They may be capable of dealing with routine operations. However, when the operation mode changes or an emergency occurs, the operator may find it difficult to contribute a quick and effective solution.

Many industrial plants have installed distributed control systems (DCS). Some of them have also installed plant-wide information systems. However, it is becoming difficult to use these advanced facilities and increase their "intelligence" to solve complex problems such as process quality control, operation optimization, troubleshooting, emergency handling, etc.

An intelligent on-line monitoring and control system (IOMCS) consists of three independent functional subsystems, the process simulator, decision support system and general information system, and is linked with an information management system. The process simulator is a graphical tool that displays predefined action variables (inputs that can be manipulated by operators) versus result (output) variables. It stores the numerical and symbolic models that relate process input variables to output variables. By using the process simulator, operators can change the action variables and observe their effects on the result variables graphically. Using expert system techniques, the decision support system helps operators improve operations through optimization and fault detection. The general information system uses the hypertext technique and provides the explanation of the IOMCS, a general overview of process operation, procedures for startup, shutdown, as well as emergency operations, and so on. In general, the IOMCS can accomplish such tasks as (1) operation optimization, (2) product quality estimation and control, (3) process alarm management, and (4) quality control.

Maintenance support systems

There are an increasing number of system elements to be monitored, due to the fast growth of equipment size and the complexity of modern industrial processes. These system elements are becoming more complex in design and generally more difficult to maintain, quickly detect the problems, diagnose and amend. The objective of intelligent maintenance support systems is to support and assist production managers and process engineers in making production procedures and scheduling maintenance strategy for process equipment. An intelligent maintenance support system can analyze the signals and data from sensors to determine if better information can be derived from the collected data using domain knowledge, thus deriving maintenance and production information from the condition monitoring data. It paves the way to implementing maintenance automation and improving the productivity, reliability, and availability of the process equipment.

For example, in order to implement the maintenance operation automation of large mining trucks and to improve the productivity, availability and reliability of the trucks, an intelligence maintenance support system (IMSS) for truck condition monitoring can be developed. It can carry out operation monitoring, fault detecting, diagnosing, alarming and production dispatching of mining trucks. Fig. 12.4.1 demonstrates the software structure of the IMSS.

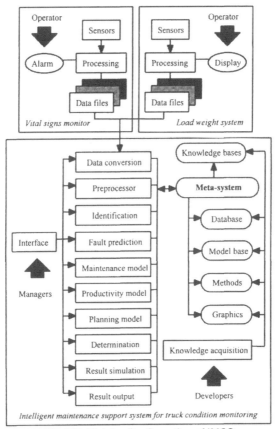

FIGURE 12.4.1. Configuration of IMSS.

Process startup/shutdown automation

Continuous manufacturing processes have four operating situations: process startup, normal operation, emergency situation, and shutdown. The majority of process control research and development focuses on normal operation that has been very well studied, and is run under fully automatic control. The other three situations, still operated manually, are more dangerous, and lack sufficient attention from academic and industrial research. Safe process operation, particularly process startup/shutdown, heavily depends on the human operator's skills. A safe and quick startup of a chemical process will reduce production cost, and bring valuable economical benefits to industrial companies. The current chemical process startup (shutdown) operation has the following difficulties:

1. The process startup is the most dangerous operation. Currently, it is operated manually and is prone to unsafe conditions.

2. There are many ill-structured problems in the startup operation that are difficult to solve by conventional control methods.

3. Since startup is an infrequent operation, most human operators are unfamiliar with such a situation. Only a very few experienced operators can safely start process operation.

4. During the startup of a process, there is little process operation information available and there exists little interaction between the operators and the process.

An intelligent system for process startup automation was developed at a refinery plant [Dong et al. 1991]. This system is linked to the process control mainframe computer, and provides such features as viewing AutoCAD™ and P&ID's (Process and Instrumentation Diagram), and maintaining a daily logbook. Mainly, it can provide appropriate advice during the startup of a section of the refinery. It was developed using an expert system shell called KnowledgePro for Windows™, which produces programs that run under Microsoft Windows™. The flexible rule structure allows this expert system to give advice for almost any possible startup state. It inquires from the user what the startup state is. Based on the response from the users, the expert system then begins stepping through each required step for the startup. The program waits at each step until the controller checks off its completion. The steps are labeled with a short descriptive title, but, more detailed information on the procedures can be obtained. As a hypertext system, it allows rapid access to the detailed information while keeping the main program section simple and easy to use.

The automatic startup/shutdown has been suggested as a new direction for future process control engineering research. Future work will focus on theoretical development and system implementation as well as industrial applications for automatic process startup/shutdown.

Integration of process design and control

The performance of a process depends not only on the soundness of its structure, but also on the efficiency of its control system. So far, the process design and control system design are two distinct activities carried out in series, which prevents the process from attaining its maximum efficiency [McAvoy 1987]. The integration of the two activities is required to improve the performance of industrial processes.

Currently, most research projects in process control systems have focused on process modeling, model identification and control algorithms. Intelligent system technologies such as qualitative analysis, heuristic knowledge and neural networks are used in process modeling and identification. Many process control algorithms associated with robust control, nonlinear control, fuzzy control and adaptive control have been used in process control system design.

With the development of manufacturing and computer techniques, industrial plants are required to respond effectively to volatile and unpredictable marketplace changes. As a result, a more flexible and safer process control design environment will be closely linked with effective process operation and production management. Since most process control facilities in industrial plants are inflexible and unchangeable, they cannot provide multiple processes and various products to satisfy customers' demands.

Process control system design can be divided into two aspects: conceptual design and detailed design. The objectives of conceptual design are to select appropriately controlled and manipulated variables, to choose a control configuration, to build control models, to study controllability, observability, reliability, operability and feasibility of control models, and to comprehensively select the best model. The objective of detailed design is to implement the process control system design in detail based on the results of the conceptual design.

Conceptual design is a creative activity. To design a process control system, we need to solve problems such as how to construct a system structure and how to choose elements (controllers, sensors and actuators) according to process dynamics, application requirements, as well as the properties of the elements. The ultimate quality of a control system design depends not only on the quality of single subsystems or components, but also on the coordination of the properties of all elements, that is, the quality of the conceptual design.

Conceptual design requires systematic coordination of a multitude of tasks. Each task has many facets dealing with data, numerical algorithms, decision making procedures, and human interaction to provide experiential knowledge. In most cases, it is a very complex and not completely understood process, particularly since it is abstract and requires creativity. It is largely a trial and error process, in which human design experts often employ heuristics or rules of thumb to solve their problems. In the past few decades, CAD techniques have been used in the detail design stage, but used less in the conceptual design stage.

To implement the integration of process design and control, a five-stage problem-solving strategy and software configuration have been proposed. The software is based on IDIS techniques. The five stages of the strategy are problem definition, conceptual design, parameter design, system analysis and comprehensive evaluation. This strategy is briefly described in Table 12.9.

At this point, a software system to implement the integration of process

TABLE 12.9.   Five stage strategy and software consideration.

No.	Stage	Software configuration
1	Problem definition stage	The system can directly access information and data from equipment selection and process planning. According to the information and knowledge stored in a functional library (a part of knowledge base), users can determine the required functions, specifications, features of a process control system. All factors about chemical processes, environmental protection and operation safety need to be fully considered when selecting functions and specifications.
2	Conceptual design stage	A control strategy to satisfy the function requirement and specifications is selected from structure library (another part of knowledge base). "Multiple to multiple" mapping between the control system functions and control strategies makes the control strategy determination more complex and diversified. A special control function can be implemented using many different control strategies, while a control strategy may possess many functions. If there is more than one control strategy, we need to select an optimal (or near optimal) one.
3	Parameter design stage	The detailed description of a selected control strategy can be completed by using the control models stored in a model-based library. The models include various process control algorithms, quantitative computation, identification, etc. All control instruments and parameters have to be selected in this stage.
4	Analysis and simulation stage	Because functions and control strategies share a "multiple to multiple" mapping configuration, numerous control strategies are usually produced. After parameters are selected, all selected control strategies are analyzed by simulation programs. Thus, the integration among different process simulation programs, control simulation packages and equipment design software is required.
5	Comprehensiv e evaluation	A target system from target knowledge library and a comprehensive mathematical model from evaluation models will be chosen to evaluate the analyzed control strategies. Techniques of fuzzy evaluations and system engineering are used during decision-making.

design and control at the early conceptual design stage is under development based on the five-stage problem-solving strategy and IDIS techniques, as described by Fig. 12.4.2.

Knowledge generation

As the new research frontier, intelligent process control has attracted much attention from academia and industry. Many new outcomes have significantly contributed to the development of artificial intelligence and process control. The experience from developing intelligent process control systems indicates that building an intelligent system is not just a translation from the existing knowledge into a computer program. It is a process in which the new expertise knowledge may be generated and acquired.

In the process of building expert systems, knowledge acquisition and representation are the two most important tasks. The methodology used in this process is different from that in the prototype problem or in the related quantitative computing programs. In the process of developing an intelligent

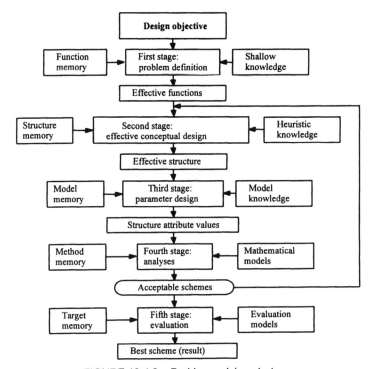

FIGURE 12.4.2. Problem-solving strategy.

system, some new methods to solve the problem may be generated which may complement the solution to the prototype problem [Rao 1991]. For example, IDSCA was developed to assist designing the controllers in multiloop control systems [Rao, Jiang and Tsai 1988]. This intelligent system seems to be relatively simple but effective for the design of industrial process control systems. The important significance behind the practical application is the investigation of both a new design criterion and an adaptive feedback testing system. The former complements the knowledge of process control, while the latter stimulates the development of intelligent systems that have high reliability and performance.

Problems

12.1. What are the relationships among KIS, IMS and DPS?

12.2. State the development stages of industry in terms of automation.

12.3. Classify the software architecture of intelligent control systems.

12.4. What are the main components for an expert system?

12.5. What are the differences between Intelligence Engineering and Knowledge Engineering?

12.6. Develop an application case study based on the 5-step procedure for developing intelligent systems.

APPENDIX A

LAPLACE TRANSFORM PAIRS

$F(s)$	$f(t)$ $t \geq 0$
1	$\delta(t)$
e^{-kTs}	$\delta(t - kT)$
$1/s$	$u(t) = 1$ $t \geq 0$ *unit step*
$1/s^2$	t
$1/s^3$	$t^2/2!$
$1/s^{n+1}$	$t^n/n!$
$1/(s+a)$	e^{-at}
$1/(s+a)^2$	te^{-at}
$s/(s+a)^2$	$(1-at)e^{at}$
$1/(s+a)^n$	$t^{n-1}e^{-at}/(n-1)!$
$1/[s(s+a)]$	$(1-e^{at})/a$

$1/[s^2(s+a)]$	$(e^{-at} + at - 1)/a^2$
$1/[(s+a)(s+b)]$	$(e^{-at} - e^{-bt})/(b-a)$
$(s+c)/[(s+a)(s+b)]$	$[(c-a)e^{-at} - (c-b)e^{-bt}]/(b-a)$
$\dfrac{\omega}{s^2 + \omega^2}$	$\sin \omega t$
$\dfrac{s}{s^2 + \omega^2}$	$\cos \omega t$
$\dfrac{\omega}{(s+a)^2 + \omega^2}$	$e^{-at} \sin \omega t$
$\dfrac{s+a}{(s+a)^2 + \omega^2}$	$e^{-at} \cos \omega t$
$\dfrac{\omega}{s^2 - \omega^2}$	$\sinh \omega t$
$\dfrac{s}{s^2 - \omega^2}$	$\cosh \omega t$
$\dfrac{\omega}{s(s^2 + \omega^2)}$	$(1 - \cos \omega t)/\omega$
$\dfrac{\omega_n^2}{s^2 + 2\zeta\omega_n s + \omega_n^2}$	$\omega_n e^{-\zeta\omega_n t} \dfrac{1}{\sqrt{1-\zeta^2}} \sin(\omega_n \sqrt{1-\zeta^2}\, t), \quad \zeta < 1$
$\dfrac{\omega_n^2}{s(s^2 + 2\zeta\omega_n s + \omega_n^2)}$	$1 - \dfrac{e^{-\zeta\omega_n t}}{\sqrt{1-\zeta^2}} \sin(\omega_n \sqrt{1-\zeta^2}\, t + \phi), \quad \phi = \cos^{-1}\zeta, \ \zeta < 1$

APPENDIX B

PROOF OF TIME DELAY APPROXIMATION FORMULA

The transfer function of a time delay element e^{-Ts} may be approximated by a rational function. Two often used approximations are as follows:

$$e^{-Ts} = \frac{1 - \tau s/2}{1 + \tau s/2} \tag{A2.1}$$

and

$$e^{-Ts} = \frac{1 - \tau s/2 + \tau^2 s^2/12}{1 + \tau s/2 + \tau^2 s^2/12} \tag{A2.2}$$

The Taylor series expansion of e^{-Ts} is

$$e^{-\tau s} = 1 - \tau s + \frac{\tau^2 s^2}{2!} - \frac{\tau^3 s^3}{3!} + \frac{\tau^4 s^4}{4!} - \frac{\tau^5 s^5}{5!} + \cdots \tag{A2.3}$$

It is easy to calculate with long division that

$$\frac{1 - \tau s/2}{1 + \tau s/2} = 1 - \tau s + \frac{\tau^2 s^2}{2} - \frac{\tau^3 s^3}{4} + \cdots \tag{A2.4}$$

A comparison of the last two equations indicates that the formula (A2.1) is correct through the first three terms, and thus is a reasonable approximation for small values of τs.

For the formula (A2.2), the right side may be expressed with long division as

$$e^{-Ts} = \frac{1 - \tau s/2 + \tau^2 s^2/12}{1 + \tau s/2 + \tau^2 s^2/12} = 1 - \tau s + \frac{\tau^2 s^2}{2} - \frac{\tau^3 s^3}{6} + \frac{\tau^4 s^4}{24} + \frac{\tau^5 s^5}{144} + \cdots \tag{A2.5}$$

394

Comparing (A2.5) with (A2.3) indicates that the formula (A2.2) is correct through the first five terms. Obviously, (A2.2) is a more accurate but more complex approximation formula than (A2.1).

APPENDIX C

COMPUTER SIMULATION LABORATORY

Instruction to *PCET*

The software *PCET* (Process Control Engineering Teachware) is developed to strengthen undergraduate process control education, and to provide an inspiring learning environment for engineering students. This microcomputer-aided education software can be useful to a wide spectrum of teaching programs, including practical experience on computer-aided design, simulation and control. This software *PCET* can be used on both color monitor computers and monochrome computers.

To run the program you need the following things:
a) IBM PC or compatible,
b) some kind of graphics card,
c) DOS, and
d) a printer and DOS 3.3 or up for printing hard screen copies (need DOS 5.0 or up for Laser printer). You may use other software to get screen hard copies.

To run *PCET*, please follow these steps:
a) Start the computer;
b) Create a directory *PCET* by typing "md *PCET*" and pressing enter;
c) Enter the new created directory *PCET* by typing "cd *PCET*" and pressing enter;
d) Put the *PCET* disk in "drive *A:*";
e) Type "copy *A:*.*" and press enter for copying software *PCET* from the floppy disc to the new created directory;
f) Type *PCET* and press enter to start the program, and you will see the introductory screen of the program;
g) Press any key (or wait about 5 seconds) to get the *"MAIN MENU"*.

Now you are ready to run the program. Follow the instruction given in the program. They should be quite explanatory.

To print a hard copy of graphic screen, graphics program (part of DOS) should be loaded into memory before *PCET* is executed. Refer to DOS manual for details.

Simulation projects

Laboratory 1. Analysis and design in the time domain
   Recommendations for proper functioning of the program:
   a) For open loop responses, keep the gain less than 100.
   b) For closed loop responses, keep the process gain times the controller gain less than 100.

Phase 1.1. Open loop system response
   Purpose: Investigate the effects of process parameters on the response of the system to a unit step input.
Project 1. Open loop responses for first order systems
   See Problem 3.10 on page 89.
   Plot the result of a) with different K on the same screen. Plot the result of b) with different T_1 on the same screen. Plot the result of c) with different τ on the same screen. Give your conclusion.
Project 2. Open loop responses for second order systems
   See Problem 3.11 on page 89.
   Plot the result of a) with different K on the same screen. Plot the result of b) with different T on the same screen. Plot the result of c) with different ζ on the same screen. Plot the result of d) with different τ on the same screen. Give your conclusion.
Project 3. Open loop responses for different order systems
   See Problem 3.12 on page 90.
   Plot four responses on the same screen, and print it out. Analyze the results.

Phase 1.2. Controller behavior analysis
   Purpose: Investigate the effects of controller parameters on the controller's output for a unit step input.
Project 4. Proportional control action (P)
   See Problem 3.13 on page 90.
   Plot the result with different K_c on the same screen.
Project 5. Proportional-integral control action (PI)
   See Problem 3.14 on page 90.

Plot the result with different K_c and the same $1/T_i$ on the same screen. Plot the result with the same K_c and different $1/T_i$ on the same screen.

Project 6. Proportional-derivative control action (*PD*)

See Problem 3.15 on page 90.

Plot the result with different K_c and the same T_d on the same screen. Plot the result with the same K_c and different T_d on the same screen.

Project 7. Proportional-integral-derivative control action (*PID*)

See Problem 3.16 on page 90.

Plot the result with different K_c , the same $1/T_i$ and the same T_d on the same screen.

Plot the result with the same K_c , different $1/T_i$ and the same T_d on the same screen.

Plot the result with the same K_c , the same $1/T_i$ and different T_d on the same screen.

Phase 1.3. Controller design

> *Purpose*: Investigate the effects of controller parameters on the response of the system to a unit step input. Learn how to use trial-and-error method to design a *P* controller, a *PI* controller or a *PID* controller to satisfy a assigned performance index.

project 8. Controller design for a first order system

See Problem 3.17 on page 91.

project 9. Controller design for a second order system

See Problem 3.18 on page 91.

Project 10. Controller design for a third order system

See Problem 3.19 on page 91. If a *PI* controller is used, can you reach the same performance criterion?

Laboratory 2. Stability and root locus technology

Phase 2.1. Routh criterion

> *Purpose*: Review Routh criterion method to determine the stability of a system. Compare three different methods to use Routh criterion in the case that there's a zero in the first column of the Routh array.

Project 11. Routh criterion and pole calculation

Use Routh array display and pole calculation in the subprogram *RSC* to solve Problem 4.9 on page 125, respectively. When you use the *RSC* to get Routh array of Problem 4.3, a zero should appear in the first column of the Routh array. In the program, we use a small positive number instead of letter ε to replace zero. Try to use other two methods to get different Routh arrays. Do you obtain the same conclusion for the system stability?

Phase 2.2. Root locus

   Purpose: Investigate the effects of open loop zeros and poles on the root locus. Learn how to use root locus technique to design controllers.

Project 12. Root locus drawing

   See Problem 4.10 and 4.11 on page 125.

Project 13. Stability determination of systems See Problem 4.12 on page 125.

Project 14. *PD* controller design

   See Problem 4.13 on page 125. (The closed loop poles should be assigned by yourself.)

Laboratory 3. Analysis and design in the frequency domain

   Purpose: Replace the time-consuming hand drawing of frequency response with a computer graphics. Learn how to draw and use frequency response. Learn to design a controller in frequency domain with a trial-and-error method.

Project 15. Nyquist plot and Bode diagram drawing of basic systems

   See Problem 5.1, 5.2, 5.3 and 5.4 on pages 164-165. In each problem, the systems have the same poles but different gain. Discuss the shapes of the Nyquist plots and Bode diagrams of the different system. Observe the effect of gain on the Nyquist plots and Bode diagrams.

project 16. System type and the Nyquist plot

   See Problem 5.5 on page 165. Find the relationship between the system type and the shape of Nyquist plots. Find the relationship between the system type and the shape of Bode diagrams. Is the system stability related to the system type?

Project 17. Stability analysis in the frequency domain

   See Problem 5.6 on page 165.

Project 18. Controller design in the frequency domain

   With subprogram *FDA*, use trial and error method to solve Problem 5.9 - 5.12 on pages 166 - 167.

Laboratory 4. Discrete time system analysis and design

   Purpose: Investigate the special problems for discrete time system, especially the selecting the sampling period and the discrete controller design.

Project 19. Effect of sampling period on the system response

   See Problem 6.10 on page 199.

Project 20. Closed loop discrete time response

See Problem 6.11 on page 199. If you increase (decrease) the gain of the process transfer function, what will happen on the system response to a unit step input?

Project 21. Discrete P PI and PID controllers design

See Problem 6.12 on page 199 for a first order system and Problem 6.13 on page 200 for a third order system. Do you have any trouble and gain experience when you select your sampling period in Problem 6.13?

Laboratory 5. Controller tuning

Phase 5.1. Controller design and tuning for a first order system
   Purpose: Investigate the effectiveness of existing controller design methods by simulation technique, and learn to tune the controller to improve the existing design.

Project 22. P controller design and tuning for a first order system
   See Problem 7.4 on page 222.

Project 23. PI controller design and tuning for a first order system
   See Problem 7.5 on page 223.

Project 24. PID controller design and tuning for a first order system
   See Problem 7.6 on page 223.

Project 25. PID controller design and tuning for a higher order system
   See Problem 7.6 on page 223.

Phase 5.2. Controller design and tuning for a higher order system
   Purpose: Learn to replace a complicated model with a simple one, and compare the responses of two models to the same unit step input to check the effectiveness of the simplification.

Project 26. PID Controller design and tuning for a higher order system with a simplified model
   See Problem 7.7 on page 220.

Project 27. PID controller tuning for a higher order system with a simplified model in frequency domain
   See Problem 7.8 on page 224. Compare the results obtained in Problems 7.7 and 7.8.

Laboratory 6. Industrial process control application
   Purpose: Study an industrial application case. Review the cascade control strategy and investigate the effects of controller parameters on the response of the system to a unit step input and to a disturbance.

Project 28. Cascade control system
   See Problem 8.6 on page 256.

REFERENCES

Allidina, A.Y. and F.M. Hughes, 1980, "Generalized Self-tuning Controller with Pole Assignment," *Proceedings of the Institution of Electrical Engineers*, Part D, 127, pp. 13-18.

Åström, K.J., J.J. Anton and K.E. Arzen, 1986, "Expert Control," *Automatica*, 22, pp. 277-286.

Åström, K.J. and B. Wittenmark, 1973, "On Self-tuning Regulators," *Automatica*, 9, pp. 185-199.

Åström, K.J. and B. Wittenmark, 1980, "Self-tuning Controllers Based on Pole-zero Placement," *Proceedings of the Institution of Electrical Engineers*, Part D, 127, pp. 120-130.

Åström, K.J. and B. Wittenmark, 1984, *Computer Controlled Systems*, Prentice-Hall, Englewood Cliffs, N.J.

Bowerman, R.G. and D.E. Glover, 1988, *Putting Expert Systems into Practices*, Van Nostrand Reinhold Company, New York.

Brogan, W.L., 1991, *Modern Control Theory*, Prentice Hall, Englewood Cliffs, N.J.

Brownston, L., R. Farrell, E. Kant and N. Martin, 1985, *Programming Expert Systems in OPS5*, Addison-Wesley, Reading, MA.

Buchanan, B.G., 1985, "Expert Systems in an Overview of Automated Reasoning and Related Fields," *Journal of Automation Reasoning*, 1, pp. 28-34.

Chen, C.T., 1984, *Linear System Theory and Design*, Holt, Rinehart and Winston, New York.

Clark, R.N., 1962, *Introduction to Automatic Control Systems*, Wiley, New York.

Clarke, D.W. and P.J. Gawthrop, 1975, "Self-tuning Controller," *Proceedings of the Institution of Electrical Engineers*, 122, pp. 929-934.

Clarke, D.W. and P.J. Gawthrop, 1979, "Self-tuning Control," *Proceedings of the Institution of Electrical Engineers*, 126, pp. 633-640.

Cohn, G.H. and G.A. Coon, 1953, "Theoretical Considerations of Retarded Control," *Trans. ASME*, 75, p. 827.

Cruise, A., R. Ennis, A. Finkel, J. Hellerstein, D. Klein, D. Loeb, M. Masullo, K. Milliken, H. Van Woerkom and N. Waite, 1987. "YES/L1: Integrating Rule-Based, Procedural, and Real-Time Programming or Industrial Applications," *Proc. IEEE Third AI Application Conf.*, pp. 134-139.

Davis, R., 1987, "Knowledge-Based Systems: The View in 1986," *AI in the 1980s and Beyond* (Grimson and Patil eds.), MIT Press, Boston, MA, pp. 16-19.

Davis, R. and D. Lenat, 1982, *Knowledge-Based Systems in Artificial Intelligence*, McGraw-Hill, Inc., New York

D'Azzo, J.J. and C.H. Houpis, 1988, *Linear Control System Analysis and Design -- Conventional and Modern*, McGraw-Hall, Inc., New York.

Dedourek, J.M., P.G. Sorenson and J.P. Tremblay, 1989, "Meta-system for Information Processing Specification Environment," *INFOR*, 27(3), pp. 311-37.

De Silva, C.W., 1989, *Control Sensors and Actuators*, Prentice-Hall, Englewood Cliffs, N.J.

De Silva, C.W. and A. MacFarlane, 1989, *Knowledge-based Control with Application to Robots*, Springer-Verlag, Berlin.

De Silva, C.W., 1993, Soft Automation of Industrial Process, *Engineering Application of Artificial Intelligence*, 6(2), pp. 87-90.

Deutsch, R., 1965, *Estimation Theory*, Prentice-Hall, Englewood Cliffs, N.J.

Donald, E.K., 1970, *Optimal Control Theory*, Prentice Hall Inc., Englewood Cliffs, N.J.

Donat, J.S., N. Bhat and T. McAvoy, 1990, "Optimizing Neural Net based Predictive Control," *Proc. American Control Conf.*, pp. 2466-2471.

Dong, R., M. Rao, H. Bacon and V. Mahalec, 1991, "An Expert Advisory System for Safe Plant Startup," *4th Intern. Symp. on Process Systems Eng.*, pp. IV4.1-4.15, Montebello, Quebec.

Dorf, R.C., 1989, *Modern Control Systems*, 5th ed., Addison-Wesley, New York.

Evans, W.R., 1954, *Control System Dynamics*, McGraw-Hill, New York.

Eykhoff, P., 1974, *System Identification - Parameter and State Estimation*, John Wiley & Sons, London.

Farquharson, F.B., 1950, "Aerodynamic Stability of Suspension Bridges, with Special Reference to the Tacoma Narrows Britge," *Bulletin 116, Part I, The Engineering Experiment Station*, University of Washington.

Franco, S., 1988, *Design with Operational Amplifiers*, McGraw-Hall, New York.

Fu, K.S., 1971, "Learning Control Systems and Intelligent Control Systems: An Intersection of Artificial Intelligence and Automatic Control," *IEEE Trans. Automat. Contr.* 16, pp. 70-72.

James, H.M., N.B. Nichols and R.S. Phillips, 1947, *Theory of Servomechanisms*, McGraw-Hill, New York.

James, J.R., D.K. Frederick, P.P. Bonissone and J.H. Taylor, 1985, "A Retrospective View of CACE-III: Considerations in Coordinating Symbolic and Numeric Computation in a Rule-Based Expert System," *Proc. IEEE 2nd Conf. AI Applic.*, Miami, FL, pp. 532-538.

Kamen, E., 1987, *Introduction to Signals and Systems*, MacMillan, New York.

Kim, H.C., X. Shen and M. Rao, 1992, "Artificial Neural Network Approach to Inferential Control of Volatility in Refinery Plants," *2nd IFAC Workshop on Algorithms and Architectures for Real-Time Control*, Seoul, Korea, pp. 90-93.

Kitzmiller, C.T. and J.S. Kowalik, 1987, "Coupling Symbolic and Numeric Computing in Knowledge-Based Systems," *AI Magazine*, Summer, pp. 85-90.

Kowalik, J.S. and C.T. Kitzmiller, 1988, *Coupling Symbolic and Numeric Computing in Expert Systems*, II (Ed.), Elsevier Science Publishers, New York.

Kuo, B.C., 1991, *Automatic Control Systems*, 6th ed., Prentice-Hall, Englewood Cliffs, N.J.

Levenson, N.G. and J.L. Stolzy, 1987, "Safety Analysis Using Petri Nets," *IEEE Software Engineering* 13, pp. 386-397.

Lirov, Y., E.Y. Rodin, B.G. McElhaney and L.W. Wilbur, 1988, "Artificial Intelligence Modeling of Control Systems," *Simulation*, 50, pp. 12-24.

Lizza, C. and C. Friedlander, 1988, "The Pilot's Associate: A Forum for the Integration of Knowledge Based Systems and Avionics," *Proc. National Aerospace and Electronics Conf. 88*, Dayton, OH, pp. 1252-1258.

Lu, S.C.-Y., 1989, "Knowledge Processing for Engineering Automation," *Proc. 15th Conf. on Production Research and Technology*, Berkeley, CA, pp. 455-468.

Luyben, W.L., 1990, *Process Modeling, Simulation, and Control for Chemical Engineers*, 2nd ed., McGraw-Hill, New York.

McAvoy, T.J., 1987, "Integration of Process Design and Control," In *Recent Development in Chemical Process and Plant Design* (Y.A. Liu, H.A. McGee Jr and W.R. Epperly, Eds) Wiley, New York, pp. 289-325.

Nagrath, I.J. and M. Gopal, 1982, *Control Systems Engineering*, 2nd ed., John Wiley & Sons, New York.

Newell, R.B. and P.L. Lee, 1988, *Applied Process Control*, Prentice-Hall of Australia, Brookvale, NSW, Austrilia.

Ogata, K., 1990, *Modern Control Engineering*, 2nd ed., Prentice-Hall, Englewood Cliffs, N.J.

Ogata, K., 1987, *Discrete-time Control Systems*, Prentice-Hall, Englewood Cliffs, N.J.

Ostroff, J.S. and W.M. Wonham, 1990, "A Framework for Real-time Discrete Event Control," *IEEE Trans. Automatic Control* 35, pp. 386-397.

Pang, G.K.H and A.G.J. MacFarlane, 1987, *An Expert System Approach to Computer-Aided Design of Multivariable Systems*, Springer-Verlag, New York.

Pang, G.K.H., M. Vidysagar and A.J. Heunis, 1990, "Development of a New Generation of Interactive CACSD Environments," *IEEE Control Systems* 10(5), pp. 40-44.

Pease, R.L., 1987, "Automation and Electric Drives," *Production Engineering*, pp. 36-40.

Rao, M., 1993, "A Graduate Course on Intelligence Engineering," *Int. J. Eng. Education.*

Rao, M., 1991. *Integrated System for Intelligent Control*, Springer Verlag, Berlin.

Rao, M. and J. Corbin, 1992. "Intelligent Operation Support System for Batch Chemical Pulping Process," *Proc. IFAC Symp. on Process Dynamics and Control*, College Park, MD, pp. 405-410.

Rao, M., Q. Wang and J. Cha, 1993, *Integrated Distributed Intelligent Systems in Manufacturing*, Chapman and Hall, Oxford.

Rao, M., T.S. Jiang and J.P. Tsai 1988, "Integrated Architecture for Intelligent Control," *Third IEEE International Symposium on Intelligent Control*, Arlington, VA, pp. 81-85.

Rao, M., T.S. Jiang and J.P. Tsai, 1988, "IDSCA: An Intelligent Direction Selector for the Controller's Action in Multiloop Control Systems," *Intern. J. Intelligent Systems*, 3, pp. 361-379.

Rao, M., J.P. Tsai and T.S. Jiang, 1987, "A Framework of Integrated Intelligent Systems," *Proc. IEEE Intern. Conf. on System, Man, and Cybernetics*, Alexandria, VA, pp. 1133-1137.

Rao, M., J.P. Tsai, and T.S. Jiang, 1988, "An Intelligent Dicisionmaker for Optimal Control," *Applied Artificial Intelligence*, 2, pp. 285-305.

Rao, M., Y. Ying and Q. Wang 1992, "Integrated Distributed Intelligent System for Process-Control System Design," *Engineering Applications of AI*, 5(6), pp. 505-518.

Raven, F.H., 1987, *Automatic Control Engineering*, McGraw-Hill, New York.

Rodd, M.G. and Q. Wu, 1990, "Knowledge-Based Computer Vision Systems for Industrial Control," *Proc. IFAC World Congress*, Tallinn, USSR, pp. 135-140.

Routh, E.J., 1877, *Dynamics of a System of Rigid Bodies*, 3rd ed., Macmillan, London.

Sage, A.P., 1977, *Optimum Systems Control*, Prentice-Hall, Englewood Cliffs, N.J.

Sanz, J.L., 1989, *Advances in Machine Vision*, Springer-Verlag, New York.

Seborg, D.E., T.F. Edgar and S.L. Shah, 1986, "Adaptive Control Strategies for Process Control: A Survey," *AIChE Journal*, 32, pp. 881-913.

Seborg, D.L., T.F. Edgar, and Mellichamp, D.A., 1989, *Process Dynamics and Control*, John Wiley & Sons, New York.

Shannon, R., 1986, "Intelligent Simulation Environment," *Proc. Intelligent Simulation Environment*, San Diego, CA, pp. 150-156.

Shinskey, F.G., 1988, *Process Control Systems -- Application, Design and Adjustment*, McGraw-Hill, New York.

Spence, M., K. Danielson, Y. Ying and M. Rao, 1993, "On-Line Advisory Expert System Development at Dashowa Peace River Pulp," *79th Annual Meeting Technical Section, Canadian Pulp and Paper Association*, Montreal, pp. A143-A147.

Stephanopoulos, G., 1984, *Chemical Process Control -- An Introduction to Theory and Practice*, Prentice-Hall, Englewood Cliffs, N.J.

Vemuri, V., 1988, *Artificial Neural Networks: Theoretical Concepts*, IEEE Computer Society Press, Washington, DC.

Venkatasubramanian, V., 1986, "A Course in Artificial Intelligence in Processing Engineering," *Chem. Eng. Ed.*, 23, p. 120.

Veuri, V., 1988, *Artificial Neural Networks: Theoretical Concepts*, IEEE Computer Society Press, Washington, DC.

Watanabe, K., 1991, *Adaptive Estimation and Control*, Prentice-Hall, Englewood Cliffs, N.J.

Wellstead, P.E., D. Prager and P. Zanker, 1979, "Pole-assignment Self-tuning Regulator," *Proceedings of the Institution of Electrical Engineers*, 126, pp. 781-787.

Zhou, M.C. and F. Dicesare, 1993, *Petri Net Synthesis for Discrete Event Control of Manufacturing Systems*, Kluwer Academic Publishers, Boston, MA.

Zadeh, L.A., 1978, "Fuzzy Sets as a Basis for a Theory of Possibility," *Fuzzy Sets and Systems*, 1, 3-28.

Ziegler, J.G. and N.B. Nichols, 1942, "Optimum Settings for Automatic Controllers," *Trans. ASME*, 64, p. 759.

INDEX

Printed and bound by CPI Group (UK) Ltd, Croydon, CR0 4YY

23/10/2024

01777667-0006

.